# OPTIMUM SIGNAL PROCESSING

# OPTIMUM SIGNAL PROCESSING:
## An Introduction

**SOPHOCLES J. ORFANIDIS**
Rutgers University

MACMILLAN PUBLISHING COMPANY
*A Division of Macmillan, Inc.*
NEW YORK

Collier Macmillan Publishers
LONDON

Macmillan Publishing Company
866 Third Avenue, New York, NY 10022

Collier Macmillan Canada, Inc.

Printed in the United States of America

printing number

1 2 3 4 5 6 7 8 9 10


**Library of Congress Cataloging in Publication Data**

Orfanidis, Sophocles J.
    Optimum signal processing.

    Bibliography: p.
    Includes index.
    1. Signal processing.    I. Title.
TK5102.5.0246    1984        621.38'043        83-24411
ISBN 0-02-949860-0

*To my parents*

*John and Cleo Orfanidis*

# Contents

# Preface

THE PURPOSE OF THIS BOOK is to provide an introduction to signal processing methods that are based on optimum Wiener filtering and least-squares estimation concepts. Such methods have a remarkably broad range of applications, ranging from the analysis and synthesis of speech, data compression, image processing and modeling, channel equalization and echo cancellation in digital data transmission, geophysical signal processing in oil exploration, linear predictive analysis of EEG signals, modern methods of high-resolution spectrum estimation, and superresolution array processing, to adaptive signal processing for sonar, radar, system identification, and adaptive control applications. The structure of the book is to present the Wiener filtering concept as the basic unifying theme that ties together the various signal processing algorithms and techniques currently used in the above applications.

The book is based on lecture notes for a second-semester graduate-level course on advanced topics in digital signal processing I have taught at Rutgers University since 1979. The book is primarily addressed to beginning graduate students in electrical engineering, but it may also be used as a reference by practicing engineers who want a conscise introduction to the subject. The prerequisites for using the book are an introductory course on digital signal processing, such as on the level of Oppenheim and Schafer's book, and some familiarity with probability and random signal concepts, such as on the level of Papoulis' book.

Chapter 1 sets many of the objectives of the book and serves both as a review of probability and random signals and as an introduction to some of the basic concepts upon which the rest of the text is built. These are the concept of correlation canceling and its connection to linear mean-squared estimation, and the concept of Gram-Schmidt orthogonalization of random variables and its connection to linear prediction and signal modeling. After a brief review of some pertinent material on random signals, such as autocorrelations, power spectra, and the periodogram and its improvements, we discuss parametric signal models in which the random signal is modeled as the output of a linear system driven by white noise and present an overview of the uses of such models in signal analysis and synthesis, spectrum estimation, signal classification, and data compression applications. A first-order autoregressive model is used to illustrate many of these ideas and to motivate some practical methods of extracting the model parameters from actual data.

Chapter 2 is also introductory, and its purpose is to present a number of straightforward applications and simulation examples that illustrate the practical usage of random signal concepts. The selected topics include simple designs for signal enhancement filters, quantization noise in digital filters, and an introduction to linear prediction based on the finite past. The last two topics are then merged into an introductory discussion of data compression by DPCM methods.

Chapter 3 introduces the concept of minimal phase signals and filters and its role in the making of parametric signal models via spectral factorization. These methods are used in Chapter 4 for the solution of the Wiener filtering problem.

The basic concept of the Wiener filter as an optimum filter for estimating one signal from another is developed in Chapter 4. The Wiener filter is also viewed as a correlation canceler and as an optimal signal separator. We consider both the stationary and nonstationary Wiener filters, as well as the more practical FIR Wiener filter. While discussing a simple first-order Wiener filter example, we take the opportunity to introduce some elementary Kalman filter concepts. We demonstrate how the steady-state Kalman filter is equivalent to the Wiener filter and how its solution may be obtained from the steady-state algebraic Riccati equation which affects the spectral factorization required in the Wiener case. We also show how the Kalman filter may be thought of as the whitening filter of the observation signal and discuss its connection to the Gram-Schmidt orthogonalization and parametric signal models of Chapter 1. This chapter is mainly theoretical in character. Practical implementations and applications of Wiener filters are discussed in Chapter 5 using block-processing methods and in Chapter 6 using real-time adaptive processing techniques.

Chapter 5 begins with a discussion of the full linear prediction problem and its connection to signal modeling and continues with the problem of linear prediction based on the finite past and its efficient solution via the Levinson recursion. We discuss the analysis and synthesis lattice filters of linear prediction, as well as the lattice realizations of more general Wiener filters that are based on the orthogonality property of the backward prediction errors. The autocorrelation, covariance, and Burg's methods of linear predictive analysis are presented, and their application to speech analysis and synthesis and to spectrum estimation is discussed. The problem of estimating the frequencies of multiple sinusoids in noise and the problem of resolving the directions of point-source emitters by spatial array processing are discussed. Four approaches to these problems are presented, namely, the classical method based on the windowed autocorrelation, the maximum entropy method based on linear prediction, Capon's maximum likelihood method, and eigenvector-based methods. We also discuss the problem of wave propagation in layered media and its connection to linear prediction, and present the dynamic predictive deconvolution procedure for deconvolving the multiple reverberation effects of a layered structure from the knowledge of its reflection or transmission response. The chapter ends with a discussion of a least-squares reformulation of the Wiener filtering problem that can be used in the design of waveshaping and spiking filters for deconvolution applications.

Real-time adaptive implementations of Wiener filters are discussed in Chapter 6. The basic operation of an adaptive filter is explained by means of the simplest possible filter, namely, the correlation canceler loop, which forms the elementary building block of higher order adaptive filters. The Widrow-Hoff LMS adaptation algorithm and its convergence properties are discussed next. Several applications of adaptive filters are presented, such as adaptive noise canceling, adaptive channel equalization and echo cancellation, adaptive signal separation and the adaptive line enhancer, adaptive spectrum estimation based on linear prediction, and adaptive array processing. We also discuss some recent developments, such as the adaptive implementation of Pisarenko's method of harmonic retrieval, and two alternative adaptation algorithms that offer very fast speed of convergence, namely, recursive least-squares, and gradient lattice adaptive filters.

The subject of Wiener filtering and linear estimation is vast. The selection of material in this book reflects my preferences and views on what should be included in an introductory course on this subject. The emphasis thoroughout the book is on the signal processing procedures that grow out of the fundamental concept of Wiener filtering. An important ingredient of the book is the inclusion of several computer experiments and assignments that demonstrate the successes and limitations of the various signal processing algorithms that are discussed. A set of FORTRAN 77 subroutines, designed to be used as a library, has been included in an appendix.

I would like to thank my colleagues Professors T. G. Marshall and P. Sannuti for their support. I am greatly indebted to Professor R. Peskin for making available his graphics system on which many of the simulation examples were run and to my graduate student Ms. L. M. Vail for her invaluable help in producing most of the computer graphs. Most of all, I would like to thank my wife Monica, without whose love and affection this book could not have been written.

Sophocles J. Orfanidis

# 1

# Random Signals

## 1.1 *Probability Density, Mean, Variance*

In this section, we present a short review of probability concepts. It is assumed that the student has had a course on the subject on the level of Papoulis' book [1].

Let $x$ be a *random variable* having *probability density $p(x)$*. Its *mean, variance,* and *second moment* are defined as the expectation values

$$m = E[x] = \int_{-\infty}^{\infty} x\, p(x)\, dx = \text{mean}$$

$$\sigma^2 = E[(x - m)^2] = \int_{-\infty}^{\infty} (x - m)^2\, p(x)\, dx = \text{variance}$$

$$E[x^2] = \int_{-\infty}^{\infty} x^2\, p(x)\, dx = \text{second moment}$$

These quantities are known as *second-order statistics* of the random variable $x$. Their importance is linked with the fact that most optimal filter design criteria require knowledge only of the second-order statistics and do not require more detailed knowledge, such as of probability densities. It is of primary importance, then, to be able to extract such quantities from the actual measured data.

1

The probability that the random variable $x$ will assume a value within an interval of values $[a,b]$ is given by

$$\text{Prob}[a \leq x \leq b] = \int_a^b p(x)\, dx = \text{shaded area}$$

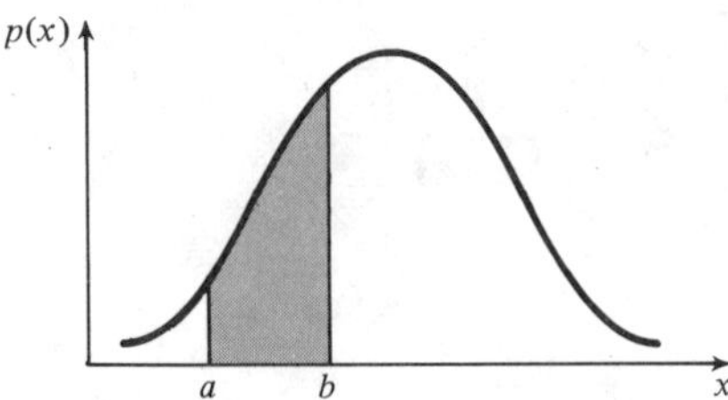

The probability density is always normalized to unity by

$$\int_{-\infty}^{\infty} p(x)\, dx = 1$$

which states that the probability of $x$ taking a value somewhere within its range of variation is unity; that is, certainty. This property also implies

$$\sigma^2 = E[(x - m)^2] = E[x^2] - (E[x])^2 = E[x^2] - m^2$$

*Example 1.1.1:*  Gaussian distribution

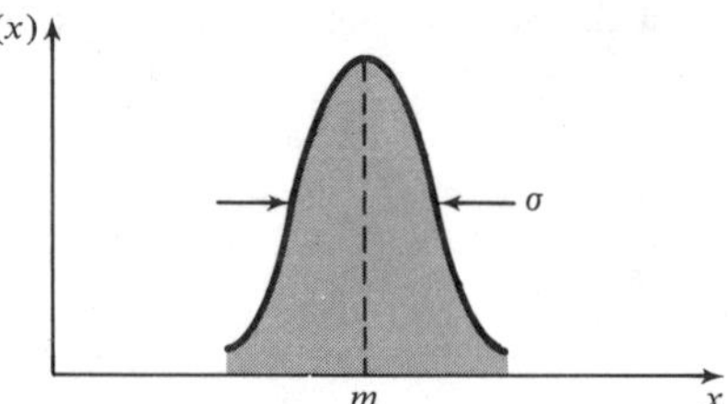

$$p(x) = \frac{1}{\sqrt{2\pi}\,\sigma} \exp[-(x - m)^2/2\sigma^2]$$

*Example 1.1.2:*  Uniform distribution

$$p(x) = \begin{cases} 1/Q, & \text{for } -Q/2 \leq x \leq Q/2 \\ 0, & \text{otherwise} \end{cases}$$

Its variance is $\sigma^2 = Q^2/12$.

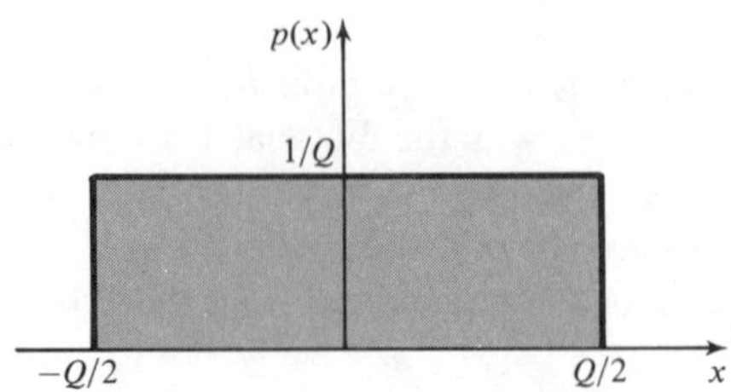

Both the gaussian and the uniform distributions will prove to be important examples. In typical signal processing problems of designing filters to remove or separate noise from signal, it is often assumed that the noise interference is gaussian. This assumption is justified on the grounds of the *central limit theorem,* provided that the noise arises from many different noise sources acting independently of each other.

The uniform distribution is also important. In digital signal processing applications, the quantization error arising from the signal quantization in the A/D converters, or the roundoff error arising from the finite accuracy of the internal arithmetic operations in digital filters, can often be assumed to be uniformly distributed.

Every computer provides system routines for the generation of random numbers. For example, the routines RANDU and GAUSS of the IBM Scientific Subroutine Package generate uniformly distributed random numbers over the interval [0,1], and gaussian-distributed numbers, respectively. GAUSS calls RANDU twelve times, thus generating twelve independent uniformly distributed random numbers $x_1, x_2, \ldots, x_{12}$. Then, their sum $x = x_1 + x_2 + \cdots + x_{12}$ will be approximately gaussian, as guaranteed by the central limit theorem. It is interesting to note that the variance of $x$ is unity, as it follows from the fact that the variance of each $x_i$ is $1/12$:

$$\sigma_x^2 = \sigma_{x_1}^2 + \sigma_{x_2}^2 + \cdots + \sigma_{x_{12}}^2 = \frac{1}{12} + \frac{1}{12} + \cdots + \frac{1}{12} = 1$$

The mean of $x$ is $12/2 = 6$. By shifting and scaling $x$, one can obtain a gaussian-distributed random number of any desired mean and variance.

## 1.2  *Chebyshev's Inequality*

The variance $\sigma^2$ of a random variable $x$ is a measure of the spread of the $x$-values about their mean. This intuitive interpretation of the variance is a direct consequence of Chebyshev's inequality, which states that the $x$-values tend to cluster about their mean in the sense that the probability of a value not occurring in the near vicinity of the mean is small; and it is smaller the smaller the variance. More precisely, for any probability density $p(x)$ and any $\Delta > 0$, the probability that $x$ will fall outside the interval of values $[m - \Delta,\ m + \Delta]$ is bounded by

$$\text{Prob}[|x - m| \geq \Delta] \leq \frac{\sigma^2}{\Delta^2}$$

(Chebyshev's inequality)

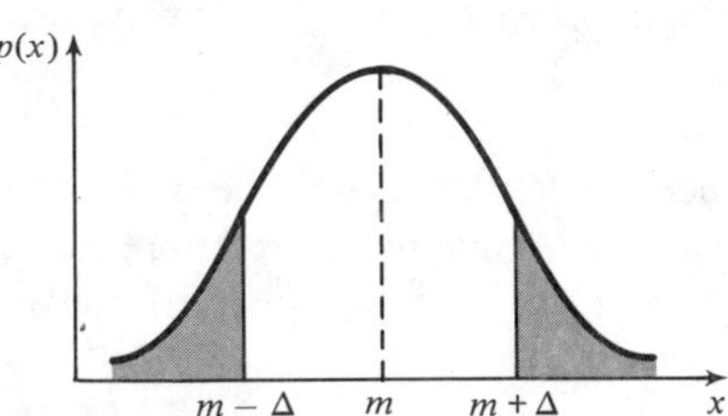

Thus, for fixed $\Delta$, as the variance $\sigma^2$ becomes smaller, the $x$-values tend to cluster more narrowly about the mean. In the extreme limiting case of a deterministic variable $x = m$, the density becomes infinitely narrow, $p(x) = \delta(x - m)$, and has zero variance.

## 1.3  *Joint and Conditional Densities, and Bayes' Rule*

Next, we discuss random vectors. A pair of two different random variables $\mathbf{x} = (x_1, x_2)$ may be thought of as a vector-valued random variable. Its statistical description is more complicated than that of a single variable and requires knowledge of the joint probability density $p(x_1, x_2)$. The two random variables may or may not have any dependence on each other. It is possible, for example, that if $x_2$ assumes a particular value, then this fact may influence, or restrict, the possible values that $x_1$ can then assume.

*Example 1.3.1:*  Suppose that $x_1$ is related to $x_2$ by

$$x_1 = 5x_2 + v$$

where $v$ is itself a random variable which is assumed to be independent (see below) of the random variable $x_2$. Then, the randomness of $x_1$ arises both from the randomness of $x_2$ and the randomness of $v$. But if a value of $x_2$ has already been realized, then the only randomness left in $x_1$ will arise only from the random variable $v$.

A quantity that provides a measure for the degree of dependency of the two variables on each other is the *conditional density* $p(x_1/x_2)$ of $x_1$ given $x_2$; and $p(x_2/x_1)$ of $x_2$ given $x_1$. These are related by *Bayes' rule*

$$p(x_1, x_2) = p(x_1/x_2)\, p(x_2) = p(x_2/x_1)\, p(x_1)$$

More generally, Bayes' rule for two events $A$ and $B$ is

$$p(A, B) = p(A/B)\, p(B) = p(B/A)\, p(A)$$

The two random variables $x_1$ and $x_2$ are *independent* of each other if they do not condition each other in any way; that is, if

$$p(x_1/x_2) = p(x_1) \qquad \text{or} \qquad p(x_2/x_1) = p(x_2)$$

In other words, the occurrence of $x_2$ does not in any way influence the variable $x_1$. When two random variables are independent, their joint density factors into the product of single (marginal) densities:

$$p(x_1, x_2) = p(x_1)\, p(x_2)$$

The converse is also true. The correlation between $x_1$ and $x_2$ is defined by the expectation value

$$E[x_1 x_2] = \int \int x_1 x_2 \, p(x_1, x_2) \, dx_1 dx_2$$

When $x_1$ and $x_2$ are independent, the correlation also factors as $E[x_1 x_2] = E[x_1] \, E[x_2]$.

The concept of a *random vector* generalizes to any dimension. A vector of $N$ random variables

$$\mathbf{x} = \begin{bmatrix} x_1 \\ x_2 \\ \vdots \\ x_N \end{bmatrix}$$

requires knowledge of the joint density

$$p(\mathbf{x}) = p(x_1, x_2, \ldots, x_N) \tag{1.3.1}$$

for its complete statistical description. The second-order statistics of $\mathbf{x}$ are its mean, its *correlation matrix*, and its *covariance matrix*, defined by

$$\mathbf{m} = E[\mathbf{x}], \quad R = E[\mathbf{x}\mathbf{x}^T], \quad \Sigma = E[(\mathbf{x} - \mathbf{m})(\mathbf{x} - \mathbf{m})^T] \tag{1.3.2}$$

where the superscript $T$ denotes transposition, and the expectation operations are defined in terms of the joint density Eq. (1.3.1); for example,

$$E[\mathbf{x}] = \int \mathbf{x} p(\mathbf{x}) d^N \mathbf{x}$$

where $d^N \mathbf{x} = dx_1 \, dx_2 \ldots dx_N$ denotes the corresponding $N$-dimensional volume element. The $ij$th element of the correlation matrix $R$ is the correlation between the $i$th random variable $x_i$ with the $j$th random variable $x_j$; that is, $R_{ij} = E[x_i x_j]$. It is easily shown that the covariance and correlation matrices are related by

$$\Sigma = R - \mathbf{m}\mathbf{m}^T$$

When the mean is zero, $R$ and $\Sigma$ coincide. Both $R$ and $\Sigma$ are *symmetric positive semi-definite* matrices.

*Example 1.3.2:* The probability density of a gaussian random vector $\mathbf{x} = (x_1, x_2, \ldots, x_N)^T$ is completely specified by its mean $\mathbf{m}$ and covariance matrix $\Sigma$; that is,

$$p(\mathbf{x}) = \frac{1}{(2\pi)^{N/2} (\det \Sigma)^{1/2}} \exp\left[ -\frac{1}{2} (\mathbf{x} - \mathbf{m})^T \Sigma^{-1} (\mathbf{x} - \mathbf{m}) \right]$$

*Example 1.3.3:*   Under a *linear* transformation, a gaussian random vector remains gaussian. Let $\mathbf{x}$ be a gaussian random vector of dimension $N$, mean $\mathbf{m}_x$, and covariance $\Sigma_x$. Show that the linearly transformed vector

$$\boldsymbol{\xi} = B\mathbf{x} \qquad \text{where } B \text{ is a nonsingular } N \times N \text{ matrix},$$

is gaussian-distributed with mean and covariance given by

$$\mathbf{m}_\xi = B\mathbf{m}_x, \; \Sigma_\xi = B\,\Sigma_x B^T \tag{1.3.3}$$

The relationships (1.3.3) are valid also for non-gaussian random vectors. They are easily derived as follows:

$$E[\boldsymbol{\xi}] = E[B\mathbf{x}] = BE[\mathbf{x}]$$

$$E[\boldsymbol{\xi}\boldsymbol{\xi}^T] = E[B\mathbf{x}(B\mathbf{x})^T] = BE[\mathbf{x}\mathbf{x}^T]\,B^T$$

The probability density $p_\xi(\boldsymbol{\xi})$ is related to the probability density $p_x(\mathbf{x})$ by the requirement that, under the above change of variables, they both yield the same elemental probabilities; that is,

$$p_\xi(\boldsymbol{\xi})\,d^N\boldsymbol{\xi} = p_x(\mathbf{x})\,d^N\mathbf{x} \tag{1.3.4}$$

Since the Jacobian of the transformation from $\mathbf{x}$ to $\boldsymbol{\xi}$ is $d^N\boldsymbol{\xi} = |\det B|d^N\mathbf{x}$, we obtain $p_\xi(\boldsymbol{\xi}) = p_x(\mathbf{x})/|\det B|$. Noting the invariance of the quadratic form

$$(\boldsymbol{\xi} - \mathbf{m}_\xi)^T\,\Sigma_\xi^{-1}\,(\boldsymbol{\xi} - \mathbf{m}_\xi) = (\mathbf{x} - \mathbf{m}_x)^T\,B^T\,(B\,\Sigma_x\,B^T)^{-1}\,B\,(\mathbf{x} - \mathbf{m}_x)$$

$$= (\mathbf{x} - \mathbf{m}_x)^T\,\Sigma_x^{-1}\,(\mathbf{x} - \mathbf{m}_x)$$

and that $\det \Sigma_\xi = \det(B\,\Sigma_x\,B^T) = (\det B)^2 \det \Sigma_x$, we obtain

$$p_\xi(\boldsymbol{\xi}) = \frac{1}{(2\pi)^{N/2}(\det \Sigma_\xi)^{1/2}}\,\exp\left[-\frac{1}{2}\,(\boldsymbol{\xi} - \mathbf{m}_\xi)^T\Sigma_\xi^{-1}(\boldsymbol{\xi} - \mathbf{m}_\xi)\right]$$

*Example 1.3.4:*   Consider two zero-mean random vectors $\mathbf{x}$ and $\mathbf{y}$ of dimensions $N$ and $M$, respectively. Show that if they are *uncorrelated* and *jointly gaussian*, then they are also *independent* of each other. That $\mathbf{x}$ and $\mathbf{y}$ are jointly gaussian means that the $(N + M)$-dimensional joint vector $\mathbf{z} = \begin{pmatrix}\mathbf{x}\\\mathbf{y}\end{pmatrix}$ is zero-mean and gaussian; that is,

$$p(\mathbf{z}) = \frac{1}{(2\pi)^{N+M/2}(\det R_{zz})^{1/2}}\,\exp\left[-\frac{1}{2}\,\mathbf{z}^T R_{zz}^{-1}\,\mathbf{z}\right]$$

where the correlation (covariance) matrix $R_{zz}$ is

$$R_{zz} = E[\mathbf{z}\mathbf{z}^T] = E\left[\begin{pmatrix}\mathbf{x}\\\mathbf{y}\end{pmatrix}\overbrace{\mathbf{x}^T,\mathbf{y}^T}\right] = \begin{bmatrix}E[\mathbf{x}\mathbf{x}^T] & E[\mathbf{x}\mathbf{y}^T]\\E[\mathbf{y}\mathbf{x}^T] & E[\mathbf{y}\mathbf{y}^T]\end{bmatrix} = \begin{bmatrix}R_{xx} & R_{xy}\\R_{yx} & R_{yy}\end{bmatrix}$$

If $\mathbf{x}$ and $\mathbf{y}$ are uncorrelated; that is, $R_{xy} = E[\mathbf{x}\mathbf{y}^T] = 0$, then the matrix $R_{zz}$ becomes block diagonal and the quadratic form of the joint vector becomes the sum of the individual quadratic forms:

$$\mathbf{z}^T R_{zz}^{-1} \mathbf{z} = [\mathbf{x}^T,\mathbf{y}^T] \begin{bmatrix} R_{xx}^{-1} & 0 \\ 0 & R_{yy}^{-1} \end{bmatrix} \begin{bmatrix} \mathbf{x} \\ \mathbf{y} \end{bmatrix} = \mathbf{x}^T R_{xx}^{-1} \mathbf{x} + \mathbf{y}^T R_{yy}^{-1} \mathbf{y}$$

Since $R_{xy} = 0$ also implies that $\det R_{zz} = (\det R_{xx})(\det R_{yy})$, it follows that the joint density $p(\mathbf{x},\mathbf{y}) = p(\mathbf{z})$ factors into the marginal densities:

$$p(\mathbf{x},\mathbf{y}) = p(\mathbf{z}) = p(\mathbf{x})\, p(\mathbf{y})$$

which shows the independence of $\mathbf{x}$ and $\mathbf{y}$.


*Example 1.3.5:*  Given a random vector $\mathbf{x}$ with mean $\mathbf{m}$ and covariance $\Sigma$, show that the best choice of a *deterministic* vector $\hat{\mathbf{x}}$ which minimizes the quantity

$$R_{ee} = E[\mathbf{e}\mathbf{e}^T] = \text{minimum}, \qquad \text{where } \mathbf{e} = \mathbf{x} - \hat{\mathbf{x}}$$

is the mean $\mathbf{m}$ itself. That is, $\hat{\mathbf{x}} = \mathbf{m}$. Also show that for this optimal choice of $\hat{\mathbf{x}}$, the actual minimum value of the quantity $R_{ee}$ is the covariance $\Sigma$.

This property is easily shown by working with the deviation of $\hat{\mathbf{x}}$ from the mean $\mathbf{m}$; that is, let

$$\hat{\mathbf{x}} = \mathbf{m} + \boldsymbol{\Delta}$$

Then, the quantity $R_{ee}$ becomes

$$\begin{aligned}
R_{ee} = E[\mathbf{e}\mathbf{e}^T] &= E[(\mathbf{x} - \mathbf{m} - \boldsymbol{\Delta})\,(\mathbf{x} - \mathbf{m} - \boldsymbol{\Delta})^T] \\
&= E[(\mathbf{x} - \mathbf{m})\,(\mathbf{x} - \mathbf{m})^T] - \boldsymbol{\Delta}\, E[(\mathbf{x} - \mathbf{m})^T] \\
&\qquad\qquad\qquad\qquad\qquad - E[(\mathbf{x} - \mathbf{m})]\,\boldsymbol{\Delta}^T + \boldsymbol{\Delta}\boldsymbol{\Delta}^T \\
&= \Sigma + \boldsymbol{\Delta}\boldsymbol{\Delta}^T
\end{aligned}$$

Where we used the fact that $E[(\mathbf{x} - \mathbf{m})] = E[\mathbf{x}] - \mathbf{m} = 0$. Since the matrix $\boldsymbol{\Delta}\boldsymbol{\Delta}^T$ is nonnegative-definite, it follows that $R_{ee}$ will be minimized when $\boldsymbol{\Delta} = 0$, and in this case the minimum value will be $\Sigma$.


## 1.4  *Correlation Canceling*

The concept of correlation canceling plays a central role in the development of many optimum signal processing algorithms, because a correlation canceler is also the *best* linear processor for estimating one signal from another.

Consider two zero-mean random vectors $\mathbf{x}$ and $\mathbf{y}$ of dimensions $N$ and $M$, respectively. If $\mathbf{x}$ and $\mathbf{y}$ are correlated with each other in the sense that

$R_{xy} = E[\mathbf{x}\mathbf{y}^T] \neq 0$, then we would like to remove such correlations by means of a linear transformation of the form

$$\mathbf{e} = \mathbf{x} - H\mathbf{y} \qquad (1.4.1)$$

where the $N \times M$ matrix $H$ must be suitably chosen such that the new pair of vectors $(\mathbf{e},\mathbf{y})$ are no longer correlated with each other; that is, we require

$$R_{ey} = E[\mathbf{e}\mathbf{y}^T] = 0 \qquad (1.4.2)$$

Using Eq. (1.4.1), we obtain

$$R_{ey} = E[\mathbf{e}\mathbf{y}^T] = E[(\mathbf{x} - H\mathbf{y})\mathbf{y}^T] = E[\mathbf{x}\mathbf{y}^T] - HE[\mathbf{y}\mathbf{y}^T] = R_{xy} - HR_{yy}$$

Then, the condition $R_{ey} = 0$ immediately implies that

$$H = R_{xy}\, R_{yy}^{-1} = E[\mathbf{x}\mathbf{y}^T]\, E[\mathbf{y}\mathbf{y}^T]^{-1} \qquad (1.4.3)$$

Using $R_{ey} = 0$, the covariance matrix of the resulting vector $\mathbf{e}$ is easily found to be

$$R_{ee} = E[\mathbf{e}\mathbf{e}^T] = E[\mathbf{e}(\mathbf{x}^T - \mathbf{y}^T H^T)] = R_{ex} - R_{ey}\, H^T = R_{ex} = E[(\mathbf{x} - H\mathbf{y})\mathbf{x}^T]$$

or

$$R_{ee} = R_{xx} - H\, R_{yx} = R_{xx} - R_{xy}\, R_{yy}^{-1}\, R_{yx} \qquad (1.4.4)$$

The vector

$$\hat{\mathbf{x}} = H\mathbf{y} = R_{xy}\, R_{yy}^{-1}\, \mathbf{y} = E[\mathbf{x}\mathbf{y}^T]\, E[\mathbf{y}\mathbf{y}^T]^{-1}\, \mathbf{y} \qquad (1.4.5)$$

obtained by linearly processing the vector $\mathbf{y}$ by the matrix $H$ is called the *linear regression*, or *orthogonal projection*, of $\mathbf{x}$ on the vector $\mathbf{y}$. In a sense to be made precise later, $\hat{\mathbf{x}}$ also represents the best "copy", or *estimate*, of $\mathbf{x}$ that can be made on the basis of the vector $\mathbf{y}$. Thus, the vector $\mathbf{e} = \mathbf{x} - H\mathbf{y} = \mathbf{x} - \hat{\mathbf{x}}$ may be thought of as the *estimation error*.

Actually, it is better to think of $\hat{\mathbf{x}} = H\mathbf{y}$ not as an estimate of $\mathbf{x}$ but rather as an estimate of *that part* of $\mathbf{x}$ which is correlated with $\mathbf{y}$. Indeed, suppose that $\mathbf{x}$ consists of two parts

$$\mathbf{x} = \mathbf{x}_1 + \mathbf{x}_2$$

such that $\mathbf{x}_1$ is correlated with $\mathbf{y}$, but $\mathbf{x}_2$ is not; that is, $R_{x_2 y} = E[\mathbf{x}_2 \mathbf{y}^T] = 0$. Then,

$$R_{xy} = E[\mathbf{x}\mathbf{y}^T] = E[(\mathbf{x}_1 + \mathbf{x}_2)\, \mathbf{y}^T] = R_{x_1 y} + R_{x_2 y} = R_{x_1 y}$$

and therefore,

$$\hat{\mathbf{x}} = R_{xy}\, R_{yy}^{-1}\, \mathbf{y} = R_{x_1 y}\, R_{yy}^{-1}\, \mathbf{y} = \hat{\mathbf{x}}_1$$

The vector $\mathbf{e} = \mathbf{x} - \hat{\mathbf{x}} = \mathbf{x}_1 + \mathbf{x}_2 - \hat{\mathbf{x}}_1 = (\mathbf{x}_1 - \hat{\mathbf{x}}_1) + \mathbf{x}_2$ consists of the estimation error $\mathbf{x}_1 - \hat{\mathbf{x}}_1$ of the $x_1$-part plus the $x_2$-part. Both of these terms are

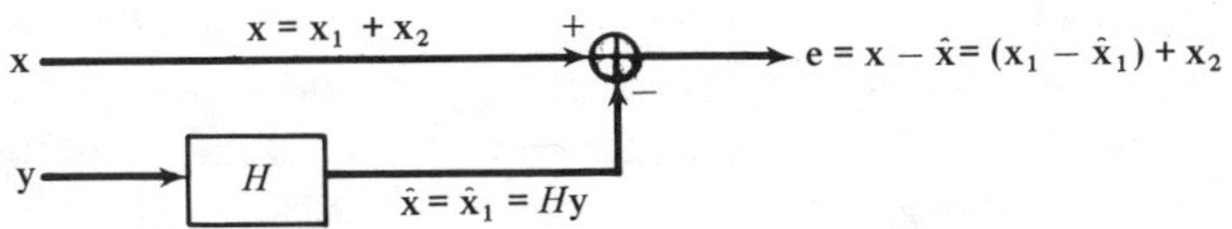


**Figure 1.1** Correlation Canceler

separately uncorrelated from **y**. These operations are summarized in block diagram form in Fig. 1.1. The most important feature of this arrangement is the *correlation cancellation* property which may be summarized as follows: If **x** has a part $\mathbf{x}_1$ which is correlated with **y**, then this part will tend to be canceled as much as possible from the output **e**. The linear processor $H$ accomplishes this by converting **y** into the *best possible copy* $\hat{\mathbf{x}}_1$ of $\mathbf{x}_1$ and then proceeding to cancel it from the output. The output vector **e** is no longer correlated with **y**. The part $\mathbf{x}_2$ of **x** which is uncorrelated with **y** remains entirely unaffected. It cannot be estimated in terms of **y**. The correlation canceler may also be thought of as an *optimal signal separator*. Indeed, the output of the processor $H$ is essentially the $\mathbf{x}_1$ component of **x**, whereas the output **e** is essentially the $\mathbf{x}_2$ component. The separation of **x** into $\mathbf{x}_1$ and $\mathbf{x}_2$ is optimal, in the sense that the $\mathbf{x}_1$ component of **x** is removed as much as possible from **e**.

Next, we discuss the *best linear estimator property* of the correlation canceler. The choice $H = R_{xy}R_{yy}^{-1}$, which guarantees correlation cancellation, is also the choice that gives the *best estimate* of **x** as a *linear* function of **y** in the form $\hat{\mathbf{x}} = H\mathbf{y}$. It is the best estimate in the sense that it produces the lowest *mean-squared* estimation error. To see this, express the covariance matrix of the estimation error in terms of $H$, as follows:

$$R_{ee} = E[\mathbf{e}\mathbf{e}^T] = E[(\mathbf{x} - H\mathbf{y})\,(\mathbf{x}^T - \mathbf{y}^T H^T)] \qquad (1.4.6)$$
$$= R_{xx} - HR_{yx} - R_{xy}H^T + HR_{yy}H^T$$

Minimizing this expression with respect to $H$ yields the optimal choice of $H$:

$$H_{\text{opt}} = R_{xy}\,R_{yy}^{-1}$$

with the minimum value for $R_{ee}$ given by

$$R_{ee}^{\min} = R_{xx} - R_{xy}\,R_{yy}^{-1}\,R_{yx}$$

Any other choice of $H$ will result in a larger value for $R_{ee}$. An alternative way to see this is to consider a deviation $\Delta H$ of $H$ from its optimal value; that is, in Eq. (1.4.6) replace $H$ by

$$H = H_{\text{opt}} + \Delta H = R_{xy}\,R_{yy}^{-1} + \Delta H$$

Then Eq. (1.4.6) may be expressed in terms of $\Delta H$ as follows:

$$R_{ee} = R_{ee}^{\min} + \Delta H R_{yy} \Delta H^T$$

Since $R_{yy}$ is positive definite, the second term always represents a nonnegative contribution above the minimum value $R_{ee}^{\min}$.

*Example 1.4.1*  If **x** and **y** are *jointly gaussian*, show that the linear estimate $\hat{\mathbf{x}} = H\mathbf{y}$ is also the *conditional mean* $E[\mathbf{x}/\mathbf{y}]$ of the vector **x** given the vector **y**. The conditional mean is defined in terms of the conditional density $p(\mathbf{x}/\mathbf{y})$ of **x** given **y** as follows:

$$E[\mathbf{x}/\mathbf{y}] = \int \mathbf{x}\, p(\mathbf{x}/\mathbf{y})\, d^N\mathbf{x}$$

Instead of computing this integral, we will use the results of Examples 1.3.3 and 1.3.4. The transformation from the jointly gaussian pair $(\mathbf{x},\mathbf{y})$ to the uncorrelated pair $(\mathbf{e},\mathbf{y})$ is linear:

$$\begin{bmatrix} \mathbf{e} \\ \mathbf{y} \end{bmatrix} = \begin{bmatrix} I_N & -H \\ O & I_M \end{bmatrix} \begin{bmatrix} \mathbf{x} \\ \mathbf{y} \end{bmatrix}$$

where $I_N$ and $I_M$ are the unit matrices of dimensions $N$ and $M$, respectively. Therefore, Example 1.3.3 implies that the transformed pair $(\mathbf{e},\mathbf{y})$ is also jointly gaussian. Furthermore, since **e** and **y** are uncorrelated, it follows from Example 1.3.4 that they must be independent of each other. The required conditional mean of **x** can be computed by writing **x** as

$$\mathbf{x} = H\mathbf{y} + \mathbf{e}$$

and noting that if **y** is given, then $H\mathbf{y}$ is no longer random. Therefore,

$$E[\mathbf{x}/\mathbf{y}] = E[(H\mathbf{y} + \mathbf{e})/\mathbf{y}] = H\mathbf{y} + E[\mathbf{e}/\mathbf{y}]$$

Since **e** and **y** are independent, the conditional mean $E[\mathbf{e}/\mathbf{y}]$ is the same as the unconditional mean $E[\mathbf{e}]$ which is zero by the zero-mean assumption. Thus,

$$E[\mathbf{x}/\mathbf{y}] = H\mathbf{y} = R_{xy}R_{yy}^{-1}\,\mathbf{y} \qquad \text{(gaussian } \mathbf{x} \text{ and } \mathbf{y}) \qquad (1.4.7)$$

*Example 1.4.2:*  Show that the conditional mean $E[\mathbf{x}/\mathbf{y}]$ is the best *unrestricted* (i.e., not necessarily linear) estimate of **x** in the *mean-squared* sense. The best linear estimate was obtained by seeking the best linear function of **y** that minimized the error criterion (1.4.6); that is, we required a priori that the estimate was to be of the form $\hat{\mathbf{x}} = H\mathbf{y}$. Here, our task is more general: Find the most general function of **y**, $\hat{\mathbf{x}} = \hat{\mathbf{x}}(\mathbf{y})$, which gives the best estimate of **x**, in the sense of producing the lowest mean-squared estimation error $\mathbf{e} = \mathbf{x} - \hat{\mathbf{x}}(\mathbf{y})$

$$R_{ee} = E[\mathbf{e}\mathbf{e}^T] = E[(\mathbf{x} - \hat{\mathbf{x}}(\mathbf{y}))\,(\mathbf{x} - \hat{\mathbf{x}}(\mathbf{y}))^T] = \min$$

The functional dependence of $\hat{\mathbf{x}}(\mathbf{y})$ on $\mathbf{y}$ is not required to be linear a priori. Using $p(\mathbf{x},\mathbf{y}) = p(\mathbf{x}/\mathbf{y})p(\mathbf{y})$, the above expectation may be written as

$$R_{ee} = \int (\mathbf{x} - \hat{\mathbf{x}}(\mathbf{y}))\,(\mathbf{x} - \hat{\mathbf{x}}(\mathbf{y}))^T\, p(\mathbf{x},\mathbf{y})\, d^N\mathbf{x}\, d^M\mathbf{y}$$

$$= \int p(\mathbf{y})\, d^M\mathbf{y} \left\{ \int (\mathbf{x} - \hat{\mathbf{x}}(\mathbf{y}))\,(\mathbf{x} - \hat{\mathbf{x}}(\mathbf{y}))^T\, p(\mathbf{x}/\mathbf{y})\, d^N\mathbf{x} \right\}$$

Since $p(\mathbf{y})$ is nonnegative for all $\mathbf{y}$, it follows that $R_{ee}$ will be minimized when the quantity

$$\int (\mathbf{x} - \hat{\mathbf{x}}(\mathbf{y}))\,(\mathbf{x} - \hat{\mathbf{x}}(\mathbf{y}))^T\, p(\mathbf{x}/\mathbf{y})\, d^N\mathbf{x}$$

is minimized with respect to $\hat{\mathbf{x}}$. But we know from Example 1.3.5 that this quantity is minimized when $\hat{\mathbf{x}}$ is chosen to be the corresponding mean; here, this is the mean with respect to the density $p(\mathbf{x}/\mathbf{y})$. Thus,

$$\hat{\mathbf{x}}(\mathbf{y}) = E[\mathbf{x}/\mathbf{y}] \tag{1.4.8}$$

To summarize, we have seen that

$$\hat{\mathbf{x}} = H\mathbf{y} = R_{xy}\, R_{yy}^{-1}\, \mathbf{y} = \text{best } \textit{linear} \text{ mean-squared estimate of } \mathbf{x}$$

$$\hat{\mathbf{x}} = E[\mathbf{x}/\mathbf{y}] = \text{best } \textit{unrestricted} \text{ mean-squared estimate of } \mathbf{x}$$

and Example 1.4.1 shows that the two are *equal* in the case of jointly gaussian vectors $\mathbf{x}$ and $\mathbf{y}$.

The concept of correlation canceling and its application to signal estimation problems will be discussed in more detail in Chapter 4. The adaptive implementation of the correlation canceler will be discussed in Chapter 6. In a typical signal processing application, the processor $H$ would represent a *linear filtering* operation and the vectors $\mathbf{x}$ and $\mathbf{y}$ would be *blocks of signal samples*. The design of such processors requires knowledge of the quantities $R_{xy} = E[\mathbf{xy}^T]$ and $R_{yy} = E[\mathbf{yy}^T]$. How does one determine these? Basically, there are two types of situations that are encountered in practice: Either both $\mathbf{x}$ and $\mathbf{y}$ are available for processing, or only $\mathbf{y}$ is available. In the first instance, one is interested in canceling the correlations between $\mathbf{x}$ and $\mathbf{y}$. Since both $\mathbf{x}$ and $\mathbf{y}$ are available, one can determine the required quantities $R_{xy}$ and $R_{yy}$ approximately; for example, by replacing ensemble averages by *time averages,* or by using *adaptive* techniques (as discussed in Chapter 6). Typical applications include noise canceling, echo canceling, channel equalization, and antenna sidelobe cancellation. In the second instance, $\mathbf{x}$ is not directly available and must be *estimated* on the basis of the available signal $\mathbf{y}$. In this case, one must have a specific *model* of the relationship between $\mathbf{x}$ and $\mathbf{y}$, from which $R_{xy}$ and $R_{yy}$ may be calculated. This is, for example, what is done in Kalman filtering.

*Example 1.4.3:* As an example of the relationship that might exist between $\mathbf{x}$ and $\mathbf{y}$, let

$$y_n = xc_n + v_n; \qquad n = 1, 2, \ldots, M$$

where $x$ and $v_n$ are zero-mean, unit-variance, random variables, and $c_n$ are known coefficients. It is further assumed that $v_n$ are mutually uncorrelated, and also uncorrelated with $x$, so that $E[v_n v_m] = \delta_{nm}$, $E[xv_n] = 0$. We would like to determine the optimal linear estimate (1.4.5) of $x$, and the corresponding estimation error (1.4.4). In obvious matrix notation we have $\mathbf{y} = x\mathbf{c} + \mathbf{v}$, with $E[x\mathbf{v}] = 0$, and $E[\mathbf{v}\mathbf{v}^T] = I$, where $I$ is the $M \times M$ unit matrix. We find

$$E[x\mathbf{y}^T] = E[x(x\mathbf{c} + \mathbf{v})^T] = \mathbf{c}^T E[x^2] + E[x\mathbf{v}^T] = \mathbf{c}^T$$

$$E[\mathbf{y}\mathbf{y}^T] = E[(x\mathbf{c} + \mathbf{v})(x\mathbf{c}^T + \mathbf{v}^T)] = \mathbf{c}\mathbf{c}^T E[x^2] + E[\mathbf{v}\mathbf{v}^T] = \mathbf{c}\mathbf{c}^T + I$$

and therefore, $H = E[x\mathbf{y}^T]\, E[\mathbf{y}\mathbf{y}^T]^{-1} = \mathbf{c}^T(I + \mathbf{c}\mathbf{c}^T)^{-1}$. Using the matrix inversion lemma we may write $(I + \mathbf{c}\mathbf{c}^T)^{-1} = I - \mathbf{c}(1 + \mathbf{c}^T\mathbf{c})^{-1}\mathbf{c}^T$, so that

$$H = \mathbf{c}^T [I - \mathbf{c}(1 + \mathbf{c}^T\mathbf{c})^{-1}\mathbf{c}^T] = (1 + \mathbf{c}^T\mathbf{c})^{-1}\mathbf{c}^T$$

The optimal estimate of $x$ is then

$$\hat{x} = H\mathbf{y} = (1 + \mathbf{c}^T\mathbf{c})^{-1}\mathbf{c}^T\mathbf{y} \qquad (1.4.9)$$

The corresponding estimation error is computed by

$$E[e^2] = R_{ee} = R_{xx} - HR_{yx} = 1 - (1 + \mathbf{c}^T\mathbf{c})^{-1}\mathbf{c}^T\mathbf{c} = (1 + \mathbf{c}^T\mathbf{c})^{-1}$$

## 1.5 *Gram-Schmidt Orthogonalization*

In the previous section, we saw that any random vector $\mathbf{x}$ may be decomposed relative to another vector $\mathbf{y}$ into two parts, $\mathbf{x} = \hat{\mathbf{x}} + \mathbf{e}$, one part which is correlated with $\mathbf{y}$, and one which is not. These two parts are uncorrelated with each other since $R_{e\hat{x}} = E[\mathbf{e}\hat{\mathbf{x}}^T] = E[\mathbf{e}\mathbf{y}^T H^T] = R_{ey} H^T = 0$. In a sense, they are orthogonal to each other. In this section, we will briefly develop such a geometrical interpretation. The usefulness of the geometrical approach is twofold: First, it provides a very simple and intuitive framework in which to formulate and understand signal estimation problems. Second, through the Gram-Schmidt orthogonalization process, it provides the basis for making *signal models*, which find themselves in a variety of signal processing applications, such as speech synthesis, data compression, and modern methods of spectrum estimation.

Geometrical ideas may be introduced by thinking of the space of random variables under consideration as a *linear vector space* [2]. For example, in the previous section we dealt with the multicomponent random variables $\mathbf{x}$ and $\mathbf{y}$ consisting, say, of the random variables $(x_1, x_2, \ldots, x_N)$ and $(y_1, y_2, \ldots, y_M)$, respectively. In this case, the space of random variables under consideration is the set

$$\{x_1, x_2, \ldots, x_N, y_1, y_2, \ldots, y_M\} \tag{1.5.1}$$

Since any linear combination of random variables from this set is itself a random variable, the above set may be enlarged by adjoining to it all such possible linear combinations. This is the linear vector space *generated* by the given set of random variables. The next step is to convert this vector space into an *inner-product space* (a Hilbert space) by defining an inner product between two random variables $u$ and $v$ as follows:

$$(u,v) = E[uv] \tag{1.5.2}$$

With this definition of an inner product, "orthogonal" means "uncorrelated." The *distance* between $u$ and $v$ is defined by the norm $\|u - v\|$ induced by the above inner product:

$$\|u - v\|^2 = E[(u - v)^2] \tag{1.5.3}$$

Mutually orthogonal (i.e., uncorrelated) random variables may be used to define *orthogonal bases*. Consider, for example, $M$ mutually orthogonal random variables $\epsilon_1, \epsilon_2, \ldots, \epsilon_M$; such that

$$(\epsilon_i, \epsilon_j) = E[\epsilon_i \epsilon_j] = 0; \qquad \text{if } i \neq j \tag{1.5.4}$$

and let $Y = \{\epsilon_1, \epsilon_2, \ldots, \epsilon_M\}$ be the linear subspace *spanned* by these $M$ random variables. Without loss of generality, we may assume that the $\epsilon_i$s are linearly independent; therefore, they form a linearly independent and orthogonal basis for the subspace $Y$.

One of the standard results on linear vector spaces is the *orthogonal decomposition* theorem [3], which in our context may be stated as follows: Any random variable $x$ may be decomposed *uniquely, with respect to* a subspace $Y$, into two mutually orthogonal parts. One part is *parallel* to the subspace $Y$ (i.e., it lies in it), and the other is *perpendicular* to it. That is,

$$x = \hat{x} + e \qquad \text{with } \hat{x} \text{ in } Y, \text{ and } e \perp Y \tag{1.5.5}$$

The component $\hat{x}$ is called the *orthogonal projection* of $x$ onto the subspace $Y$. This decomposition is depicted in Fig. 1.2. The orthogonality condition $e \perp Y$ means that $e$ must be orthogonal to every vector in $Y$; or equivalently, to every basis vector $\epsilon_i$; that is,

$$(e, \epsilon_i) = E[e\epsilon_i] = 0, \qquad \text{for } i = 1, 2, \ldots, M \tag{1.5.6}$$

Since the component $\hat{x}$ lies in $Y$, it may be expanded in terms of the orthogonal basis in the form

$$\hat{x} = \sum_{i=1}^{M} a_i \epsilon_i$$

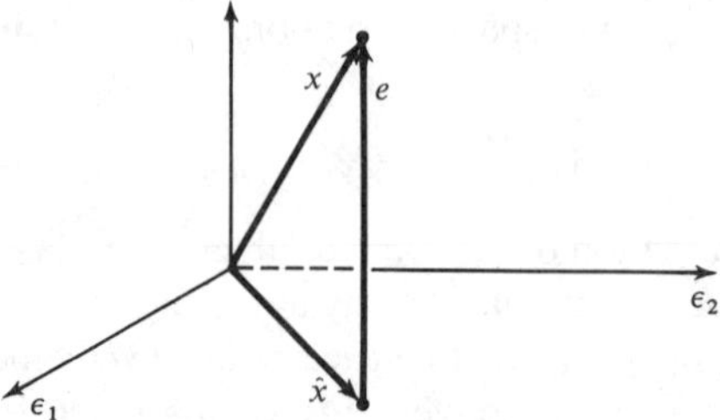

**Figure 1.2** Orthogonal Decomposition with Respect to $Y = \{\epsilon_1, \epsilon_2\}$

The coefficients $a_i$ can be determined using the orthogonality equations (1.5.6), as follows,

$$(x, \epsilon_i) = (\hat{x} + e, \epsilon_i) = (\hat{x}, \epsilon_i) + (e, \epsilon_i) = (\hat{x}, \epsilon_i)$$

$$= \left( \sum_{j=1}^{M} a_j \epsilon_j, \epsilon_i \right) = \sum_{j=1}^{M} a_j (\epsilon_j, \epsilon_i) = a_i (\epsilon_i, \epsilon_i)$$

where in the last equality we used Eq. (1.5.4). Thus, $a_i = (x, \epsilon_i)(\epsilon_i, \epsilon_i)^{-1}$, or $a_i = E[x\epsilon_i]E[\epsilon_i \epsilon_i]^{-1}$, and we can write Eq. (1.5.5) as

$$x = \hat{x} + e = \sum_{i=1}^{M} E[x\epsilon_i]E[\epsilon_i \epsilon_i]^{-1}\epsilon_i + e \qquad (1.5.7)$$

Equation (1.5.7) may also be written in a compact matrix form by introducing the $M$-vector

$$\boldsymbol{\epsilon} = \begin{bmatrix} \epsilon_1 \\ \epsilon_2 \\ \vdots \\ \epsilon_M \end{bmatrix}$$

the corresponding cross-correlation $M$-vector

$$E[x\boldsymbol{\epsilon}] = \begin{bmatrix} E[x\epsilon_1] \\ E[x\epsilon_2] \\ \vdots \\ E[x\epsilon_M] \end{bmatrix}$$

and the correlation matrix $R_{\epsilon\epsilon} = E[\boldsymbol{\epsilon}\boldsymbol{\epsilon}^T]$, which is *diagonal* because of the conditions (1.5.4):

$$R_{\epsilon\epsilon} = E[\boldsymbol{\epsilon}\boldsymbol{\epsilon}^T] = \text{diag}\{E[\epsilon_1^2], E[\epsilon_2^2], \ldots, E[\epsilon_M^2]\}$$

Then, Eq. (1.5.7) may be written as

$$x = \hat{x} + e = E[x\epsilon^T]\, E[\epsilon\epsilon^T\,]^{-1}\epsilon + e \tag{1.5.8}$$

The orthogonality equations (1.5.6) can be written as

$$R_{e\epsilon} = E[e\epsilon^T] = 0 \tag{1.5.9}$$

Equations (1.5.8) and (1.5.9) represent the unique orthogonal decomposition of any random variable $x$ relative to a linear subspace $Y$ of random variables. If one has a collection of $N$ random variables $x_1, x_2, \ldots, x_N$, then each one may be orthogonally decomposed with respect to the same subspace $Y$, giving $x_i = \hat{x}_i + e_i$, $i = 1, 2, \ldots, N$. All these may be grouped together into a compact matrix from as

$$\mathbf{x} = \hat{\mathbf{x}} + \mathbf{e} = E[\mathbf{x}\epsilon^T]\, E[\epsilon\epsilon^T]^{-1}\epsilon + \mathbf{e} \tag{1.5.10}$$

with

$$R_{e\epsilon} = E[\mathbf{e}\epsilon^T] = 0$$

where $\mathbf{x}$ stands for the column $N$-vector $\mathbf{x} = [x_1, x_2, \ldots, x_N]^T$, and so on. This is identical to the correlation canceler decomposition of the previous section.

Next, we briefly discuss the orthogonal projection theorem. In Section 1.4, we noted the best linear estimator property of the correlation canceler decomposition. The same result may be understood geometrically by means of the orthogonal projection theorem, which states: The orthogonal projection $\hat{x}$ of a vector $x$ onto a linear subspace $Y$ is that vector in $Y$ which lies *closest* to $x$ with respect to the distance induced by the inner product of the vector space.

The theorem is a simple consequence of the orthogonal decomposition theorem and the Pythagorean theorem. Indeed, let $x = \hat{x} + e$ be the unique orthogonal decomposition of $x$ with respect to $Y$, so that $\hat{x} \in Y$ and $e \perp Y$, and let $y$ be an arbitrary vector in $Y$; noting that $(\hat{x} - y) \in Y$ and therefore $e \perp (\hat{x} - y)$, we have

$$\|x - y\|^2 = \|(\hat{x} - y) + e\|^2 = \|\hat{x} - y\|^2 + \|e\|^2$$

or, in terms of Eq. (1.5.3),

$$E[(x - y)^2] = E[(\hat{x} - y)^2] + E[e^2]$$

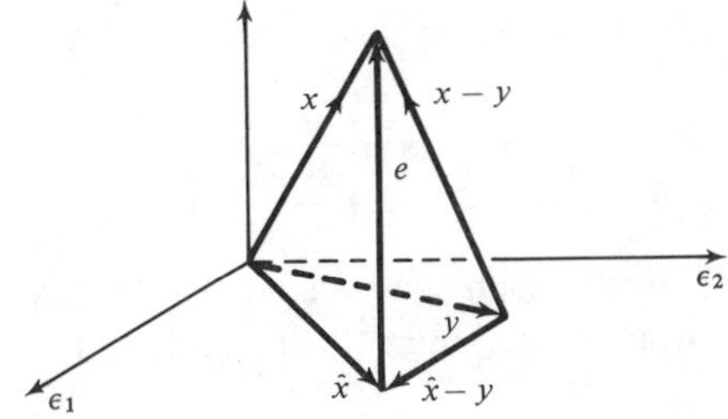

Since the vector $y$ varies over the subspace $Y$, it follows that the above quantity will be minimized when $y = \hat{x}$. In summary, $\hat{x}$ represents the *best approximation* of $x$ that can be made as a *linear* function of the random variables in $Y$ in the minimum mean-squared sense.

Above, we developed the orthogonal decomposition of a random variable relative to a linear subspace $Y$ which was generated by means of an orthogonal basis $\epsilon_1, \epsilon_2, \ldots, \epsilon_M$. In practice, the subspace $Y$ is almost always defined by means of a nonorthogonal basis, such as a collection of random variables

$$Y = \{y_1, y_2, \ldots, y_M\}$$

which may be mutually correlated. The subspace $Y$ is defined again as the *linear span* of this basis. The *Gram-Schmidt* orthogonalization process is a recursive procedure of generating an orthogonal basis $\epsilon_1, \epsilon_2, \ldots, \epsilon_M$ from $y_1, y_2, \ldots, y_M$. The basic idea of the method is this: Initialize the procedure by selecting $\epsilon_1 = y_1$. Next, consider $y_2$ and decompose it relative to $\epsilon_1$. Then, the component of $y_2$ which is perpendicular to $\epsilon_1$ is selected as $\epsilon_2$; so that $(\epsilon_1, \epsilon_2) = 0$. Next, take $y_3$ and decompose it relative to the subspace spanned by $\{\epsilon_1, \epsilon_2\}$ and take the corresponding perpendicular component to be $\epsilon_3$, and so on. For example, the first three steps of the procedure are

$$\epsilon_1 = y_1$$

$$\epsilon_2 = y_2 - E[y_2\epsilon_1]\, E[\epsilon_1\epsilon_1]^{-1}\epsilon_1$$

$$\epsilon_3 = y_3 - E[y_3\epsilon_1]\, E[\epsilon_1\epsilon_1]^{-1}\epsilon_1 - E[y_3\epsilon_2]\, E[\epsilon_2\epsilon_2]^{-1}\epsilon_2$$

and at the $n$th iteration step

$$\epsilon_n = y_n - \sum_{i=1}^{n-1} E[y_n\epsilon_i]\, E[\epsilon_i\epsilon_i]^{-1}\epsilon_i, \qquad 2 \leq n \leq M \qquad (1.5.11)$$

The basis $\{\epsilon_1, \epsilon_2, \ldots, \epsilon_M\}$ generated in this way is orthogonal by construction. The Gram-Schmidt process may be understood in terms of a hierarchy of subspaces defined by

$$Y_1 = \{\epsilon_1\} = \{y_1\}$$

$$Y_2 = \{\epsilon_1, \epsilon_2\} = \{y_1, y_2\}$$

$$Y_3 = \{\epsilon_1, \epsilon_2, \epsilon_3\} = \{y_1, y_2, y_3\}$$

$$\vdots$$

$$Y_n = \{\epsilon_1, \epsilon_2, \ldots, \epsilon_n\} = \{y_1, y_2, \ldots, y_n\}$$

for $n = 1, 2, \ldots, M$, where each is a subspace of the next one and differs from the next by the addition of one more basis vector. The second term in Eq.

(1.5.11) may be recognized now as the component of $y_n$ parallel to the subspace $Y_{n-1}$. We may denote this as

$$\hat{y}_{n/n-1} = \sum_{i=1}^{n-1} E[y_n \epsilon_i] \, E[\epsilon_i \epsilon_i]^{-1} \epsilon_i \qquad (1.5.12)$$

Then Eq. (1.5.11) may be written as

$$\epsilon_n = y_n - \hat{y}_{n/n-1} \qquad \text{or} \qquad y_n = \hat{y}_{n/n-1} + \epsilon_n \qquad (1.5.13)$$

which represents the *orthogonal decomposition* of $y_n$ relative to the subspace $Y_{n-1}$. Since, the term $\hat{y}_{n/n-1}$ already lies in $Y_{n-1}$, we have the direct sum decomposition

$$Y_n = Y_{n-1} \oplus \{y_n\} = Y_{n-1} \oplus \{\epsilon_n\}$$

Introducing the notation

$$b_{ni} = E[y_n \epsilon_i] \, E[\epsilon_i \epsilon_i]^{-1} \qquad \text{for } 1 \leq i \leq n - 1 \qquad (1.5.14)$$

and

$$b_{nn} = 1$$

we may write Eq. (1.5.13) in the form

$$y_n = \sum_{i=1}^{n} b_{ni} \epsilon_i = \epsilon_n + \sum_{i=1}^{n-1} b_{ni} \epsilon_i = \epsilon_n + \hat{y}_{n/n-1} \qquad (1.5.15)$$

for $1 \leq n \leq M$. And in matrix form,

$$\mathbf{y} = B \boldsymbol{\epsilon} \qquad \text{where } \mathbf{y} = \begin{bmatrix} y_1 \\ y_2 \\ \vdots \\ y_M \end{bmatrix} \text{ and } \boldsymbol{\epsilon} = \begin{bmatrix} \epsilon_1 \\ \epsilon_2 \\ \vdots \\ \epsilon_M \end{bmatrix} \qquad (1.5.16)$$

and $B$ is a *lower-triangular* matrix with matrix elements given by (1.5.14). Its main diagonal is *unity*. For example, for $M = 4$ we have

$$\begin{bmatrix} y_1 \\ y_2 \\ y_3 \\ y_4 \end{bmatrix} = \begin{bmatrix} 1 & 0 & 0 & 0 \\ b_{21} & 1 & 0 & 0 \\ b_{31} & b_{32} & 1 & 0 \\ b_{41} & b_{42} & b_{43} & 1 \end{bmatrix} \begin{bmatrix} \epsilon_1 \\ \epsilon_2 \\ \epsilon_3 \\ \epsilon_4 \end{bmatrix}$$

Both the matrix $B$ and its inverse $B^{-1}$ are unit lower-triangular matrices. The information contained in the two bases $\mathbf{y}$ and $\boldsymbol{\epsilon}$ is the *same*. Going from the basis $\mathbf{y}$ to the basis $\boldsymbol{\epsilon}$ removes all the *redundant correlations* that may exist in $\mathbf{y}$ and "distills" the essential information contained in $\mathbf{y}$ to its most basic form.

Because the basis $\boldsymbol{\epsilon}$ is uncorrelated, every basis vector $\epsilon_i$; $i = 1, \ldots ,M$ will represent something different, or new. Therefore, the random variables $\epsilon_i$ are sometimes called the *innovations*, and the representation (1.5.16) of $\mathbf{y}$ in terms of $\boldsymbol{\epsilon}$, is called the *innovations representation* of $\mathbf{y}$.

Since the correlation matrix $R_{\epsilon\epsilon} = E[\boldsymbol{\epsilon}\boldsymbol{\epsilon}^T]$ is diagonal, the transformation (1.5.16) corresponds to a *LU (lower-upper)-Cholesky factorization* of the correlation matrix of $\mathbf{y}$; that is,

$$R_{yy} = E[\mathbf{y}\mathbf{y}^T] = BE[\boldsymbol{\epsilon}\boldsymbol{\epsilon}^T] B^T = BR_{\epsilon\epsilon}B^T \tag{1.5.17}$$

We note also the *invariance* of the projected vector $\hat{\mathbf{x}}$ of Eq. (1.5.10) under such *change of basis*:

$$\hat{\mathbf{x}} = E[\mathbf{x}\boldsymbol{\epsilon}^T] E[\boldsymbol{\epsilon}\boldsymbol{\epsilon}^T]^{-1}\boldsymbol{\epsilon} = E[\mathbf{x}\mathbf{y}^T] E[\mathbf{y}\mathbf{y}^T]^{-1}\mathbf{y} \tag{1.5.18}$$

This shows the *equivalence* of the orthogonal decompositions (1.5.10) to the correlation canceler decompositions (1.4.1). We may also apply the property (1.5.18) to $y_n$ itself. Defining the vectors

$$\boldsymbol{\epsilon}_{n-1} = \begin{bmatrix} \epsilon_1 \\ \epsilon_2 \\ \vdots \\ \epsilon_{n-1} \end{bmatrix} \qquad \mathbf{y}_{n-1} = \begin{bmatrix} y_1 \\ y_2 \\ \vdots \\ y_{n-1} \end{bmatrix}$$

we may write the projection $\hat{y}_{n/n-1}$ of $y_n$ on the subspace $Y_{n-1}$ given by Eq. (1.5.12) as follows:

$$\hat{y}_{n/n-1} = E[y_n\boldsymbol{\epsilon}_{n-1}^T] E[\boldsymbol{\epsilon}_{n-1}\boldsymbol{\epsilon}_{n-1}^T]^{-1}\boldsymbol{\epsilon}_{n-1}$$

$$= E[y_n \mathbf{y}_{n-1}^T] E[\mathbf{y}_{n-1}\mathbf{y}_{n-1}^T]^{-1}\mathbf{y}_{n-1} \tag{1.5.19}$$

Equation (1.5.13) is then written as

$$\epsilon_n = y_n - \hat{y}_{n/n-1} = y_n - E[y_n\mathbf{y}_{n-1}^T] E[\mathbf{y}_{n-1}\mathbf{y}_{n-1}^T]^{-1}\mathbf{y}_{n-1} \tag{1.5.20}$$

which provides a construction of $\epsilon_n$ directly in terms of the $y_n$s. We note that the quantity $\hat{y}_{n/n-1}$ is also the *best linear estimate* of $y_n$ that can be made on the basis of the *previous* $y_n$s, $Y_{n-1} = \{y_1, y_2, \ldots, y_{n-1}\}$. If the index $n$ represents the time index, as it does for random signals, then $\hat{y}_{n/n-1}$ is the best linear *prediction* of $y_n$ on the basis of *its past*; and $\epsilon_n$ is the corresponding *prediction error*.

*Example 1.5.1:* Consider the three random variables $\{y_1, y_2, y_3\}$, and let $R_{ij} = E[y_i y_j]$ for $i,j = 1,2,3$, denote their correlation matrix. Then, the explicit construction indicated in Eq. (1.5.20) can be carried out as follows: The required vectors $\mathbf{y}_{n-1}$ are

$$\mathbf{y}_1 = [y_1], \qquad \mathbf{y}_2 = \begin{bmatrix} y_1 \\ y_2 \end{bmatrix}$$

and hence

$$E[\mathbf{y}_2\,\mathbf{y}_1^T] = E[\mathbf{y}_2\,y_1] = R_{21}$$

$$E[\mathbf{y}_1\,\mathbf{y}_1^T] = E[\mathbf{y}_1\,y_1] = R_{11}$$

$$E[\mathbf{y}_3\,\mathbf{y}_2^T] = E[\mathbf{y}_3\,\overbrace{y_1,y_2}] = [R_{31},R_{32}]$$

$$E[\mathbf{y}_2\,\mathbf{y}_2^T] = E\left[\binom{y_1}{y_2}\overbrace{y_1,y_2}\right] = \begin{bmatrix} R_{11} & R_{12} \\ R_{21} & R_{22} \end{bmatrix}$$

Therefore, Eq. (1.5.20) becomes

$$\epsilon_1 = y_1$$

$$\epsilon_2 = y_2 - R_{21}\,R_{11}^{-1}\,y_1$$

$$\epsilon_3 = y_3 - [R_{31},R_{32}] \begin{bmatrix} R_{11} & R_{12} \\ R_{21} & R_{22} \end{bmatrix}^{-1} \begin{bmatrix} y_1 \\ y_2 \end{bmatrix}$$

## 1.6  *Partial Correlations*

A concept intimately connected to the Gram-Schmidt orthogonalization is that of the partial correlation. It plays a central role in linear prediction applications.

Consider the Gram-Schmidt orthogonalization of a random vector $\mathbf{y}$ in the form $\mathbf{y} = B\boldsymbol{\epsilon}$, where $B$ is a unit lower-triangular matrix, and $\boldsymbol{\epsilon}$ is a vector of mutually uncorrelated components. Inverting, we have

$$\boldsymbol{\epsilon} = A\mathbf{y} \tag{1.6.1}$$

where $A = B^{-1}$. Now, suppose the vector $\mathbf{y}$ is arbitrarily subdivided into three subvectors as follows:

$$\mathbf{y} = \begin{bmatrix} \mathbf{y}_0 \\ \mathbf{y}_1 \\ \mathbf{y}_2 \end{bmatrix}$$

where $\mathbf{y}_0$, $\mathbf{y}_1$, $\mathbf{y}_2$ do not necessarily have the same dimension. Then, the matrix equation (1.6.1) may also be decomposed in a block-compatible form:

$$\begin{bmatrix} \boldsymbol{\epsilon}_0 \\ \boldsymbol{\epsilon}_1 \\ \boldsymbol{\epsilon}_2 \end{bmatrix} = \begin{bmatrix} A_{00} & 0 & 0 \\ A_{11} & A_{10} & 0 \\ A_{22} & A_{21} & A_{20} \end{bmatrix} \begin{bmatrix} \mathbf{y}_0 \\ \mathbf{y}_1 \\ \mathbf{y}_2 \end{bmatrix} \tag{1.6.2}$$

where $A_{00}$, $A_{10}$, $A_{20}$ are *unit lower-triangular* matrices. Since $\mathbf{y}$ has components which are generally correlated with each other, it follows that $\mathbf{y}_0$ will be correlated with $\mathbf{y}_1$, and $\mathbf{y}_1$ will be correlated with $\mathbf{y}_2$. Thus, through the intermediate action of $\mathbf{y}_1$, the vector $\mathbf{y}_0$ will be indirectly coupled with the vector $\mathbf{y}_2$. The question we would like to ask is this: Suppose the effect of the intermediate vector $\mathbf{y}_1$

were to be removed, then what would be the correlation that is left between $\mathbf{y}_0$ and $\mathbf{y}_2$? This is the *partial correlation*. It represents the "true", or "direct" influence of $\mathbf{y}_0$ on $\mathbf{y}_2$, when the indirect influence via $\mathbf{y}_1$ is removed. To remove the effect of $\mathbf{y}_1$, we project both $\mathbf{y}_0$ and $\mathbf{y}_2$ on the subspace spanned by $\mathbf{y}_1$ and then subtract these parts from both; that is, let

$$\mathbf{e}_0 = \mathbf{y}_0 - (\text{projection of } \mathbf{y}_0 \text{ on } \mathbf{y}_1)$$

$$\mathbf{e}_2 = \mathbf{y}_2 - (\text{projection of } \mathbf{y}_2 \text{ on } \mathbf{y}_1)$$

or

$$\mathbf{e}_0 = \mathbf{y}_0 - R_{01} R_{11}^{-1} \mathbf{y}_1 \tag{1.6.3}$$

$$\mathbf{e}_2 = \mathbf{y}_2 - R_{21} R_{11}^{-1} \mathbf{y}_1$$

where we defined $R_{ij} = E[\mathbf{y}_i \, \mathbf{y}_j^T]$; for $i,j = 0,1,2$. We define the *partial correlation* (PARCOR) *coefficient* between $\mathbf{y}_0$ and $\mathbf{y}_2$, with the effect of the intermediate $\mathbf{y}_1$ removed, as follows:

$$\Gamma = E[\mathbf{e}_2 \, \mathbf{e}_0^T] \, E[\mathbf{e}_0 \, \mathbf{e}_0^T]^{-1} \tag{1.6.4}$$

Then, $\Gamma$ may be expressed in terms of the entries of the matrix $A$ as follows:

$$\Gamma = -A_{20}^{-1} A_{22} \tag{1.6.5}$$

To prove this result, we consider the last equation of (1.6.2):

$$\boldsymbol{\epsilon}_2 = A_{22} \, \mathbf{y}_0 + A_{21} \, \mathbf{y}_1 + A_{20} \, \mathbf{y}_2 \tag{1.6.6}$$

By construction, $\boldsymbol{\epsilon}_2$ is orthogonal to $\mathbf{y}_1$, so that $E[\boldsymbol{\epsilon}_2 \, \mathbf{y}_1^T] = 0$. Thus we obtain the relationship:

$$\begin{aligned} E[\boldsymbol{\epsilon}_2 \, \mathbf{y}_1^T] &= A_{22} \, E[\mathbf{y}_0 \, \mathbf{y}_1^T] + A_{21} \, E[\mathbf{y}_1 \, \mathbf{y}_1^T] + A_{20} \, E[\mathbf{y}_2 \, \mathbf{y}_1^T] \\ &= A_{22} \, R_{01} + A_{21} \, R_{11} + A_{20} \, R_{21} = 0 \end{aligned} \tag{1.6.7}$$

Using Eqs. (1.6.3) and (1.6.7), we may express $\boldsymbol{\epsilon}_2$ in terms of $\mathbf{e}_0$ and $\mathbf{e}_2$, as follows:

$$\begin{aligned} \boldsymbol{\epsilon}_2 &= A_{22} \, (\mathbf{e}_0 + R_{01} R_{11}^{-1} \mathbf{y}_1) + A_{21} \, \mathbf{y}_1 + A_{20} \, (\mathbf{e}_2 + R_{21} R_{11}^{-1} \mathbf{y}_1) \\ &= A_{22} \, \mathbf{e}_0 + A_{20} \, \mathbf{e}_2 + (A_{22} \, R_{01} + A_{21} \, R_{11} + A_{20} \, R_{21}) \, R_{11}^{-1} \mathbf{y}_1 \\ &= A_{22} \, \mathbf{e}_0 + A_{20} \, \mathbf{e}_2 \end{aligned} \tag{1.6.8}$$

Now, by construction, $\boldsymbol{\epsilon}_2$ is orthogonal to both $\mathbf{y}_0$ and $\mathbf{y}_1$, and hence also to $\mathbf{e}_0$; that is, $E[\boldsymbol{\epsilon}_2 \, \mathbf{e}_0^T] = 0$. Using Eq. (1.6.8) we obtain

$$E[\boldsymbol{\epsilon}_2 \, \mathbf{e}_0^T] = A_{22} \, E[\mathbf{e}_0 \, \mathbf{e}_0^T] + A_{20} \, E[\mathbf{e}_2 \, \mathbf{e}_0^T] = 0$$

from which Eq. (1.6.5) follows. It is interesting also to note that Eq. (1.6.8) may be written as

$$\boldsymbol{\epsilon}_2 = A_{20} \, \mathbf{e}$$

where $\mathbf{e} = \mathbf{e}_2 - \Gamma \, \mathbf{e}_0$ is the orthogonal complement of $\mathbf{e}_2$ relative to $\mathbf{e}_0$.

*Example 1.6.1:* An important special case of Eq. (1.6.5) is when $\mathbf{y}_0$ and $\mathbf{y}_2$ are selected as the first and last components of $\mathbf{y}$, and therefore $\mathbf{y}_1$ consists of *all the intermediate* components. For example, suppose $\mathbf{y} = [y_0, y_1, y_2, y_3, y_4]^T$. Then, the decomposition (1.6.2) can be written as follows:

$$
\begin{bmatrix} \epsilon_0 \\ \epsilon_1 \\ \epsilon_2 \\ \epsilon_3 \\ \epsilon_4 \end{bmatrix} =
\left[ \begin{array}{c|cccc|c} 1 & 0 & 0 & 0 & 0 \\ \hline a_{11} & 1 & 0 & 0 & 0 \\ a_{22} & a_{21} & 1 & 0 & 0 \\ a_{33} & a_{32} & a_{31} & 1 & 0 \\ \hline a_{44} & a_{43} & a_{42} & a_{41} & 1 \end{array} \right]
\begin{bmatrix} y_0 \\ y_1 \\ y_2 \\ y_3 \\ y_4 \end{bmatrix}
\tag{1.6.9}
$$

where $\mathbf{y}_0$, $\mathbf{y}_1$, and $\mathbf{y}_2$ are chosen as the vectors

$$
\mathbf{y}_0 = [y_0], \qquad \mathbf{y}_1 = \begin{bmatrix} y_1 \\ y_2 \\ y_3 \end{bmatrix}, \qquad \mathbf{y}_2 = [y_4]
$$

The matrices $A_{20}$ and $A_{22}$ are in this case the scalars $A_{20} = [1]$ and $A_{22} = [a_{44}]$. Therefore, the corresponding PARCOR coefficient (1.6.5) is

$$
\Gamma = -a_{44}
$$

Clearly, the first column $[1, a_{11}, a_{22}, a_{33}, a_{44}]^T$ of $A$ contains all the *lower order* PARCOR coefficients; that is, the quantity

$$
\gamma_p = -a_{pp} \qquad p = 1, 2, 3, \ldots
$$

represents the partial correlation coefficient between $y_0$ and $y_p$, with the effect of all the intermediate variables $y_1, y_2, \ldots, y_{p-1}$ removed.

We note, in passing, the "backward" indexing of the entries of the matrix $A$ in Eqs. (1.6.2) and (1.6.9). It corresponds to writing $\epsilon_n$ in a convolutional form

$$
\epsilon_n = \sum_{i=0}^{n} a_{ni}\, y_{n-i} = \sum_{i=0}^{n} a_{n,n-i}\, y_i
\tag{1.6.10}
$$
$$
= y_n + a_{n1}\, y_{n-1} + a_{n2}\, y_{n-2} + \cdots + a_{nn}\, y_0
$$

and conforms to standard notation in linear prediction applications. Comparing Eq. (1.6.10) with Eq. (1.5.13), we note that the projection of $y_n$ onto the subspace $Y_{n-1}$ may also be expressed directly in terms of the correlated basis $Y_{n-1} = \{y_0, y_1, \ldots, y_{n-1}\}$ as follows:

$$
\hat{y}_{n/n-1} = -[a_{n1}\, y_{n-1} + a_{n2}\, y_{n-2} + \cdots + a_{nn}\, y_0]
\tag{1.6.11}
$$

An alternative expression was given in Eq. (1.5.19).

The ideas discussed in the last three sections are basic in the development of optimum signal processing algorithms, and will be pursued further in subsequent chapters. However, taking a brief look ahead, we point out how some of these concepts fit into the signal processing context:

1. The correlation canceling/orthogonal decompositions of Eqs. (1.4.1) and (1.5.10) form the basis of optimum Wiener and Kalman filtering.
2. The Gram-Schmidt process expressed by Eqs. (1.5.13) and (1.5.20) forms the basis of linear prediction and is also used in the development of the Kalman filter.
3. The representation $\mathbf{y} = B\boldsymbol{\epsilon}$ may be thought of as a signal model for synthesizing $\mathbf{y}$ by processing the uncorrelated (white noise) vector $\boldsymbol{\epsilon}$ through the linear filter $B$. The lower-triangular nature of $B$ is equivalent to causality. Such signal models have a very broad range of applications, among which are speech synthesis and modern methods of spectrum estimation.
4. The inverse representation $\boldsymbol{\epsilon} = A\mathbf{y}$ of Eqs. (1.6.1) and (1.6.10) corresponds to the analysis filters of linear prediction. The PARCOR coefficients will turn out to be the reflection coefficients of the lattice filter realizations of linear prediction.
5. The Cholesky factorization (1.5.17) is the matrix analog of the spectral factorization theorem. It not only facilitates the solution of optimum Wiener filtering problems, but also the making of signal models of the type of Eq. (1.5.16).

## 1.7  *Random Signals*

A random signal (random process, or stochastic process) is defined as a sequence of random variables $\{x_0, x_1, x_2, \ldots, x_n, \ldots\}$ where the index $n$ is taken to be the time. The statistical description of so many random variables is very complicated since it requires knowledge of all the joint densities

$$p(x_0, x_1, \ldots, x_n)$$

for each $n = 0, 1, 2, \ldots$.

If the mean $E[x_n]$ of the random signal is not zero, it can be removed by redefining a new signal $x_n - E[x_n]$. From now on, we will assume that this has been done, and shall work with *zero-mean* random signals. The *autocorrelation function* is defined as

$$R_{xx}(n,m) = E[x_n x_m]; \qquad n,m = 0,1,2, \ldots$$

Sometimes it will be convenient to think of the random signal as a (possibly infinite) random vector $\mathbf{x} = [x_0, x_1, \ldots, x_n, \ldots]^T$, and of the autocorrelation function as a (possibly infinite) matrix $R_{xx} = E[\mathbf{x}\mathbf{x}^T]$. $R_{xx}$ is positive semi-definite and symmetric. The autocorrelation function may also be written as

$$R_{xx}(n + k,n) = E[x_{n+k} x_n] \tag{1.7.1}$$

It provides a *measure* of the influence of the sample $x_n$ on the sample $x_{n+k}$, which lies in the future (if $k > 0$) by $k$ units of time. The relative time separation $k$ of the two samples is called *the lag*.

If the signal $x_n$ is *stationary* (or wide-sense stationary), then the above average is *independent* of the absolute time $n$, and is a function only of the *relative lag* $k$; abusing somewhat the above notation, we may write in this case:

$$R_{xx}(k) = E[x_{n+k} \, x_n] = E[x_{n'+k} \, x_{n'}] \qquad \text{(autocorrelation)} \qquad (1.7.2)$$

In other words, the self-correlation properties of a stationary signal $x_n$ are the same on the average, regardless of when this average is computed. In a way, the stationary random signal $x_n$ looks the same for all times. In this sense, if we take two different blocks of data of length $N$, as shown in Fig. 1.3, we should expect the *average properties*, such as means and autocorrelations, extracted from these blocks of data to be roughly the same. The relative time separation of the two blocks as a whole should not matter.

A direct consequence of stationarity is the *reflection-invariance* of the autocorrelation function $R_{xx}(k)$ of Eq. (1.7.2):

$$R_{xx}(k) = E[x_{n+k} \, x_n] = R_{xx}(-k) \qquad (1.7.3)$$

One way to introduce a systematization of the various types of random signals is the Markov classification into zeroth order Markov, first order Markov, and so on. The simplest possible random signal is the zeroth order Markov, or *purely random signal*, defined by the requirement that all the (zero-mean) random variables $x_0, x_1, \ldots, x_n, \ldots$ be independent of each other and arise from a common density $p(x)$; this implies

$$p(x_0, x_1, \ldots, x_n, \ldots) = p(x_0) \, p(x_1) \ldots p(x_n) \ldots$$
$$R_{xx}(n,m) = E[x_n \, x_m] = 0; \qquad \text{for } n \neq m$$

Such a random signal is stationary. The quantity $R_{xx}(n,n)$ is independent of $n$, and represents the variance of each sample:

$$R_{xx}(0) = E[x_n^2] = \sigma_x^2$$

In this case, the autocorrelation function $R_{xx}(k)$ may be expressed compactly as

$$R_{xx}(k) = E[x_{n+k} \, x_n] = \sigma_x^2 \, \delta(k) \qquad (1.7.4)$$

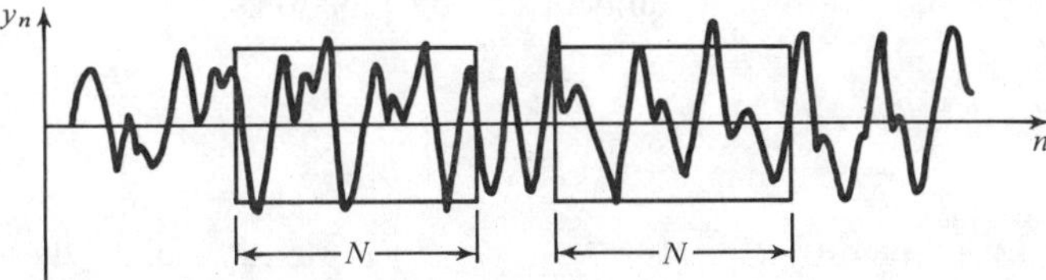

**Figure 1.3**  Blocks of Data from a Stationary Signal

A purely random signal has no memory, as can be seen from the property

$$p(x_n,x_{n-1}) = p(x_n)\,p(x_{n-1}) \qquad \text{or} \qquad p(x_n/x_{n-1}) = p(x_n)$$

That is, the occurrence of $x_{n-1}$ at time instant $n-1$ does not in any way affect, or restrict, the values of $x_n$ at the next time instant. Successive signal values are entirely independent of each other. Past values do not influence future values. No memory is retained from sample to sample; the next sample will take a value regardless of the value that the previous sample has already taken. Since successive samples are random, such a signal will exhibit very rapid time variations. But it will also exhibit slow time variations. Such time variations are best discussed in the frequency domain. This will lead directly to frequency concepts, power spectra, periodograms, and the like. It is expected that a purely random signal will contain all frequencies, from the very low to the very high, in equal proportions (white noise).

The next least complicated signal is the first order Markov signal, which has memory only of one sampling instant. Such a signal "remembers" only the previous sample. It is defined by the requirement that

$$p(x_n/x_{n-1},x_{n-2}, \ldots ,x_1, x_0) = p(x_n/x_{n-1})$$

which states that $x_n$ may be influenced directly only by the previous sample value $x_{n-1}$, and not by the samples $x_{n-2}, \ldots ,x_0$ that are further in the past. The complete statistical description of such random signal is considerably simplified. It is sufficient to know only the marginal densities $p(x_n)$ and the conditional densities $p(x_n/x_{n-1})$. Any other joint density may be constructed in terms of these. For instance,

$$
\begin{aligned}
p(x_3,x_2,x_1,x_0) &= p(x_3/x_2,x_1,x_0)\,p(x_2,x_1,x_0) && \text{(by Bayes' rule)} \\
&= p(x_3/x_2)\,p(x_2,x_1,x_0) && \text{(by the Markov property)} \\
&= p(x_3/x_2)\,p(x_2/x_1,x_0)\,p(x_1,x_0) \\
&= p(x_3/x_2)\,p(x_2/x_1)\,p(x_1,x_0) \\
&= p(x_3/x_2)\,p(x_2/x_1)\,p(x_1/x_0)\,p(x_0)
\end{aligned}
$$

## 1.8 *Power Spectrum and Its Interpretation*

The *power spectral density* of a *stationary* random signal $x_n$ is defined as the double-sided $z$-transform of its autocorrelation function

$$S_{xx}(z) = \sum_{k=-\infty}^{\infty} R_{xx}(k)\,z^{-k} \tag{1.8.1}$$

where $R_{xx}(k)$ is given by Eq. (1.7.2). If $R_{xx}(k)$ is strictly stable, the region of convergence of $S_{xx}(z)$ will include the unit circle in the complex $z$-plane. This allows us to define the *power spectrum* $S_{xx}(\omega)$ of the random signal $x_n$ by setting $z = \exp(j\omega)$ in Eq. (1.8.1); abusing the notation somewhat, we have in this case

$$S_{xx}(\omega) = \sum_{k=-\infty}^{\infty} R_{xx}(k)\, e^{-j\omega k} \qquad \text{(power spectrum)} \qquad (1.8.2)$$

This quantity conveys very useful information. It is a measure of the frequency content of the signal $x_n$ and of the distribution of the power of $x_n$ over frequency. To see this, consider the inverse $z$-transform

$$R_{xx}(k) = \oint_{u.c.} S_{xx}(z)\, z^k\, \frac{dz}{2\pi j z} \qquad (1.8.3)$$

where, since $R_{xx}(k)$ is stable, the integration contour may be taken to be the unit circle. Using $z = \exp(j\omega)$, we find for the integration measure

$$\frac{dz}{2\pi j z} = \frac{d\omega}{2\pi}$$

Thus, Eq. (1.8.3) may also be written as an inverse Fourier transform

$$R_{xx}(k) = \int_{-\pi}^{\pi} S_{xx}(\omega)\, e^{j\omega k}\, \frac{d\omega}{2\pi} \qquad (1.8.4)$$

In particular, the variance of $x_n$ can be written as

$$R_{xx}(0) = \sigma_x^2 = E[x_n^2] = \int_{-\pi}^{\pi} S_{xx}(\omega)\, \frac{d\omega}{2\pi} \qquad (1.8.5)$$

Since the quantity $E[x_n^2]$ represents the average *total power* contained in $x_n$, it follows that $S_{xx}(\omega)$ will represent the *power per unit frequency interval*. A typical power spectrum is depicted in Fig. 1.4. As suggested by this figure, it is possible for the power to be mostly concentrated about some frequencies and not about others. The area under the curve represents the total power of the signal $x_n$.

If $x_n$ is an uncorrelated (white-noise) random signal with a delta-function

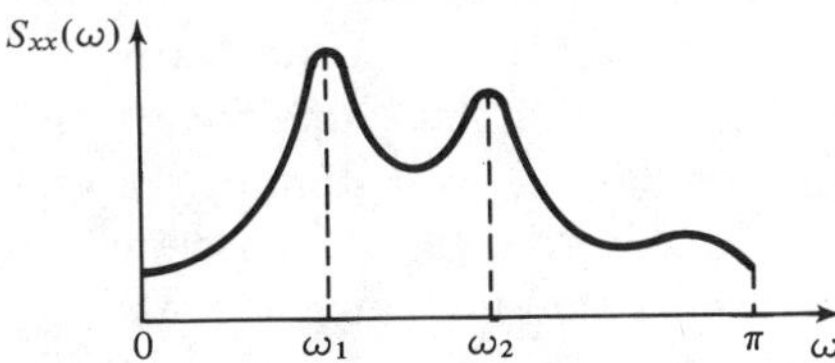


**Figure 1.4**  Typical Power Spectrum

autocorrelation, given by Eq. (1.7.4), it will have a *flat* power spectrum with power level equal to the variance $\sigma_x^2$:

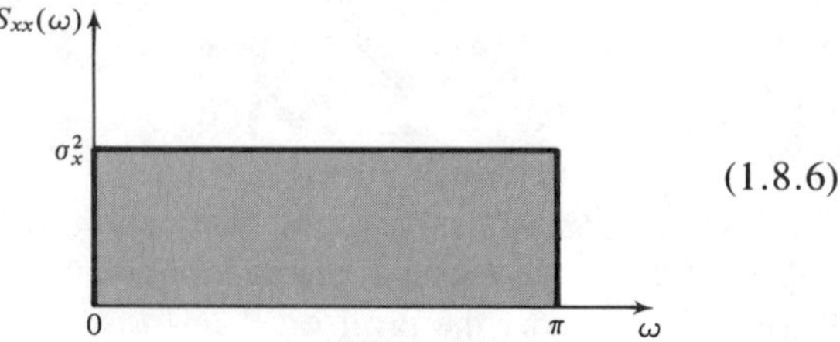

$$(1.8.6)$$

Another useful concept is that of the *cross-correlation* and *cross-spectrum* between two stationary random sequences $x_n$ and $y_n$. These are defined by

$$R_{yx}(k) = E[y_{n+k} x_n] \qquad S_{yx}(z) = \sum_{k=-\infty}^{\infty} R_{yx}(k) z^{-k} \qquad (1.8.7)$$

Using stationarity, it is easy to show the reflection symmetry property

$$R_{yx}(k) = R_{xy}(-k) \qquad (1.8.8)$$

that is analogous to Eq. (1.7.3). In the $z$-domain, the reflection symmetry properties (1.7.3) and (1.8.8) are translated into:

$$S_{xx}(z) = S_{xx}(z^{-1}) \qquad S_{yx}(z) = S_{xy}(z^{-1}) \qquad (1.8.9)$$

respectively; and also

$$S_{xx}(\omega) = S_{xx}(-\omega) \qquad S_{yx}(\omega) = S_{xy}(-\omega) \qquad (1.8.10)$$

## 1.9 *Sample Autocorrelation and the Periodogram*

From now on we will work mostly with stationary random signals. If a block of $N$ signal samples is available, we will assume that it is a segment from a stationary signal. The length $N$ of the available data segment is an important consideration. For example, in computing frequency spectra, we know that high resolution in frequency requires a long record of data. However, if the record is too long the assumption of stationarity may no longer be justified. This is the case in many applications, as for example in speech and EEG signal processing. The speech waveform does not remain stationary for long time intervals. It may be assumed to be stationary only for short time intervals. Such a signal may be called *piece-wise stationary*. If it is divided into short segments of duration of 20–30 milliseconds, then the portion of speech within each segment may be assumed to be a segment from a stationary signal. A typical piece-wise stationary signal is depicted in Fig. 1.5.

The main reason for assuming stationarity, or piece-wise stationarity, is that

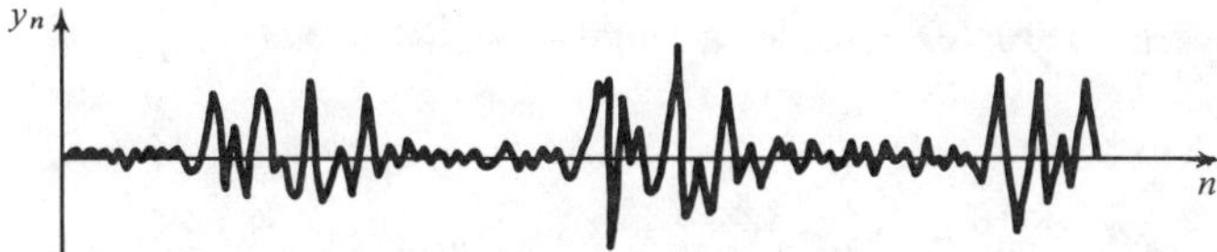

**Figure 1.5**  Piece-Wise Stationary Signal

most of our methods of handling random signals depend heavily on this assumption. For example, the statistical autocorrelations based on the ensemble averages (1.7.2) may be replaced in practice by *time averages*. This can be justified only if the signals are stationary (actually, they must be ergodic). If the underlying signal processes are not stationary (and therefore definitely not ergodic) we cannot use time averages. If a signal is piece-wise stationary and divided into stationary blocks, then for each such block, ensemble averages may be replaced by time averages. The time average approximation of an autocorrelation function is called the *sample autocorrelation* and is defined as follows: Given a block of length $N$ of measured signal samples

$$y_0, y_1, y_2, \cdots, y_{N-1}$$

define

$$\hat{R}_{yy}(k) = \frac{1}{N} \sum_{n=0}^{N-1-k} y_{n+k}\, y_n \qquad \text{for } 0 \leqslant k \leqslant (N-1) \qquad (1.9.1)$$

and

$$\hat{R}_{yy}(k) = \hat{R}_{yy}(-k) \qquad \text{for } -(N-1) \leqslant k \leqslant -1$$

The *periodogram* of this record of data is defined as the (double-sided) $z$-transform of the sample autocorrelation

$$\hat{S}_{yy}(z) = \sum_{k=-(N-1)}^{N-1} \hat{R}_{yy}(k)\, z^{-k} \qquad (1.9.2)$$

It may be thought of as an approximation (estimate) of the true power spectral density $S_{yy}(z)$. It is easily shown that the periodogram may be expressed in terms of the $z$-transform of the data sequence itself, as

$$\hat{S}_{yy}(z) = \frac{1}{N}\, Y(z)\, Y(z^{-1}) \qquad (1.9.3)$$

where

$$Y(z) = \sum_{n=0}^{N-1} y_n\, z^{-n} \qquad (1.9.4)$$

As a concrete example, consider a length-3 signal $\mathbf{y} = (y_0, y_1, y_2)$. Then,

$$
\begin{aligned}
Y(z)\, Y(z^{-1}) &= (y_0 + y_1\, z^{-1} + y_2\, z^{-2})\, (y_0 + y_1\, z + y_2\, z^2) \\
&= (y_0^2 + y_1^2 + y_2^2) + (y_0\, y_1 + y_1\, y_2)z^{-1} \\
&\quad + (y_0\, y_1 + y_1\, y_2)z + (y_0\, y_2)z^{-2} + (y_0\, y_2)z^2
\end{aligned}
$$

from which we extract the inverse $z$-transform

$$
\hat{R}_{yy}(0) = [y_0^2 + y_1^2 + y_2^2]/3
$$

$$
\hat{R}_{yy}(-1) = \hat{R}_{yy}(1) = [y_0\, y_1 + y_1\, y_2]/3
$$

$$
\hat{R}_{yy}(-2) = \hat{R}_{yy}(2) = [y_0\, y_2]/3
$$

These equations may also be written in a nice matrix form, as follows

$$
\underbrace{
\begin{bmatrix}
\hat{R}_{yy}(0) & \hat{R}_{yy}(1) & \hat{R}_{yy}(2) \\
\hat{R}_{yy}(1) & \hat{R}_{yy}(0) & \hat{R}_{yy}(1) \\
\hat{R}_{yy}(2) & \hat{R}_{yy}(1) & \hat{R}_{yy}(0)
\end{bmatrix}
}_{\hat{R}_{yy}}
$$

$$
= \underbrace{
\begin{bmatrix}
y_0 & y_1 & y_2 & 0 & 0 \\
0 & y_0 & y_1 & y_2 & 0 \\
0 & 0 & y_0 & y_1 & y_2
\end{bmatrix}
}_{Y^T}
\underbrace{
\begin{bmatrix}
y_0 & 0 & 0 \\
y_1 & y_0 & 0 \\
y_2 & y_1 & y_0 \\
0 & y_2 & y_1 \\
0 & 0 & y_2
\end{bmatrix}
}_{Y}
\frac{1}{3}; \quad \hat{R}_{yy} = Y^T Y/3
$$

The matrix $\hat{R}_{yy}$ on the left is called the sample autocorrelation matrix. It is a *Toeplitz matrix*; that is, it has the same entry in each diagonal. The right hand side also shows that the autocorrelation matrix is positive definite. In the general case of a length-$N$ sequence $y_n$, the matrix $Y$ has $N$ columns, each a down-shifted (delayed) version of the previous one, corresponding to a total of $N - 1$ delays. This requires the length of each column to be $N + (N - 1)$; that is, there are $2N - 1$ rows. We will encounter again this matrix factorization of the covariance matrix $\hat{R}_{yy}$ in the least-squares design of waveshaping filters.

The sample autocorrelation may also be thought of as *ordinary convolution*. Note that $Y(z^{-1})$ represents the $z$-transform of a signal which is the original signal $\mathbf{y} = (y_0, y_1, \ldots, y_{N-1})$ *reflected* about the time origin. The reflected signal may be made causal by a *delay* of $N - 1$ units of time. The reflected-delayed signal has some significance, and is known as the *reversed signal*. Its $z$-transform is the *reverse polynomial* of $Y(z)$

$$Y^R(z) = z^{-(N-1)}Y(z^{-1})$$

$$(0 \ldots 0 \qquad 0 \; y_0 \quad y_1 \ldots \quad y_{N-2}y_{N-1}) \to Y(z) \qquad\qquad = \text{original}$$

$$(y_{N-1} \, y_{N-2} \ldots y_1 \, y_0 \quad 0 \ldots \quad 0 \quad 0) \quad \to Y(z^{-1}) \qquad\qquad = \text{reflected}$$

$$(0 \ldots 0 \qquad 0 \; y_{N-1} \, y_{N-2} \ldots y_1 \quad y_0) \quad \to z^{-(N-1)}Y(z^{-1}) = \text{reversed}$$

The periodogram is expressed then in the form

$$\hat{S}_{yy}(z) = \frac{1}{N} Y(z)Y(z^{-1}) = \frac{1}{N} Y(z)Y^R(z) \, z^{(N-1)}$$

which implies that $\hat{R}_{yy}(k)$ may be obtained by convolving the original data sequence with the reversed sequence and then advancing the result in time by $N-1$ time units. This is seen by the following convolution table.

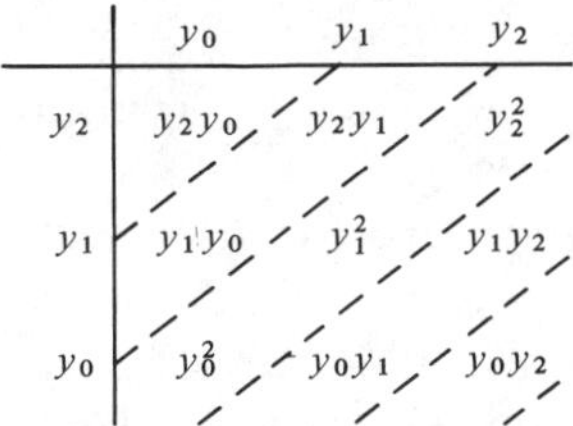

The *periodogram spectrum* is obtained by substituting $z = e^{j\omega}$

$$\hat{S}_{yy}(\omega) = \frac{1}{N} \, | \, Y(\omega) \, |^2 = \frac{1}{N} \left| \sum_{n=0}^{N-1} y_n \, e^{-j\omega n} \right|^2 \qquad (1.9.5)$$

The periodogram spectrum (1.9.5) may be computed efficiently using FFT methods. The digital frequency $\omega$ [in radians/sample] is related to the physical frequency $f$ [in Hz] by

$$\omega = 2\pi f T = \frac{2\pi f}{f_s}$$

where $f_s$ is the sampling rate, and $T = 1/f_s$ the time interval between samples. The frequency resolution afforded by a length-$N$ sequence is

$$\Delta\omega = \frac{2\pi}{N} \qquad \text{or} \qquad \Delta f = \frac{f_s}{N} = \frac{1}{NT} = \frac{1}{T_R} \text{ [in Hz]}$$

where $T_R = NT$ is the *duration* of the data record. The periodogram spectrum suffers from two major drawbacks. First, the rectangular windowing of the data segment introduces significant *sidelobe leakage*. This can cause misinterpretation of sidelobe spectral peaks as being part of the true spectrum. And second, it is well-known that the periodogram spectrum is *not* a good (consistent) estimator of the true power spectrum $S_{yy}(\omega)$. The development of methods to *improve* on the periodogram is the subject of *classical spectral analysis* [4–12]. We just mention, in passing, one of the most popular of such methods; namely, *Welch's method* [13]. The given data record of length $N$ is subdivided into $K$ shorter segments which may be overlapping or nonoverlapping. If they are nonoverlapping then each will have length $M = N/K$; if they are 50% overlapping then $M = 2N/(K + 1)$. Each such segment is then windowed by a length-$M$ data window, such as a Hamming window. The window reduces the sidelobe frequency leakage at the expense of resolution. The window $w(n)$ is typically normalized to have unit average energy; that is, $(1/M)\sum_{n=0}^{M-1} w(n)^2 = 1$. The periodogram of each windowed segment is then computed by FFT methods and the $K$ periodograms are averaged together to obtain the spectrum estimate

$$S(\omega) = \frac{1}{K} \sum_{i=1}^{K} S_i(\omega)$$

where $S_i(\omega)$ is the periodogram of the $i$th segment. The above subdivision into segments imitates ensemble averaging, and therefore, it results in a spectrum estimate of improved statistical stability. However, since each periodogram is computed from a length-$M$ sequence, the frequency resolution is reduced from $\Delta\omega = 2\pi/N$ to roughly $\Delta\omega = 2\pi/M$ (for a well-designed window). Therefore, to maintain high frequency resolution (large $M$), as well as improved statistical stability of the spectrum estimate (large $K$), a long data record $N$ is required; a condition that can easily come into conflict with stationarity. The so-called "modern methods" of spectrum estimation, which are based on *parametric signal models*, can provide high resolution spectrum estimates from *short* data records $N$.

## 1.10 *Random Signal Models and Their Uses*

Models that provide a characterization of the properties and nature of random signals are of primary importance in the design of optimum signal processing systems. This section offers an overview of such models and outlines their major applications. Many of the ideas presented here will be developed in greater detail in later chapters.

One of the most useful ways to model a random signal [14] is to consider it as being the *output* of a *causal and stable* linear filter $B(z)$ which is driven by a *stationary uncorrelated* (white-noise) sequence $\{\epsilon_0, \epsilon_1, \ldots, \epsilon_n, \ldots\}$:

$$\epsilon_n \longrightarrow \boxed{B(z)} \longrightarrow y_n \qquad B(z) = \sum_{n=0}^{\infty} b_n z^{-n}$$

where $R_{\epsilon\epsilon}(k) = E[\epsilon_{n+k}\epsilon_n] = \sigma_\epsilon^2 \delta(k)$. The output random signal $y_n$ is obtained by convolving the input sequence $\epsilon_n$ with the filter's impulse response $b_n$:

$$y_n = \sum_{i=0}^{n} b_{n-i}\epsilon_i \qquad n = 0,1,2, \ldots \tag{1.10.1}$$

The stability of the filter $B(z)$ is essential as it guarantees the stationarity of the sequence $y_n$. This point will be discussed later on. By readjusting, if necessary, the value of $\sigma_\epsilon^2$, we may assume that $b_0 = 1$. Then Eq. (1.10.1) corresponds exactly to the Gram-Schmidt form of Eqs. (1.5.15) and (1.5.16), where the matrix elements $b_{ni}$ are given in terms of the impulse response of the filter $B(z)$:

$$b_{ni} = b_{n-i} \tag{1.10.2}$$

In this case, the structure of the matrix $B$ is considerably simplified. Writing the convolutional equation (1.10.1) in matrix form

$$\begin{bmatrix} y_0 \\ y_1 \\ y_2 \\ y_3 \\ y_4 \end{bmatrix} = \begin{bmatrix} 1 & 0 & 0 & 0 & 0 \\ b_1 & 1 & 0 & 0 & 0 \\ b_2 & b_1 & 1 & 0 & 0 \\ b_3 & b_2 & b_1 & 1 & 0 \\ b_4 & b_3 & b_2 & b_1 & 1 \end{bmatrix} \begin{bmatrix} \epsilon_0 \\ \epsilon_1 \\ \epsilon_2 \\ \epsilon_3 \\ \epsilon_4 \end{bmatrix}$$

we observe that the first column of $B$ is the impulse response $b_n$ of the filter. Each subsequent column is a down-shifted (delayed) version of the previous one, and each diagonal has the same entry (i.e., $B$ is a Toeplitz matrix). The lower-triangular nature of $B$ is equivalent to the assumed *causality* of the filter $B(z)$.

Such signal models are quite general. In fact, there is a general theorem by Wold that essentially guarantees the *existence* of such models for any stationary signal $y_n$ [15]. Wold's construction of $B(z)$ is none other than the Gram-Schmidt construction of the orthogonalized basis $\epsilon_n$. However, the practical usage of such models requires further that the transfer function $B(z)$ be *rational*; that is, the ratio of two polynomials in $z$. In this case, the I/O equation (1.10.1) is most conveniently expressed as a *difference equation*.

*Example 1.10.1:* Suppose

$$B(z) = \frac{1 + c_1 z^{-1} + c_2 z^{-2}}{1 + d_1 z^{-1} + d_2 z^{-2}}$$

Then Eq. (1.10.1) is equivalent to the difference equation

$$y_n = -d_1 y_{n-1} - d_2 y_{n-2} + \epsilon_n + c_1 \epsilon_{n-1} + c_2 \epsilon_{n-2}$$

which may be realized as follows.

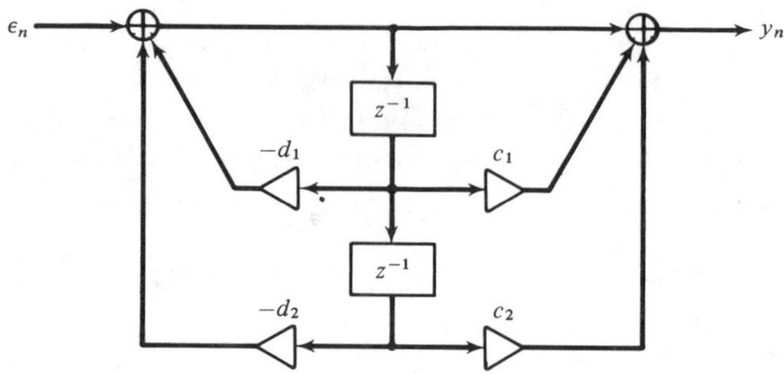


The filter $B(z)$ is called a *synthesis filter* and may be thought of as a random signal generator, or a signal model for the random signal $y_n$. The numerator and denominator coefficients of the filter $B(z)$, and the variance $\sigma_\epsilon^2$ of the input white noise, are referred to as the *model parameters*. For instance, in Example 1.10.1 the model parameters are $\{c_1, c_2, d_1, d_2; \sigma_\epsilon^2\}$. Such parametric models have received a lot of attention in recent years. They are very common in speech and geophysical signal processing, image processing, EEG signal processing, spectrum estimation, data compression, and other time series analysis applications. How are such models used? One of the main objectives in such applications has been to develop appropriate *analysis procedures* for extracting the model parameters (filter coefficients and $\sigma_\epsilon^2$) on the basis of a given set of actual signal samples $y_n$. This is a *system identification* problem. The analysis procedures are designed to provide effectively the *best fit* of the data samples to a particular model. The procedures typically begin with a measured block of signal samples $\{y_0, y_1, \ldots, y_{N-1}\}$—also referred to as an *analysis frame*—and through an appropriate analysis algorithm extract *estimates* of the model parameters. This is depicted in Fig. 1.6.

The given frame of samples $\{y_0, y_1, \ldots, y_{N-1}\}$ is *represented* now by the set of model parameters extracted from it. Following the analysis procedure, the resulting model may be used in a variety of ways. The four major uses of such models are in:

1. Signal synthesis
2. Spectrum estimation
3. Signal classification
4. Data compression

We will discuss each of these briefly. To synthesize a particular realization of the random signal $y_n$, it is only necessary to recall from memory the appropriate

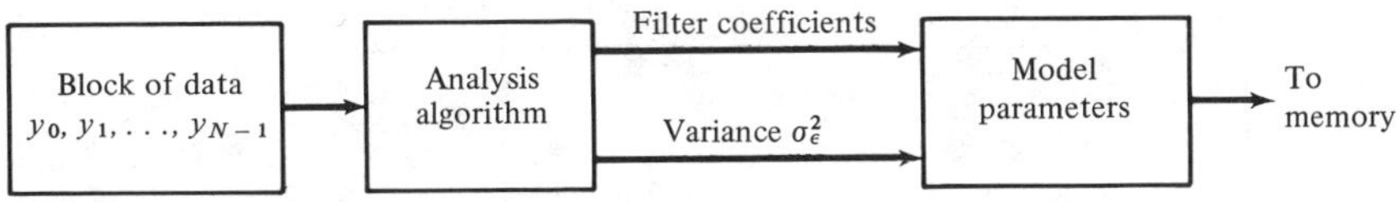

**Figure 1.6**  Analysis Procedure

model parameters, to generate a random uncorrelated sequence $\epsilon_n$ having variance $\sigma_\epsilon^2$, and send it through the filter $B(z)$. Such uncorrelated sequence may be *computer-generated* using a standard random number generator routine. The synthetic signal will appear at the output of the filter. This is shown in Fig. 1.7. This is the basic principle behind most speech synthesis systems. In speech, the synthesis filter $B(z)$ represents a model of the transfer characteristics of the *vocal tract* considered as an acoustic tube. A typical analysis frame of speech has duration of 20 msec. If sampled at a 10-kHz sampling rate, it will consist of $N$ = 200 samples. To synthesize a particular frame of 200 samples, the model parameters representing that frame are recalled from memory, and the synthesis filter is run for 200 sampling instances generating 200 output speech samples, which may be sent to a D/A converter. The next frame of 200 samples can be synthesized by recalling from memory *its* model parameters, and so on. Entire words or sentences can be synthesized in such a piece-wise, or frame-wise, manner. A realistic representation of each speech frame requires the specification of two additional parameters besides the filter coefficients and $\sigma_\epsilon^2$; namely, the *pitch* period and a *voiced/unvoiced* (V/UV) decision. Unvoiced sounds, such as the "sh" in the word "should," have a white-noise sounding nature, and are generated by the turbulent flow of air through constrictions of the vocal tract. Such sounds may be represented adequately by the above random signal models. On the other hand, voiced sounds, such as vowels, are pitched sounds, and have a pitch period associated with them. They may be assumed to be generated by the periodic excitation of the vocal tract by a train of impulses separated by the pitch period. The vocal tract responds to each of these impulses by producing its impulse response, resulting therefore in a quasi-periodic output which is characteristic of such sounds. Thus, depending on the type of sound, the nature of the generator of the excitation input to the synthesis filter will be different; namely, it will be a *random noise generator* for unvoiced sounds, and a *pulse train generator* for voiced sounds. A typical speech synthesis system that incorporates the above features is shown in Fig. 1.8.

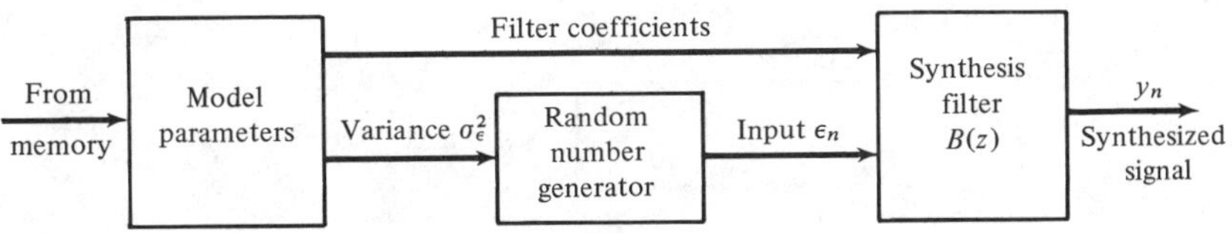

**Figure 1.7**  Signal Synthesis

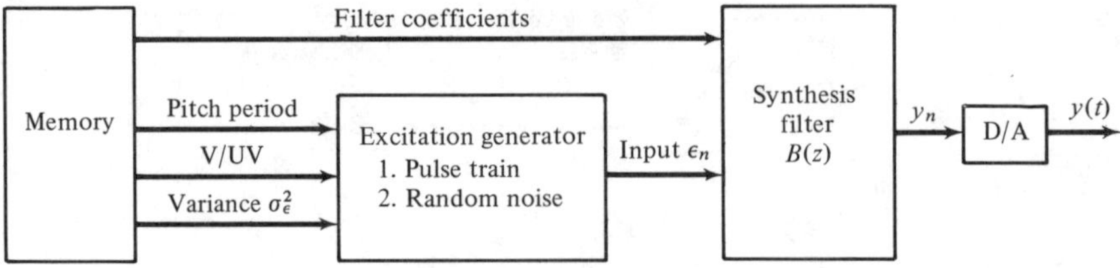

**Figure 1.8**  Typical Speech Synthesis System

Another major application of parametric models is to *spectrum estimation*. This is based on the property that

$$S_{yy}(\omega) = \sigma_\epsilon^2 \, |B(\omega)|^2 \tag{1.10.6}$$

which will be proved later. It states that the spectral shape of the power spectrum $S_{yy}(\omega)$ of the signal $y_n$ arises only from the spectral shape of the model filter $B(\omega)$. For example, the signal $y_n$ generated by the model of Example 1.10.1 will have

$$S_{yy}(\omega) = \sigma_\epsilon^2 \left| \frac{1 + c_1 e^{-j\omega} + c_2 e^{-2j\omega}}{1 + d_1 e^{-j\omega} + d_2 e^{-2j\omega}} \right|^2$$

This approach to spectrum estimation is depicted in Fig. 1.9. The parametric approach to spectrum estimation must be contrasted with the classical approach which is based on *direct computation* of the Fourier transform of the available data record; that is, the periodogram spectrum, or its improvements. The classical periodogram method is shown in Fig. 1.10. As we mentioned in the previous section, spectrum estimates based on such parametric models tend to have much better frequency resolution properties than the classical methods, especially when the length $N$ of the available data record is short.

In *signal classification* applications, such as speech recognition, speaker verification, or EEG pattern classification, the basic problem is to compare two available blocks of data samples and decide whether they belong to the same class or not. One of the two blocks might be a prestored and preanalyzed *reference template* against which the other block is to be compared. Instead of comparing the data records sample by sample, what are compared are the corresponding model parameters extracted from these blocks. In pattern recognition nomenclature, the vector of model parameters is the "feature vector". The closeness of the two sets of model parameters to each other is decided on the basis of an

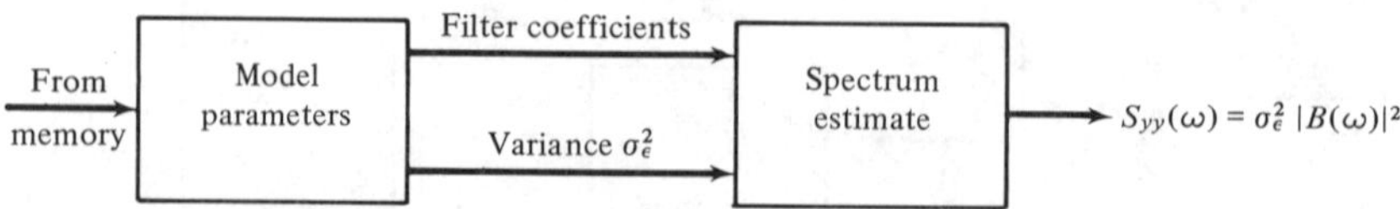

**Figure 1.9**  Spectrum Estimation with Parametric Models

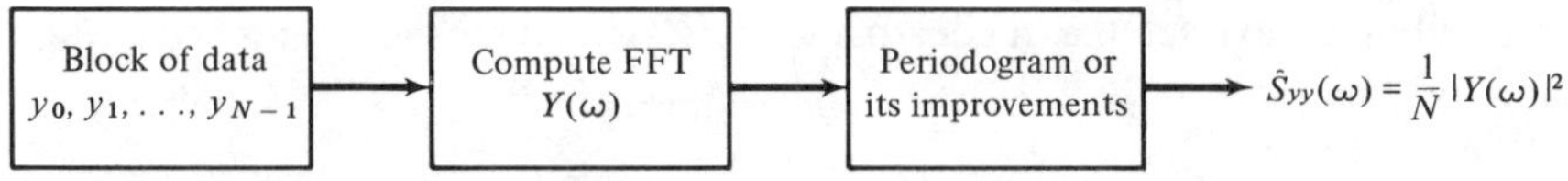

**Figure 1.10**   Classical Spectrum Estimation

appropriate *distance measure*. We will discuss examples of distance measures for speech and EEG signals in Chapter 5. This approach to signal classification is depicted in Fig. 1.11.

Next, we discuss the application of such models to *data compression*. The signal synthesis method described above is a form of data compression because instead of saving the $N$ data samples $y_n$ as such, what are saved are the corresponding model parameters which are typically much *fewer* in number than $N$. For example, in speech synthesis systems a savings of about a factor of 20 in memory may be achieved with this approach. Indeed, as we discussed above, a typical frame of speech consists of 200 samples, whereas the number of model parameters typically needed to represent this frame is about 10 to 15. The main limitation of this approach is that the reproduction of the original signal segment is not exact but depends on the particular realization of the computer-generated input sequence $\epsilon_n$ that drives the model. Speech synthesized in such manner is still intelligible, but it has lost some of its naturalness. Such signal synthesis methods are not necessarily as successful or appropriate in all applications. For example, in image processing, if one makes a parametric model of an image and attempts to "synthesize" it by driving the model with a computer-generated uncorrelated sequence, the reproduced image will bear no resemblance to the original image; not even vaguely so.

For *exact* reproduction, both the model parameters and the entire sequence $\epsilon_n$ must be stored. This would still provide some form of data compression, as will be explained below. Such an approach to data compression is widely used in digital data transmission or digital data storage applications for all types of data, including speech and image data. The method may be described as follows: the given data record $\{y_0, y_1, \ldots, y_{N-1}\}$ is subjected to the appropriate analysis

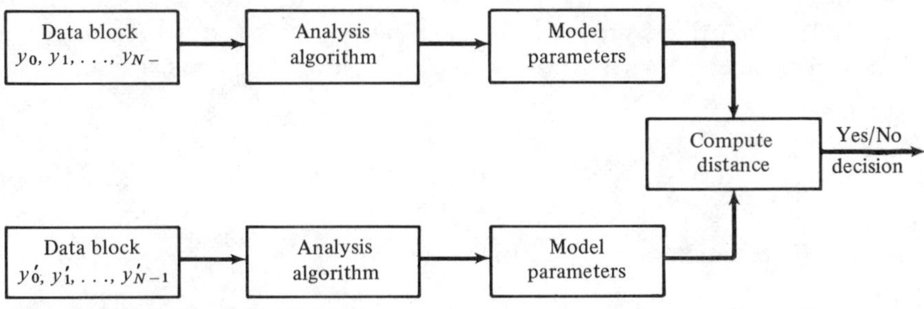

**Figure 1.11**   Signal Classification

algorithm to extract the model parameters, and then the segment is filtered through the *inverse filter*

$$A(z) = \frac{1}{B(z)} \qquad y_n \longrightarrow \boxed{A(z)} \longrightarrow \epsilon_n \qquad (1.10.7)$$

to provide the sequence $\epsilon_n$. The inverse filter $A(z)$ is also known as the *whitening filter*, the *prediction-error filter*, or the *analysis filter*. The resulting sequence $\epsilon_n$ has a compressed dynamic range relative to $y_n$ and therefore it requires fewer number of bits for the representation of each sample $\epsilon_n$. A quantitative measure for the *data compression gain* is given by the ratio $G = \sigma_y^2/\sigma_\epsilon^2$ which is always greater than one. This can be seen easily using Eqs. (1.10.6) and (1.8.5)

$$\sigma_y^2 = \int_{-\pi}^{\pi} S_{yy}(\omega)\, \frac{d\omega}{2\pi} = \sigma_\epsilon^2 \int_{-\pi}^{\pi} |B(\omega)|^2\, \frac{d\omega}{2\pi} = \sigma_\epsilon^2 \sum_{n=0}^{\infty} b_n^2$$

Since $b_0 = 1$, we find

$$G = \frac{\sigma_y^2}{\sigma_\epsilon^2} = \sum_{n=0}^{\infty} b_n^2 = 1 + b_1^2 + b_2^2 + \cdots \qquad (1.10.8)$$

The entire sequence $\epsilon_n$ and the model parameters are then transmitted over the data link, or stored in memory. At the receiving end, the original sequence $y_n$ may be reconstructed exactly using the synthesis filter $B(z)$ driven by $\epsilon_n$. This approach to data compression is depicted in Fig. 1.12. Not shown in Fig. 1.12 are the *quantization* and *encoding* operations that must be performed on $\epsilon_n$ in order to transmit it over the digital channel. An example that properly takes into account the quantization effects will be discussed in more detail in Chapter 2.

Filtering the sequence $y_n$ through the inverse filter $A(z)$ requires that $A(z)$ be stable and causal. If we write $B(z)$ as the ratio of two polynomials

$$B(z) = \frac{N(z)}{D(z)} \qquad (1.10.9)$$

then the stability and causality of $B(z)$ requires that the zeros of the polynomial $D(z)$ lie inside the unit circle in the complex $z$-plane; whereas the stability and causality of the inverse $A(z) = D(z)/N(z)$ requires the zeros of $N(z)$ to be inside the unit circle. Thus, both the poles and the zeros of $B(z)$ must be inside the unit circle. Such filters are called *minimal phase* filters and will be discussed further in Chapter 3. When $A(z)$ is stable and causal it may be expanded in the form

$$A(z) = \sum_{n=0}^{\infty} a_n z^{-n} = 1 + a_1 z^{-1} + a_2 z^{-2} + \cdots \qquad (1.10.10)$$

and the I/O equation of Eq. (1.10.7) becomes

$$\epsilon_n = \sum_{i=0}^{n} a_i y_{n-i} = y_n + a_1 y_{n-1} + a_2 y_{n-2} + \cdots + a_n y_0 \qquad (1.10.11)$$

**Figure 1.12**   Data Compression

for $n = 0,1,2, \ldots$ It may be written in matrix form $\boldsymbol{\epsilon} = A\mathbf{y}$ as

$$
\begin{bmatrix} \epsilon_0 \\ \epsilon_1 \\ \epsilon_2 \\ \epsilon_3 \\ \epsilon_4 \end{bmatrix} = \begin{bmatrix} 1 & 0 & 0 & 0 & 0 \\ a_1 & 1 & 0 & 0 & 0 \\ a_2 & a_1 & 1 & 0 & 0 \\ a_3 & a_2 & a_1 & 1 & 0 \\ a_4 & a_3 & a_2 & a_1 & 1 \end{bmatrix} \begin{bmatrix} y_0 \\ y_1 \\ y_2 \\ y_3 \\ y_4 \end{bmatrix}
$$

Both this matrix form and Eq. (1.10.11) are recognized as special cases of Eqs. (1.6.1) and (1.6.10). According to Eq. (1.6.11), the quantity

$$\hat{y}_{n/n-1} = -[a_1 y_{n-1} + a_2 y_{n-2} + \cdots a_n y_0] \tag{1.10.12}$$

is the projection of $y_n$ on the subspace spanned by $Y_{n-1} = \{y_{n-1}, y_{n-2}, \ldots, y_0\}$. Therefore, it represents the best linear estimate of $y_n$ on the basis of (all) its past values $Y_{n-1}$; that is, $\hat{y}_{n/n-1}$ is the *best prediction* of $y_n$ from its (entire) past. Equation (1.10.11) gives the corresponding prediction error $\epsilon_n = y_n - \hat{y}_{n/n-1}$. We note here an interesting connection between linear prediction concepts and signal modeling concepts [15–17]; namely, that the best linear predictor (1.10.12) determines the whitening filter $A(z)$ which, in turn, determines the generator model $B(z) = 1/A(z)$ of $y_n$. In other words, solving the prediction problem also solves the modeling problem.

The above modeling approach to the representation of stationary time series, and its relationship to the Gram-Schmidt construction and linear prediction was initiated by Wold and developed further by Kolmogorov [15,16].

The most general model filter $B(z)$ given in Eq. (1.10.9) is called an *auto-regressive moving average* (ARMA), or a pole-zero model. Two special cases of interest are the *moving average* (MA), or all-zero models, and the *autoregressive* (AR), or all-pole models. The MA model has a nontrivial numerator only, $B(z) = N(z)$, so that $B(z)$ is a finite polynomial:

$$B(z) = 1 + b_1 z^{-1} + b_2 z^{-2} + \cdots + b_M z^{-M} \qquad \text{(MA model)}$$

The AR model has a nontrivial denominator only, $B(z) = 1/D(z)$, so that its inverse $A(z) = D(z)$ is a polynomial:

$$B(z) = \frac{1}{1 + a_1 z^{-1} + a_2 z^{-2} + \cdots + a_M z^{-M}} \qquad \text{(AR model)}$$

$$A(z) = 1 + a_1 z^{-1} + a_2 z^{-2} + \cdots + a_M z^{-M}$$

Autoregressive models are the most widely used models, because the analysis algorithms for extracting the model parameters $\{a_1, a_2, \ldots, a_M; \sigma_\epsilon^2\}$ are fairly simple. In the sequel, we will concentrate mainly on such models.

## 1.11  *Filter Model of First Order Markov Process*

To gain some understanding of filter models of the above type, we consider now a very simple example of a first order recursive filter $B(z)$ driven by a purely random sequence of variance $\sigma_\epsilon^2$:

$$\epsilon_n \longrightarrow \boxed{B(z)} \longrightarrow y_n \qquad B(z) = \frac{1}{1 - az^{-1}}$$

This serves also as a simple model for generating a first order Markov signal. The signal $y_n$ is generated by the difference equation of the filter:

$$y_n = ay_{n-1} + \epsilon_n \tag{1.11.1}$$

Let the probability of the $n$th sample $\epsilon_n$ be $f(\epsilon_n)$. We would like to show that

$$p(y_n/y_{n-1}, y_{n-2}, \ldots, y_1, y_0) = p(y_n/y_{n-1}) = f(\epsilon_n) = f(y_n - ay_{n-1})$$

which not only shows the Markov property of $y_n$, but also how to compute the related conditional density. Perhaps the best way to see this is to start at $n = 0$:

$$y_0 = \epsilon_0 \qquad \text{(assuming zero initial conditions)}$$

$$y_1 = ay_0 + \epsilon_1$$

$$y_2 = ay_1 + \epsilon_2, \qquad \text{etc.}$$

To compute $p(y_2/y_1, y_0)$, suppose that $y_1$ and $y_0$ are both given. Since $y_1$ is given, the third equation above shows that the randomness left in $y_2$ arises from $\epsilon_2$ only. Thus, $p(y_2/y_1) = f(\epsilon_2)$. From the first two equations it follows that specifying $y_0$ and $y_1$ is equivalent to specifying $\epsilon_0$ and $\epsilon_1$. Therefore, $p(y_2/y_1, y_0) = f(\epsilon_2/\epsilon_1, \epsilon_0) = f(\epsilon_2)$, the last equation following from the purely random nature of the sequence $\epsilon_n$. We have shown that

$$p(y_2/y_1, y_0) = p(y_2/y_1) = f(\epsilon_2) = f(y_2 - ay_1)$$

Using the results of Section 1.7, we also note

$$p(y_2, y_1, y_0) = p(y_2/y_1)\, p(y_1/y_0)\, p(y_0)$$

$$= f(\epsilon_2)\, f(\epsilon_1)\, f(\epsilon_0)$$

$$= f(y_2 - ay_1)\, f(y_1 - ay_0)\, f(y_0)$$

The solution of the difference equation (1.11.1) is obtained by convolving the impulse response of the filter $B(z)$

$$b_n = a^n u(n) \qquad u(n) = \text{unit step}$$

with the input sequence $\epsilon_n$ as follows:

$$y_n = \sum_{i=0}^{n} b_i \, \epsilon_{n-i} = \sum_{i=0}^{n} a^i \, \epsilon_{n-i} \tag{1.11.2}$$

for $n = 0,1,2, \ldots$ . This is the innovations representation of $y_n$ given by Eqs. (1.5.15), (1.5.16), and (1.10.1). In matrix form it reads:

$$\begin{bmatrix} y_0 \\ y_1 \\ y_2 \\ y_3 \end{bmatrix} = \begin{bmatrix} 1 & 0 & 0 & 0 \\ a & 1 & 0 & 0 \\ a^2 & a & 1 & 0 \\ a^3 & a^2 & a & 1 \end{bmatrix} \begin{bmatrix} \epsilon_0 \\ \epsilon_1 \\ \epsilon_2 \\ \epsilon_3 \end{bmatrix} \tag{1.11.3}$$

The inverse equation $\epsilon = B^{-1}y = Ay$, is obtained by writing Eq. (1.11.1) as $\epsilon_n = y_n - ay_{n-1}$. In matrix form, this reads

$$\begin{bmatrix} \epsilon_0 \\ \epsilon_1 \\ \epsilon_2 \\ \epsilon_3 \end{bmatrix} = \begin{bmatrix} 1 & 0 & 0 & 0 \\ -a & 1 & 0 & 0 \\ 0 & -a & 1 & 0 \\ 0 & 0 & -a & 1 \end{bmatrix} \begin{bmatrix} y_0 \\ y_1 \\ y_2 \\ y_3 \end{bmatrix} \tag{1.11.4}$$

According to the discussion of Example 1.6.1, the partial correlation coefficients can be read off from the *first column* of this matrix. We conclude, therefore, that all partial correlation coefficients of order greater than two are zero. This property is in accordance with our intuition about first order Markov processes; due to the recursive nature of Eq. (1.11.1) a given sample, say $y_n$, will have an indirect influence on all future samples. However, the *only direct* influence is to the next sample.

Higher order Markov random signals can be generated by sending white noise through higher order filters. For example, the second order difference equation

$$y_n = a_1 y_{n-1} + a_2 y_{n-2} + \epsilon_n \tag{1.11.5}$$

will generate a second order Markov signal. In this case, the difference equation directly couples two successive samples, but not more than two. Therefore, all the partial correlations of order greater than three will be zero. This may be seen also by writing Eq. (1.11.5) in matrix form and inspecting the first column of the matrix $A$:

$$\begin{bmatrix} \epsilon_0 \\ \epsilon_1 \\ \epsilon_2 \\ \epsilon_3 \\ \epsilon_4 \end{bmatrix} = \begin{bmatrix} 1 & 0 & 0 & 0 & 0 \\ -a_1 & 1 & 0 & 0 & 0 \\ -a_2 & -a_1 & 1 & 0 & 0 \\ 0 & -a_2 & -a_1 & 1 & 0 \\ 0 & 0 & -a_2 & -a_1 & 1 \end{bmatrix} \begin{bmatrix} y_0 \\ y_1 \\ y_2 \\ y_3 \\ y_4 \end{bmatrix}$$

## 1.12  *Stability and Stationarity*

In this section we discuss the importance of stability of the signal generator filter $B(z)$. We demonstrate that the generated signal $y_n$ will be stationary *only* if the generating filter is stable. And in this case, the sequence $y_n$ will become stationary only after the *transient effects* introduced by the filter have died out.

To demonstrate these ideas, consider the lag $= 0$ autocorrelation of our first order Markov signal

$$
\begin{aligned}
R_{yy}(n,n) = E[y_n^2] &= E[(ay_{n-1} + \epsilon_n)^2] \\
&= a^2 E[y_{n-1}^2] + 2aE[y_{n-1}\epsilon_n] + E[\epsilon_n^2] \qquad (1.12.1) \\
&= a^2 R_{yy}(n-1,n-1) + \sigma_\epsilon^2
\end{aligned}
$$

where we set $\sigma_\epsilon^2 = E[\epsilon_n^2]$, and $E[y_{n-1}\,\epsilon_n] = 0$, which follows by using Eq.(1.11.2) to get

$$
y_{n-1} = \epsilon_{n-1} + a\epsilon_{n-2} + a^2\epsilon_{n-3} + \cdots + a^{n-1}\epsilon_0
$$

and noting that $\epsilon_n$ is uncorrelated with all these terms, due to its white-noise nature. The above difference equation for $R_{yy}(n,n)$ can now be solved to get

$$
R_{yy}(n,n) = E[y_n^2] = \frac{\sigma_\epsilon^2}{1 - a^2} + \sigma_\epsilon^2\left(1 - \frac{1}{1 - a^2}\right)a^{2n} \qquad (1.12.2)
$$

where the initial condition was taken to be $E[y_0^2] = E[\epsilon_0^2] = \sigma_\epsilon^2$.

If the filter is stable; that is, $|a| < 1$, then the second term above tends to zero exponentially, and $R_{yy}(n,n)$ eventually "loses" its dependence on the absolute time $n$. For large $n$, it tends to the steady-state value

$$
R_{yy}(0) = E[y_n^2] = \sigma_y^2 = \frac{\sigma_\epsilon^2}{1 - a^2} \qquad (1.12.3)
$$

The same result is obtained, of course, by assuming stationarity from the start. The difference equation (1.12.1) can be written as

$$
E[y_n^2] = a^2 E[y_{n-1}^2] + \sigma_\epsilon^2
$$

If $y_n$ is assumed to be already stationary, then $E[y_{n-1}^2] = E[y_n^2]$. This implies the same steady-state solution as Eq. (1.12.3).

If the filter is unstable; that is, $|a| > 1$, then the second term of Eq. (1.12.2) blows up exponentially. The marginal case $a = 1$ is also unacceptable, but is of historical interest being the famous Wiener process, or random walk. In this case, the signal model is

$$
y_n = y_{n-1} + \epsilon_n
$$

and the difference equation for the variance becomes

$$
R_{yy}(n,n) = R_{yy}(n-1,n-1) + \sigma_\epsilon^2
$$

with solution

$$R_{yy}(n,n) = E[y_n^2] = (n + 1) \sigma_\epsilon^2$$

In summary, for true stationarity to set in, the signal generator filter $B(z)$ must be strictly stable (all its poles must be strictly inside the unit circle).

## 1.13 *Parameter Identification by the Maximum Likelihood Method*

One of the most important practical questions is how to extract the model parameters, such as the above filter parameter $a$, from the actual data values. As an introduction to the analysis methods used to answer this question, let us suppose that the white noise input sequence $\epsilon_n$ is gaussian

$$f(\epsilon_n) = \frac{1}{\sqrt{2\pi}\,\sigma_\epsilon} \exp(-\epsilon_n^2/2\sigma_\epsilon^2)$$

and assume that a block of $N$ measured values of the signal $y_n$ is available

$$y_0, y_1, y_2, \ldots, y_{N-1}$$

Can we extract the filter parameter $a$ from this block of data? Can we also extract the variance $\sigma_\epsilon^2$ of the driving white noise $\epsilon_n$? If so, then instead of saving the $N$ measured values $y_0, y_1, \ldots, y_{N-1}$, we can save the extracted filter parameter $a$ and the variance $\sigma_\epsilon^2$. Whenever we want to ''synthesize'' our original sequence $y_n$, we will simply generate a white-noise input sequence $\epsilon_n$ of variance $\sigma_\epsilon^2$, using a pseudorandom number generator routine, and then drive with it the signal model whose parameter $a$ was previously extracted from the original data. Somehow, all the significant information contained in the original samples, has now been packed or compressed into the two numbers $a$ and $\sigma_\epsilon^2$.

One possible criterion for extracting the filter parameter $a$ is the maximum likelihood (ML) criterion: The parameter $a$ is selected so as to *maximize* the joint density

$$p(y_{N-1}, \ldots, y_1, y_0) = f(\epsilon_{N-1}) \ldots f(\epsilon_1) f(\epsilon_0)$$

$$= \left(\frac{1}{\sqrt{2\pi}\,\sigma_\epsilon}\right)^N \exp\left[-\sum_{n=1}^{N-1}(y_n - ay_{n-1})^2/2\sigma_\epsilon^2\right]\exp[-y_0^2/2\sigma_\epsilon^2]$$

that is, the parameter $a$ is selected so as to render the actual measured values $y_0, y_1, \ldots, y_{N-1}$ most likely. The criterion is equivalent to minimizing the exponent with respect to $a$:

$$\mathcal{E}(a) = \sum_{n=1}^{N-1}(y_n - ay_{n-1})^2 + y_0^2 = \sum_{n=0}^{N-1} e_n^2 = \min \qquad (1.13.1)$$

where we set $e_n = y_n - a y_{n-1}$, and $e_0 = y_0$. The minimization of Eq. (1.13.1) gives

$$\frac{\partial \mathcal{E}(a)}{\partial a} = -2 \sum_{n=1}^{N-1} (y_n - a y_{n-1}) y_{n-1} = 0$$

or

$$a = \frac{\displaystyle\sum_{n=1}^{N-1} y_n \, y_{n-1}}{\displaystyle\sum_{n=1}^{N-1} y_{n-1} \, y_{n-1}} = \frac{y_0 y_1 + y_1 y_2 + \cdots + y_{N-2} y_{N-1}}{y_0^2 + y_1^2 + \cdots + y_{N-2}^2} \qquad (1.13.2)$$

There is a potential problem with the above ML criterion for extracting the filter parameter $a$; namely, the parameter may turn out to have magnitude greater than one, which will correspond to an unstable filter generating the sequence $y_n$. This is easily seen from Eq. (1.13.2). Whereas the numerator has dependence on the last sample $y_{N-1}$, the denominator does not. Therefore it is possible, for sufficiently large values of $y_{N-1}$, for the parameter $a$ to be greater than one. There are other criteria for extracting the Markov model parameters that guarantee the stability of the resulting synthesis filters, such as the so-called autocorrelation method, or Burg's method. These will be discussed later on.

## 1.14  *Parameter Identification by the Yule-Walker Method*

In this section, we introduce the *autocorrelation* or *Yule-Walker method* of extracting the model parameters from a block of data. We begin by expressing the model parameters in terms of output statistical quantities and then replace ensemble averages by time averages. Assuming stationarity has set in, we find

$$R_{yy}(1) = E[y_n \, y_{n-1}] = E[(a y_{n-1} + \epsilon_n) y_{n-1}] = a E[y_{n-1}^2] + E[\epsilon_n \, y_{n-1}]$$

$$= a \, R_{yy}(0)$$

from which

$$a = \frac{R_{yy}(1)}{R_{yy}(0)}$$

The input parameter $\sigma_\epsilon^2$ can be expressed as

$$\sigma_\epsilon^2 = (1 - a^2) \, \sigma_y^2 = (1 - a^2) \, R_{yy}(0)$$

These two equations may be written in matrix form as

$$\begin{bmatrix} R_{yy}(0) & R_{yy}(1) \\ R_{yy}(1) & R_{yy}(0) \end{bmatrix} \begin{bmatrix} 1 \\ -a \end{bmatrix} \begin{bmatrix} \sigma_\epsilon^2 \\ 0 \end{bmatrix}$$

These are called the *normal equations* of linear prediction. Their generalization will be considered later on. These results are important because they allow the extraction of the signal model parameters directly in terms of *output* quantities; that is, from experimentally accessible quantities.

We may obtain estimates of the model parameters by replacing the theoretical autocorrelations by the corresponding *sample autocorrelations,* defined by Eq. (1.9.1):

$$\hat{a} = \frac{\hat{R}_{yy}(1)}{\hat{R}_{yy}(0)} = \frac{\displaystyle\sum_{n=0}^{N-1-1} y_{n+1}\, y_n}{\displaystyle\sum_{n=0}^{N-1} y_n\, y_n} = \frac{y_0 y_1 + y_1 y_2 + \cdots + y_{N-2} y_{N-1}}{y_0^2 + y_1^2 + \cdots + y_{N-2}^2 + y_{N-1}^2}$$

$$\hat{\sigma}_\epsilon^2 = (1 - \hat{a}^2)\, \hat{R}_{yy}(0)$$

It is easily checked that the parameter $\hat{a}$, defined as above, is always of magnitude less than one; thus, the stability of the synthesis filter is guaranteed. Note the difference with the ML expression. The numerators are the same, but the denominators differ by an extra term. It is also interesting to note that the above expressions may be obtained by a *minimization criterion;* known as the auto-correlation method, or the Yule-Walker method:

$$\mathcal{E}(a) = \sum_{n=0}^{N} e_n^2 = \sum_{n=0}^{N} (y_n - a y_{n-1})^2 = \min \qquad (1.14.1)$$

This differs from the ML criterion (1.13.1) only in the range of summation for $n$. Whereas in the ML criterion the summation index $n$ does not run off the ends of the data block, it does so in the Yule-Walker case. We may think of the block of data as having been extended to both directions by padding it with zeros

$$0, \ldots, 0, y_0, y_1, \ldots, y_{N-1}, 0, 0, \ldots 0, \ldots$$

The difference between this and the ML criterion arises from the last term in the sum

$$\mathcal{E}(a) = \sum_{n=0}^{N} e_n^2 = \sum_{n=0}^{N-1} e_n^2 + e_N^2 = \sum_{n=0}^{N-1} (y_n - a y_{n-1})^2 + (0 - a y_{N-1})^2$$

The Yule-Walker analysis algorithm for this first order example is summarized in Fig. 1.13.

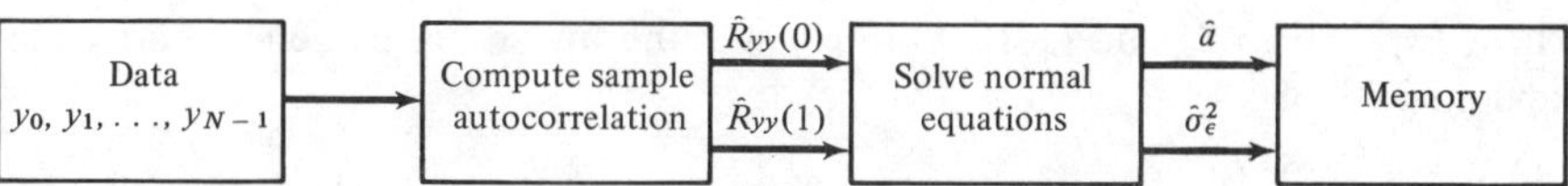

**Figure 1.13**   Yule-Walker Analysis Method

## 1.15  *Linear Prediction and Signal Modeling*

Linear prediction ideas are introduced in the context of our simple example by noting that the least-squares minimization criteria (1.13.1) and (1.14.1)

$$\mathcal{E}(a) = \sum_n e_n^2 = \text{minimum} \qquad (1.15.1)$$

essentially force each $e_n$ to be small. Thus, if we reinterpret

$$\hat{y}_n = ay_{n-1}$$

as the linear prediction of the sample $y_n$ made on the basis of just the previous sample $y_{n-1}$, then $e_n = y_n - ay_{n-1} = y_n - \hat{y}_n$ may be thought of as the prediction error. The minimization criterion (1.15.1) essentially minimizes the prediction error in an average least-squares sense, thus attempting to make the best prediction possible.

As we mentioned in Section 1.10, the solution of the linear prediction problem provides the corresponding random signal generator model for $y_n$, which can be used, in turn, in a number of ways as outlined in Section 1.10. This is the main reason for our interest in linear prediction.

A more intuitive way to understand the connection between linear prediction and signal models is as follows: Suppose we have a predictor $\hat{y}_n$ of $y_n$ which is not necessarily the best predictor. The predictor $\hat{y}_n$ is given as a linear combination of the past values $\{y_{n-1}, y_{n-2}, \ldots\}$:

$$\hat{y}_n = -[a_1 y_{n-1} + a_2 y_{n-2} + \cdots] \qquad (1.15.2)$$

The corresponding prediction error will be

$$e_n = y_n - \hat{y}_n = y_n + a_1 y_{n-1} + a_2 y_{n-2} + \cdots \qquad (1.15.3)$$

and it may be considered as the output of the prediction-error filter $A(z)$ (which is assumed to be stable and causal):

$$y_n \longrightarrow \boxed{A(z)} \longrightarrow e_n \qquad A(z) = 1 + a_1 z^{-1} + a_2 z^{-2} + \cdots$$

Suppose further that $A(z)$ has a stable and causal inverse filter

$$e_n \longrightarrow \boxed{B(z)} \longrightarrow y_n \qquad B(z) = \frac{1}{A(z)} = 1 + b_1 z^{-1} + b_2 z^{-2} + \cdots$$

so that $y_n$ may be expressed *causally* in terms of $e_n$; that is,

$$y_n = e_n + b_1 e_{n-1} + b_2 e_{n-2} + \cdots \qquad (1.15.4)$$

Then, Eqs. (1.15.3) and (1.15.4) imply that the linear spaces generated by the random variables

$$\{y_{n-1}, y_{n-2}, \ldots\} \qquad \text{and} \qquad \{e_{n-1}, e_{n-2}, \ldots\}$$

are the same space. One can pass from one set to the other by a causal and causally invertible linear filtering operation.

Now, if the prediction $\hat{y}_n$ of $y_n$ is the best possible prediction, then what remains after the prediction is made—namely, the error signal $e_n$—should be entirely *unpredictable* on the basis of the past values $\{y_{n-1}, y_{n-2}, \ldots\}$. That is, $e_n$ must be uncorrelated with all of these. But this implies that $e_n$ must be uncorrelated with all $\{e_{n-1}, e_{n-2}, \ldots\}$, and therefore $e_n$ must be a white-noise sequence. It follows that $A(z)$ and $B(z)$ are the analysis and synthesis filters for the sequence $y_n$.

The *least-squares* minimization criteria of the type (1.15.1) that are based on time averages, provide a *practical* way to solve the *linear prediction problem* and hence also the *modeling problem*. Their generalization to higher order predictors will be discussed in Chapter 5.

## Problems

### Problem 1.1:

Two dice are available for throwing. One is fair, but the other bears only sixes. One die is selected as follows: A coin is tossed. If the outcome is tails then the fair die is selected, but if the outcome is heads, the biased die is selected. The coin itself is not fair, and the probability of bearing heads or tails is 1/3 or 2/3, respectively. A die is now selected according to this procedure and tossed twice and the number of sixes is noted.

Let $x$ be a random variable that takes on the value 0 when the fair die is selected or 1 if the biased die is selected. Let $y$ be a random variable denoting the number of sixes obtained in the two tosses; thus, the possible values of $y$ are 0,1,2.

(a) For all possible values of $x$ and $y$, compute $p(y/x)$; that is, the probability that the number of sixes will be $y$, given that the $x$ die was selected.

(b) For each $y$, compute $p(y)$; that is, the probability that the number of sixes will be $y$, regardless of which die was selected.

(c) Compute the mean number of sixes $E[y]$.

(d) For all values of $x$ and $y$, compute $p(x/y)$; that is, the probability that we selected die $x$, given that we already observed a $y$ number of sixes.

### Problem 1.2:

(a) Following the notation of Section 1.4, show the matrix identity

$$\begin{bmatrix} I_N & -H \\ 0 & I_M \end{bmatrix} \begin{bmatrix} R_{xx} & R_{xy} \\ R_{yx} & R_{yy} \end{bmatrix} \begin{bmatrix} I_N & -H \\ 0 & I_M \end{bmatrix}^T = \begin{bmatrix} R_{xx} - R_{xy}R_{yy}^{-1}R_{yx} & 0 \\ 0 & R_{yy} \end{bmatrix}$$

where $H = R_{xy}R_{yy}^{-1}$.

(b) Rederive the correlation canceling results (1.4.3) and (1.4.4) using this identity.

***Problem 1.3:***
Using the matrix identity of Problem 1.2, derive directly the result of Example 1.4.1; that is, $E[\mathbf{x}/\mathbf{y}] = R_{xy}R_{yy}^{-1}\mathbf{y}$. Work directly with probability densities; do not use the results of Examples 1.3.3 and 1.3.4.

***Problem 1.4:***
Show that the orthogonal projection $\hat{\mathbf{x}}$ of a vector $\mathbf{x}$ onto another vector $\mathbf{y}$, defined by Eq. (1.4.5) or Eq. (1.5.18), is a linear function of $\mathbf{x}$; that is, show

$$\widehat{A_1\,\mathbf{x}_1 + A_2\,\mathbf{x}_2} = A_1\,\hat{\mathbf{x}}_1 + A_2\,\hat{\mathbf{x}}_2$$

***Problem 1.5:*** Noise Canceling
Suppose $x$ consists of two components $x = s + n_1$, a desired component $s$, and a noise component $n_1$. Suppose that $y$ is a related noise component $n_2$ to which we have access, $y = n_2$. The relationship between $n_1$ and $n_2$ is assumed to be linear, $n_1 = Fn_2$. For example, $s$ might represent an electrocardiogram signal which is contaminated by 60 Hz power frequency pick-up noise $n_1$; then, a reference 60 Hz noise $y = n_2$ can be obtained from the wall outlet. (a) Show that the correlation canceler is $H = F$, and that complete cancellation of $n_1$ takes place. (b) If $n_1 = Fn_2 + v$, where $v$ is uncorrelated with $n_2$ and $s$, show that $H = F$ still, and $n_1$ is canceled completely. The part $v$ remains unaffected.

***Problem 1.6:*** Signal Cancellation Effects
In the previous problem, we assumed that the reference signal $y$ did not contain any part related to the desired component $s$. There are applications, however, where both the signal and the noise components contribute to both $x$ and $y$, as for example in antenna sidelobe cancellation. Since the reference signal $y$ contains part of $s$, the correlation canceler will act also to cancel part of the useful signal $s$ from the output. To see this effect, consider a simple one-dimensional example

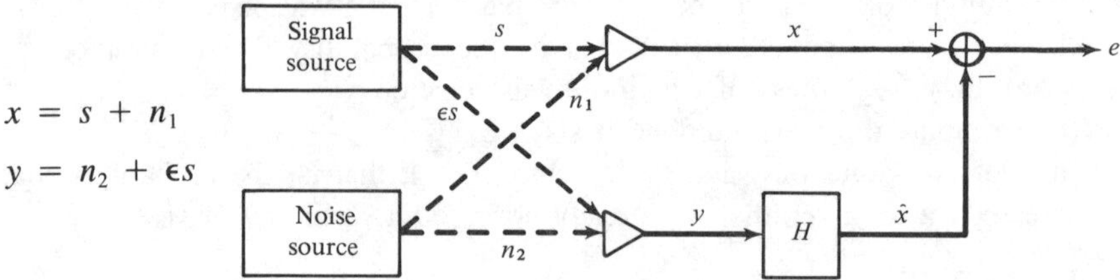


with $n_1 = Fn_2$, where we assume that $y$ contains a small part proportional to the desired signal $s$. Assume that $n_2$ and $s$ are uncorrelated. Show that the output $e$ of the correlation canceler will contain a reduced noise component $n_1$ as well as a partially canceled signal $s$, as follows:

$$e = as + bn_1 \qquad \text{where} \qquad a = 1 - \frac{F\epsilon(1 + F\epsilon G)}{1 + F^2\epsilon^2 G}, \ b = -\epsilon FGa$$

and $G$ is a signal to noise ratio $G = E[s^2]/E[n_1^2]$. Note that when $\epsilon = 0$, then $a = 1$, and $b = 0$, as it should.

### Problem 1.7:

Consider a special case of Example 1.4.3 defined by $c_n = 1$, so that $y_n = x + v_n$, $n = 1,2, \ldots ,M$. This represents the noisy measurement of a constant $x$. By comparing the corresponding mean-squared estimation errors $E[e^2]$, show that the optimal estimate of $x$ given in Eq. (1.4.9) is indeed better than the straight average estimate

$$\hat{x}_{av} = \frac{1}{M} (y_1 + y_2 + \cdots + y_M)$$

### Problem 1.8: Recursive Estimation

Consider the subspaces $Y_n = \{y_1,y_2, \ldots ,y_n\}$ for $n = 1,2, \ldots ,M$, as defined in Section 1.5. Equation (1.5.18) defines the estimate $\hat{x}$ of a random vector $\mathbf{x}$ based on the largest one of these subspaces; namely, $Y_M$.

(a) Show that this estimate can also be generated recursively as follows

$$\hat{\mathbf{x}}_n = \hat{\mathbf{x}}_{n-1} + \mathbf{G}_n(y_n - \hat{y}_{n/n-1})$$

for $n = 1,2, \ldots ,M$; and initialized by $\hat{\mathbf{x}}_0 = 0$, and $\hat{y}_{1/0} = 0$, where $\hat{\mathbf{x}}_n$ denotes the best estimate of $\mathbf{x}$ based on the subspace $Y_n$ and $\mathbf{G}_n$ is a gain coefficient given by $\mathbf{G}_n = E[\mathbf{x}\epsilon_n] E[\epsilon_n\epsilon_n]^{-1}$. (*Hint:* Note $\hat{\mathbf{x}}_n = \Sigma_{i=1}^n E[\mathbf{x}\epsilon_i] E[\epsilon_i\epsilon_i]^{-1}\epsilon_i$.)

(b) Show that the innovations $\epsilon_n = y_n - \hat{y}_{n/n-1}$ is orthogonal to $\hat{\mathbf{x}}_{n-1}$; that is, show that $E[\hat{\mathbf{x}}_{n-1}\epsilon_n] = 0$ for $n = 1,2, \ldots .M$.

(c) Let $\mathbf{e}_n = \mathbf{x} - \hat{\mathbf{x}}_n$ be the corresponding estimation error of $\mathbf{x}$ with respect to the subspace $Y_n$. Using Eq. (1.4.4), show that its covariance matrix can be expressed in the $\epsilon$-basis as

$$R_{e_n e_n} = R_{xx} - \sum_{i=1}^n E[\mathbf{x}\epsilon_i] E[\epsilon_i\epsilon_i]^{-1} E[\epsilon_i\mathbf{x}^T]$$

(d) The above recursive construction represents a successive improvement of the estimate of $\mathbf{x}$, as more and more $y$s are taken into account; that is, as the subspaces $Y_n$ are successively enlarged. Verify that $\hat{\mathbf{x}}_n$ is indeed a better estimate than $\hat{\mathbf{x}}_{n-1}$ by showing that the mean-squared estimation error $R_{e_n e_n}$ of $\hat{\mathbf{x}}_n$ is smaller than the mean-squared error $R_{e_{n-1} e_{n-1}}$ of $\hat{\mathbf{x}}_{n-1}$. This is a very intuitive result; the more information we use the better the estimate.

Such recursive updating schemes are the essence of Kalman filtering. In that context, $\mathbf{G}_n$ is referred to as the "Kalman gain."

### Problem 1.9:

The recursive updating procedure given in Problem 1.8 is useful only if the gain coefficient $\mathbf{G}_n$ can be computed at each iteration $n$. For that, a knowledge of the relationship between $\mathbf{x}$ and $y_n$ is required. Consider the case of Example 1.4.3 where $y_n = c_n x + v_n$; define the vectors

$$\mathbf{c}_n = [c_1, c_2, \ldots, c_n]^T \qquad \text{and} \qquad \mathbf{y}_n = [y_1, y_2, \ldots, y_n]^T$$

$$\text{for } n = 1, 2, \ldots, M$$

and let $\hat{x}_n$ and $e_n = x - \hat{x}_n$ denote the estimate of $x$ on the basis of $Y_n$ and the corresponding estimation error.

(a) Using Eq. (1.4.9), show that

$$\hat{x}_n = (1 + \mathbf{c}_n^T \mathbf{c}_n)^{-1} \mathbf{c}_n^T \mathbf{y}_n \qquad \text{and} \qquad E[e_n^2] = E[x e_n] = (1 + \mathbf{c}_n^T \mathbf{c}_n)^{-1}$$

(b) Using Eq. (1.5.19), compute $\hat{y}_{n/n-1}$ and show that it may be expressed in the form

$$\hat{y}_{n/n-1} = c_n \hat{x}_{n-1} = c_n (1 + \mathbf{c}_{n-1}^T \mathbf{c}_{n-1})^{-1} \mathbf{c}_{n-1}^T \mathbf{y}_{n-1}$$

(c) Let $e_{n-1} = x - \hat{x}_{n-1}$ be the estimation error based on $Y_{n-1}$. Writing

$$\epsilon_n = y_n - \hat{y}_{n/n-1} = (c_n x + v_n) - c_n \hat{x}_{n-1} = c_n e_{n-1} + v_n$$

show that

$$E[\epsilon_n \epsilon_n] = (1 + \mathbf{c}_n^T \mathbf{c}_n)(1 + \mathbf{c}_{n-1}^T \mathbf{c}_{n-1})^{-1}$$

$$E[x \epsilon_n] = c_n (1 + \mathbf{c}_{n-1}^T \mathbf{c}_{n-1})^{-1}$$

(d) Show that the estimate $\hat{x}_n$ of $x$ can be computed recursively by

$$\hat{x}_n = \hat{x}_{n-1} + G_n(y_n - c_n \hat{x}_{n-1}), \qquad \text{where} \qquad G_n = c_n (1 + \mathbf{c}_n^T \mathbf{c}_n)^{-1}$$


***Problem 1.10:***

Rederive the recursive updating equation given in Problem 1.9 (d), without any reference to innovations or projections, by simply manipulating Eq. (1.4.9) algebraically, and writing it in recursive form.


***Problem 1.11:***

The Gram-Schmidt orthogonalization procedure for a subspace $Y = \{y_1, y_2, \ldots, y_M\}$ is initialized at the leftmost random variable $y_1$, $\epsilon_1 = y_1$, and progresses to the right by successively orthogonalizing $y_2$, $y_3$, and so on. It results in the lower triangular representation $\mathbf{y} = B\boldsymbol{\epsilon}$. The procedure can just as well be started at the rightmost variable $y_M$ and proceed backwards as follows:

$$\eta_M = y_M$$

$$\eta_{M-1} = y_{M-1} - (\text{projection of } y_{M-1} \text{ on } \eta_M)$$

$$\eta_{M-2} = y_{M-2} - (\text{projection of } y_{M-2} \text{ on } \{\eta_M, \eta_{M-1}\})$$

and so on. Show that the resulting uncorrelated vector $\boldsymbol{\eta} = [\eta_1, \eta_2, \ldots, \eta_M]^T$ is related to $\mathbf{y} = [y_1, y_2, \ldots, y_M]^T$ by a linear transformation

$$\mathbf{y} = U\boldsymbol{\eta}$$

where $U$ is a unit *upper-triangular* matrix. Show also that this corresponds to a $UL$ (rather than $LU$) Cholesky factorization of the covariance matrix $R_{yy}$.

### Problem 1.12:

Since "orthogonal" means "uncorrelated," the Gram-Schmidt orthogonalization procedure can also be understood as a correlation canceling operation. Explain how Eq. (1.5.20) may be thought of as a special case of the correlation canceler defined by Eqs. (1.4.1) and (1.4.2). What are $\mathbf{x}$, $\mathbf{y}$, $\mathbf{e}$, and $H$, in this case? Draw the correlation canceler diagram of Fig. 1.1 as it applies here, showing explicitly the components of all the vectors.

### Problem 1.13:

Using Eq. (1.6.11), show that the vector of coefficients $[a_{n1}, a_{n2}, \ldots, a_{nn}]^T$ can be expressed explicitly in terms of the $\mathbf{y}$-basis as follows:

$$\begin{bmatrix} a_{n1} \\ a_{n2} \\ \vdots \\ a_{nn} \end{bmatrix} = -E[\mathbf{y}_{n-1}\,\mathbf{y}_{n-1}^T]^{-1}\,E[y_n\,\mathbf{y}_{n-1}], \qquad \text{where} \qquad \mathbf{y}_{n-1} = \begin{bmatrix} y_{n-1} \\ y_{n-2} \\ \vdots \\ y_0 \end{bmatrix}$$

### Problem 1.14:

Show that the mean-squared estimation error of $y_n$ on the basis of $Y_{n-1}$—that is, $E[\epsilon_n^2]$, where $\epsilon_n = y_n - \hat{y}_{n/n-1}$—can be expressed as

$$E[\epsilon_n^2] = E[\epsilon_n\,y_n] = E[y_n^2] - E[y_n\,\mathbf{y}_{n-1}^T]\,E[\mathbf{y}_{n-1}\,\mathbf{y}_{n-1}^T]^{-1}\,E[y_n\,\mathbf{y}_{n-1}]$$

### Problem 1.15:

Let $\mathbf{a}_n = [1, a_{n1}, a_{n2}, \ldots, a_{nn}]^T$ for $n = 1, 2, \ldots, M$. Show that the results of the last two problems can be combined into one enlarged matrix equation

$$E[\mathbf{y}_n\,\mathbf{y}_n^T]\,\mathbf{a}_n = E[\epsilon_n^2]\,\mathbf{u}_n$$

where $\mathbf{u}_n$ is the unit-vector $\mathbf{u}_n = [1, 0, 0, \ldots, 0]^T$ consisting of one followed by $n$ zeros, and $\mathbf{y}_n = [y_n, y_{n-1}, \ldots, y_1, y_0]^T = [y_n, \mathbf{y}_{n-1}^T]^T$.

### Problem 1.16:

The quantity $\hat{y}_{n/n-1}$ of Eq. (1.5.19) is the best estimate of $y_n$ based on all the previous $y$s; namely, $Y_{n-1} = \{y_0, y_1, \ldots, y_{n-1}\}$. This can be understood in three ways: First, in terms of the orthogonal projection theorem as we demonstrated in the text. Second, in terms of the correlation canceler interpretation as suggested in Problem 1.12. And third, it may be proved directly as follows: Let $\hat{y}_{n/n-1}$ be given as a linear combination of the previous $y$s as in Eq. (1.6.11); the coefficients $[a_{n1}, a_{n2}, \ldots, a_{nn}]^T$ are to be chosen optimally to minimize the estimation error $\epsilon_n$ given by Eq. (1.6.10) in the mean-squared sense. In terms of the notation of Problem 1.15, Eq. (1.6.10) and $E[\epsilon_n^2]$ can be written in the compact vectorial form

$$\epsilon_n = \mathbf{a}_n^T \mathbf{y}_n \quad \text{and} \quad \mathcal{E}(\mathbf{a}_n) = E[\epsilon_n^2] = \mathbf{a}_n^T E[\mathbf{y}_n \mathbf{y}_n^T] \, \mathbf{a}_n$$

The quantity $\mathcal{E}(\mathbf{a}_n)$ is to be minimized with respect to $\mathbf{a}_n$. The minimization must be subject to the constraint that the first entry of the vector $\mathbf{a}_n$ be unity. This constraint can be expressed in vector form as

$$\mathbf{a}_n^T \mathbf{u}_n = 1$$

where $\mathbf{u}_n$ is the unit vector defined in Problem 1.15. Incorporate this constraint with a Lagrange multiplier and minimize the performance index

$$\mathcal{E}(\mathbf{a}_n) = \mathbf{a}_n^T E[\mathbf{y}_n \mathbf{y}_n^T] \, \mathbf{a}_n + \lambda \, (1 - \mathbf{a}_n^T \mathbf{u}_n)$$

with respect to $\mathbf{a}_n$, then fix $\lambda$ by enforcing the constraint, and finally show that the resulting solution of the minimization problem is identical to that given in Problem 1.15.

### Problem 1.17:
A random signal $x(n)$ is defined as a linear function of time by

$$x(n) = a + bn$$

where $a$ and $b$ are independent zero-mean gaussian random variables of variances $\sigma_a^2$ and $\sigma_b^2$, respectively.

   (a) Compute $E[x(n)^2]$.
   (b) Is $x(n)$ a stationary process? Is it ergodic? Explain.
   (c) For each fixed $n$, compute the probability density $p(x(n))$.
   (d) For each fixed $n$ and $m$ $(n \neq m)$, compute the conditional probability density $p(x(n)/x(m))$ of $x(n)$ given $x(m)$.
(*Hint:* $x(n) - x(m) = b(n - m)$.)

### Problem 1.18:
Compute the sample autocorrelation of the sequences
   (a) $y_n = 1$, for $0 \leq n \leq 10$
   (b) $y_n = (-1)^n$, for $0 \leq n \leq 10$
in two ways: First in the time domain, using Eq. (1.9.1), and then in the $z$-domain, using Eq. (1.9.3) and computing its inverse $z$-transform.

### Problem 1.19: FFT Computation of Autocorrelations
In many applications, a fast computation of sample autocorrelations or cross-correlations is required, as in the matched filtering operations in radar data

processors. A fast way to compute the sample autocorrelation $\hat{R}_{yy}(k)$ of a length-$N$ data segment $\mathbf{y} = [y_0, y_1, \ldots , y_{N-1}]^T$ is based on Eq. (1.9.5) which can be computed using FFTs. Performing an inverse FFT on Eq. (1.9.5), we find the computationally efficient formula

$$\hat{R}_{yy}(k) = \frac{1}{N} \, \mathrm{IFFT}[|\mathrm{FFT}(\mathbf{y})|^2] \qquad (\mathrm{P.1})$$

To avoid wrap-around errors introduced by the FFT, the length $N'$ of the FFT must be selected to be greater than the length of the function $\hat{R}_{yy}(k)$. Since $\hat{R}_{yy}(k)$ is double-sided with an extent $-(N - 1) \leqslant k \leqslant (N - 1)$, it will have length equal to $2N - 1$. Thus, we must select $N' \geqslant 2N - 1$. To see the wrap-around effects, consider a length-4 signal $\mathbf{y} = [1,2,2,1]^T$.

    (a) Compute $\hat{R}_{yy}(k)$ using the time-domain definition.

    (b) Compute $\hat{R}_{yy}(k)$ according to Eq. (P.1) using 4-point FFTs.

    (c) Repeat using 8-point FFTs.

***Problem 1.20:*** Computer Experiment

    (a) Generate 1000 samples $x(n)$, $n = 0,1, \ldots ,999$, of a zero-mean, unit-variance, white gaussian noise sequence.

    (b) Compute and plot the first 100 lags of its sample autocorrelation; that is, $\hat{R}_{yy}(k)$, for $k = 0,1, \ldots ,99$. Does $\hat{R}_{xx}(k)$ look like a delta function, $\delta(k)$?

    (c) Generate 10 different realizations of the length-1000 sequence $x(n)$, and compute 100 lags of the corresponding sample autocorrelations. Define an average autocorrelation by

$$\hat{R}(k) = \frac{1}{10} \sum_{i=1}^{10} \hat{R}_i(k), \qquad k = 0,1, \ldots ,99,$$

where $\hat{R}_i(k)$ is the sample autocorrelation of the $i$th realization of $x(n)$. Plot $\hat{R}(k)$ versus $k$. Do you notice any improvement?

***Problem 1.21:***

A 500-millisecond record of a stationary random signal is sampled at a rate of 2 kHz and the resulting $N$ samples are recorded for further processing. What is $N$? The record of $N$ samples is then divided into $K$ contiguous segments, each of length $M$, so that $M = N/K$. The periodograms from each segment are computed and averaged together to obtain an estimate of the power spectrum of the signal. A frequency resolution of $\Delta f = 20$ Hz is required. What is the shortest length $M$ that will guarantee such resolution? (Larger $M$s will have better resolution than required but will result in a poorer power spectrum estimate because $K$ will be smaller.) What is $K$ in this case?

**Problem 1.22:**

A random signal $y_n$ is generated by sending unit-variance zero-mean white noise $\epsilon_n$ through the filters defined by the following difference equations:

1. $y_n = -0.9\, y_{n-1} + \epsilon_n$
2. $y_n = 0.9\, y_{n-1} + \epsilon_n + \epsilon_{n-1}$
3. $y_n = \epsilon_n + 2\, \epsilon_{n-1} + \epsilon_{n-2}$
4. $y_n = -0.81\, y_{n-2} + \epsilon_n$
5. $y_n = 0.1\, y_{n-1} + 0.72\, y_{n-2} + \epsilon_n - 2\, \epsilon_{n-1} + \epsilon_{n-2}$

(a) For each case, determine the transfer function $B(z)$ of the filter and draw its canonical implementation form, identify the set of model parameters, and decide whether the model is ARMA, MA, or AR.

(b) Write explicitly the power spectrum $S_{yy}(\omega)$ using Eq. (1.10.6).

(c) Based on the pole/zero pattern of the filter $B(z)$, draw a rough sketch of the power spectrum $S_{yy}(\omega)$ for each case.

**Problem 1.23:** Computer Experiment

Two different realizations of a stationary random signal $y(n)$, $n = 0,1, \ldots ,19$ are given. It is known that this signal has been generated by a model of the form

$$y(n) = ay(n - 1) + \epsilon(n)$$

where $\epsilon(n)$ is gaussian zero-mean white noise of variance $\sigma_\epsilon^2$.

(a) Estimate the model parameters $a$ and $\sigma_\epsilon^2$, using the maximum likelihood criterion for both realizations. (The exact, simulated, values were $a = 0.95$ and $\sigma_\epsilon^2 = 1$.)

(b) Repeat, using the Yule-Walker method.

This situation might, for example, arise in speech analysis where $y(n)$ might represent a short segment of sampled unvoiced speech from which the filter parameters (model parameters) are to be extracted and stored for future regeneration of that segment. A realistic speech model would of course require a higher-order filter; typically, order 10 to 15.

| $n$ | $y(n)$ | $y(n)$ |
|---|---|---|
| 0 | 3.848 | 5.431 |
| 1 | 3.025 | 5.550 |
| 2 | 5.055 | 4.873 |
| 3 | 4.976 | 5.122 |
| 4 | 6.599 | 5.722 |
| 5 | 6.217 | 5.860 |
| 6 | 6.572 | 6.133 |
| 7 | 6.388 | 5.628 |
| 8 | 6.500 | 6.479 |
| 9 | 5.564 | 4.321 |
| 10 | 5.683 | 5.181 |
| 11 | 5.255 | 4.279 |
| 12 | 4.523 | 5.469 |
| 13 | 3.952 | 5.087 |
| 14 | 3.668 | 3.819 |
| 15 | 3.668 | 2.968 |
| 16 | 3.602 | 2.751 |
| 17 | 1.945 | 3.306 |
| 18 | 2.420 | 3.103 |
| 19 | 2.104 | 3.694 |

***Problem 1.24:*** Computer Experiment

(a) Using the Yule-Walker estimates $\{\hat{a};\ \hat{\sigma}_\epsilon^2\}$ of the model parameters extracted from the first realization of $y_n$ given in Problem 1.23, make a plot of the estimate of the power spectrum following Eq. (1.10.6); that is,

$$\hat{S}_{yy}(\omega) = \frac{\hat{\sigma}_\epsilon^2}{|1 - \hat{a}\ e^{-j\omega}|^2}$$

versus frequency $\omega$ in the interval $0 \leqslant \omega \leqslant \pi$.

(b) Also, plot the true power spectrum

$$S_{yy}(\omega) = \frac{\sigma_\epsilon^2}{|1 - a\ e^{-j\omega}|^2}$$

defined by the true model parameters $\{a;\ \sigma_\epsilon^2\} = \{0.95;\ 1.0\}$.

(c) Using the given data values $y_n$ for the first realization, compute and plot the corresponding periodogram spectrum of Eq. (1.9.5). Preferably, plot all three spectra on the same graph. Compute the spectra at 100 or 200 equally-spaced frequency points in the interval $[0, \pi]$.

(d) Repeat parts (a) through (c) using the second realization of $y_n$.

Better agreement between estimated and true spectra can be obtained using Burg's analysis procedure instead of the Yule-Walker method. Burg's method performs remarkably well on the basis of very short data records. The Yule-Walker method also performs well but it requires somewhat longer records. These methods will be compared in Chapter 5.

# References


1. A. Papoulis. *Probability, Random Variables, and Stochastic Processes*, New York, McGraw-Hill, 1965.

2. J. L. Doob, *Stochastic Processes*, New York, Wiley, 1953.

3. P. R. Halmos, *Finite-Dimensional Vector Spaces*, New York, Van Nostrand, 1958.

4. R. B. Blackman and J. W. Tukey, *The Measurement of Power Spectra*, New York, Dover, 1958.

5. C. Bingham, M. D. Godfrey, and J. W. Tukey, Modern Techniques of Power Spectrum Estimation, *IEEE Trans. Audio Electroacoust.*, **AU-15,** 56–66 (1967).

6. G. M. Jenkins and D. G. Watts, *Spectral Analysis and its Applications*, San Francisco, Holden-Day, 1968.

7. A. V. Oppenheim and R. W. Schafer, *Digital Signal Processing*, Englewood Cliffs, Prentice-Hall, 1975.

8. R. K. Otnes and L. Enochson, *Digital Time Series Analysis*, New York, Wiley, 1972.

9. W. Davenport and W. Root, *Introduction to the Theory of Random Signals and Noise*, New York, McGraw-Hill, 1958.

10. D. Childers, Ed., *Modern Spectrum Analysis*, New York, Wiley, 1978.

11. F. J. Harris, On the Use of Windows for Harmonic Analysis with the Discrete Fourier Transform, *Proc. IEEE*, **66**, 51–83 (1978).

12. A. H. Nuttal and G. C. Carter, A Generalized Framework for Power Spectral Estimation, *IEEE Trans. Acoust., Speech, Signal Process.*, **ASSP-28**, 334–335 (1980).

13. P. D. Welch, The Use of Fast Fourier Transform for the Estimation of Power Spectra: A Method Based on Time Averaging Over Short, Modified Periodograms, *IEEE Trans. Audio Electroacoust.*, **AU-15**, 70–73 (1967).

14. G. P. Box and G. M. Jenkins, *Time Series Analysis Forecasting and Control*, New York, Holden-Day, 1970.

15. H. Wold, *A Study in the Analysis of Time Series*, Uppsala, Sweden, Almqvist and Wiksell, 1931 and 1954.

16. A. N. Kolmogorov, Sur l'Interpolation et Extrapolation des Suites Stationnaires, *C. R. Acad. Sci.*, **208**, 2043–2045 (1939). See also Interpolation and Extrapolation of Stationary Random Sequences, and Stationary Sequences in Hilbert Space, reprinted in T. Kailath, Ed., *Linear Least-Squares Estimation*, Stroudsburg, PA, Dowden, Hutchinson, and Ross, 1977.

17. E. A. Robinson, *Time Series Analysis and Applications*, Houston, TX, Goose Pond Press, 1981.

# 2

# Some Signal Processing Applications

IN THE NEXT FEW SECTIONS, we shall present some applications of the random signal concepts that we introduced in the previous chapter. We shall discuss system identification by cross-correlation techniques, design simple filters to remove noise from noisy measurements, apply these concepts to the problem of quantization effects in digital filter structures, introduce the problem of linear prediction and its iterative solution through Levinson's algorithm, and discuss a data compression example.

## 2.1  *Filtering of Stationary Random Signals*

In this section, we discuss the effect of linear filtering on random signals. The results are very basic and of importance in suggesting guidelines for the design of signal processing systems for many applications of random signals [1–3].

Suppose a stationary random signal $x_n$ is sent into a linear filter defined by a transfer function $H(z)$. Let $y_n$ be the output random signal. Our objective is to derive relationships between the autocorrelation functions of the input and output signals, and also between the corresponding power spectra.

$$x_n \longrightarrow \boxed{H(z)} \longrightarrow y_n \qquad H(z) = \sum_{n=0}^{\infty} h_n z^{-n}$$

Using the input/output filtering equation in the $z$-domain

$$Y(z) = H(z)\, X(z) \tag{2.1.1}$$

we determine first a relationship between the periodograms of the input and output signals. Using the factorization (1.9.3) and dropping the factor $1/N$ for convenience, we find

$$\hat{S}_{yy}(z) = Y(z)\, Y(z^{-1})$$
$$= H(z)\, X(z)\, H(z^{-1})\, X(z^{-1}) = H(z)\, H(z^{-1})\, X(z)\, X(z^{-1}) \quad (2.1.2)$$
$$= H(z)\, H(z^{-1})\, \hat{S}_{xx}(z) = S_{hh}(z)\, \hat{S}_{xx}(z)$$

where we used the notation $S_{hh}(z) = H(z)\,H(z^{-1})$. This quantity is the $z$-transform of the sample autocorrelation of the filter; that is,

$$S_{hh}(z) = H(z)\, H(z^{-1}) = \sum_{k=-\infty}^{\infty} R_{hh}(k)\, z^{-k} \quad (2.1.3)$$

where $R_{hh}(k)$ is the filter's autocorrelation function

$$R_{hh}(k) = \sum_{n} h_{n+k}\, h_{n} \quad (2.1.4)$$

Equation (2.1.3) is easily verified by writing

$$R_{hh}(k) = \sum_{i,j=0}^{\infty} h_i h_j\, \delta(k - (i - j))$$

Taking inverse $z$-transforms of Eq. (2.1.2), we find the time-domain equivalent relationships between input and output sample autocorrelations

$$\hat{R}_{yy}(k) = \sum_{m=-\infty}^{\infty} R_{hh}(m)\, \hat{R}_{xx}(k - m) = \text{convolution of } R_{hh} \text{ with } \hat{R}_{xx} \quad (2.1.5)$$

Similarly, we find for the cross-periodograms

$$\hat{S}_{yx}(z) = Y(z)\, X(z^{-1}) = H(z)\, X(z)\, X(z^{-1}) = H(z)\, \hat{S}_{xx}(z) \quad (2.1.6)$$

and also, replacing $z$ by $z^{-1}$,

$$\hat{S}_{xy}(z) = \hat{S}_{xx}(z)\, H(z^{-1}) \quad (2.1.7)$$

The same relationships hold for the statistical autocorrelations and power spectra. We summarize these as follows: In the $z$-domain the power spectral densities are related by

$$S_{yy}(z) = H(z)\, H(z^{-1})\, S_{xx}(z)$$
$$S_{yx}(z) = H(z)\, S_{xx}(z) \quad (2.1.8)$$
$$S_{xy}(z) = S_{xx}(z)\, H(z^{-1})$$

Setting $z = \exp(j\omega)$, we may also write Eq. (2.1.8) in terms of the corresponding power spectra, as follows

$$S_{yy}(\omega) = |H(\omega)|^2 \, S_{xx}(\omega)$$

$$S_{yx}(\omega) = H(\omega) \, S_{xx}(\omega) \tag{2.1.9}$$

$$S_{xy}(\omega) = S_{xx}(\omega) \, H(\omega)^*$$

In the time domain, the correlation functions are related by

$$R_{yy}(k) = \sum_{m=-\infty}^{\infty} R_{hh}(m) \, R_{xx}(k-m)$$

$$R_{yx}(k) = \sum_{m=0}^{\infty} h_m R_{xx}(k-m) \tag{2.1.10}$$

The proof of these is straightforward; for example, to prove Eq. (2.1.10), use stationarity and the I/O convolutional equation

$$y_n = \sum_{m=0}^{\infty} h_m x_{n-m}$$

to find

$$R_{yy}(k) = E[y_{n+k}\, y_n] = E\left[ \sum_{i=0}^{\infty} h_i x_{n+k-i} \sum_{j=0}^{\infty} h_j x_{n-j} \right]$$

$$= \sum_{i,j=0}^{\infty} h_i h_j \, E[x_{n+k-i}\, x_{n-j}] = \sum_{i,j=0}^{\infty} h_i h_j R_{xx}(k - (i - j))$$

$$= \sum_{i,j,m} h_i h_j \, \delta(m - (i - j)) \, R_{xx}(k - m) = \sum_{m} R_{hh}(m) \, R_{xx}(k - m)$$

An important special case is when the input signal is white with variance $\sigma_x^2$:

$$R_{xx}(k) = E[x_{n+k}\, x_n] = \sigma_x^2 \, \delta(k) \qquad S_{xx}(z) = \sigma_x^2 \tag{2.1.11}$$

Then, Eqs. (2.1.8)–(2.1.10) simplify to

$$S_{yy}(z) = H(z) \, H(z^{-1}) \, \sigma_x^2$$

$$S_{yx}(z) = H(z) \, \sigma_x^2 \tag{2.1.12}$$

and

$$S_{yy}(\omega) = |H(\omega)|^2 \, \sigma_x^2$$

$$S_{yx}(\omega) = H(\omega) \, \sigma_x^2 \tag{2.1.13}$$

and

$$R_{yy}(k) = \sigma_x^2 \sum_{n=0}^{\infty} h_{n+k} \, h_n$$

$$R_{yx}(k) = \sigma_x^2 \, h_k \tag{2.1.14}$$

These results show how the filtering operation reshapes the flat white-noise spectrum of the input signal into a shape defined by the magnitude response $|H(\omega)|^2$ of the filter, and how the filtering operation introduces self-correlations in the output signal. Equation (2.1.13) is also the proof of the previously stated result (1.10.6).

As an example, consider the first order Markov signal $y_n$ defined previously as the output of the filter

$$y_n = ay_{n-1} + \epsilon_n \qquad H(z) = \frac{1}{1 - az^{-1}}$$

driven by white noise $\epsilon_n$ of variance $\sigma_\epsilon^2$. The impulse response of the filter is

$$h_n = a^n u(n), \qquad u(n) = \text{unit step}$$

The output autocorrelation $R_{yy}(k)$ may be computed in two ways. In the time-domain (assuming first that $k \geq 0$):

$$R_{yy}(k) = \sigma_\epsilon^2 \sum_{n=0}^{\infty} h_{n+k}\, h_n = \sigma_x^2 \sum_{n=0}^{\infty} a^{n+k}\, a^n = \sigma_x^2\, a^k \sum_{n=0}^{\infty} a^{2n} = \frac{\sigma_x^2\, a^k}{1 - a^2}$$

And second, in the $z$-domain using power spectral densities and inverse $z$-transforms (again take $k \geq 0$):

$$S_{yy}(z) = H(z)\, H(z^{-1})\, \sigma_\epsilon^2 = \frac{\sigma_\epsilon^2}{(1 - az^{-1})(1 - az)}$$

$$R_{yy}(k) = \oint_{u.c.} S_{yy}(z)\, z^k\, \frac{dz}{2\pi jz} = \oint_{u.c.} \frac{\sigma_\epsilon^2 z^k}{(z - a)(1 - az)}\, \frac{dz}{2\pi j}$$

$$= (\text{Residue at } z = a) = \frac{\sigma_\epsilon^2 a^k}{1 - a^2}$$

In particular, we verify the results of Section 1.14:

$$R_{yy}(0) = \frac{\sigma_\epsilon^2}{1 - a^2} \qquad R_{yy}(1) = \frac{\sigma_\epsilon^2\, a}{1 - a^2}$$

$$a = \frac{R_{yy}(1)}{R_{yy}(0)} \qquad \sigma_\epsilon^2 = (1 - a^2)\, R_{yy}(0)$$

It is interesting to note the exponentially decaying nature of $R_{yy}(k)$ with increasing lag $k$, as shown in Fig. 2.1. We noted earlier that direct correlations exist only between samples separated by lag one; and that indirect correlations also exist due to the indirect influence of a given sample $y_n$ on all future samples, as propagated by the difference equation. In going from one sampling instant to the next, the difference equation scales $y_n$ by a factor $a$; therefore, we expect these indirect correlations to decrease fast (exponentially) with increasing lag.

Whenever the autocorrelation drops off very fast with increasing lag, this

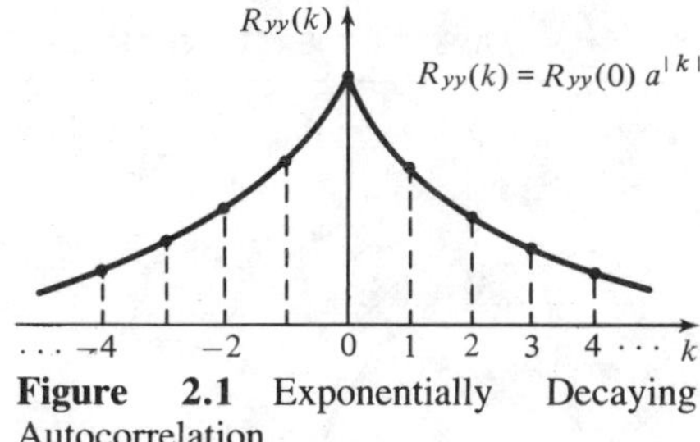



**Figure 2.1** Exponentially Decaying Autocorrelation

can be taken as an indication that there exists a stable difference equation model for the random signal.

However, not all random signals have exponentially decaying autocorrelations. For example, a pure sinusoid with random phase

$$y_n = A \cos(\omega_0 n + \phi)$$

where $\phi$ is a uniformly distributed random phase, has autocorrelation

$$R_{yy}(k) = \frac{1}{2} A^2 \cos(\omega_0 k)$$

which never dies out. A particular realization of the random variable $\phi$ defines the entire realization of the time series $y_n$. Thus, as soon as $\phi$ is fixed, the entire $y_n$ is fixed. Such random signals are called *deterministic*, since a few past values—e.g., a complete cycle—of $y_n$ are sufficient to determine all future values of $y_n$.

## 2.2 *System Identification by Cross-Correlation Methods*

The filtering results derived in Section 2.1 suggest a system identification procedure to identify an unknown system $H(z)$ on the basis of input/output measurements: Generate pseudorandom white noise $x_n$, send it through the unknown linear system, and compute the cross-correlation of the resulting output sequence $y_n$ with the known sequence $x_n$. According to Eq. (2.1.14), this cross-correlation is proportional to the impulse response of the unknown system. This identification scheme is shown in Fig. 2.2. A simulated example is shown in Fig. 2.3. The

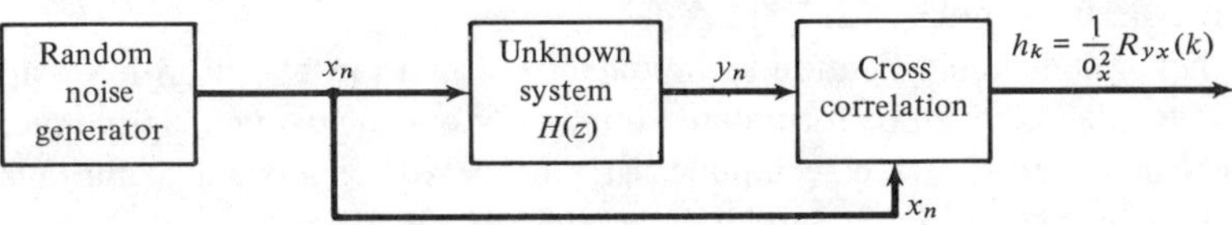



**Figure 2.2** System Identification

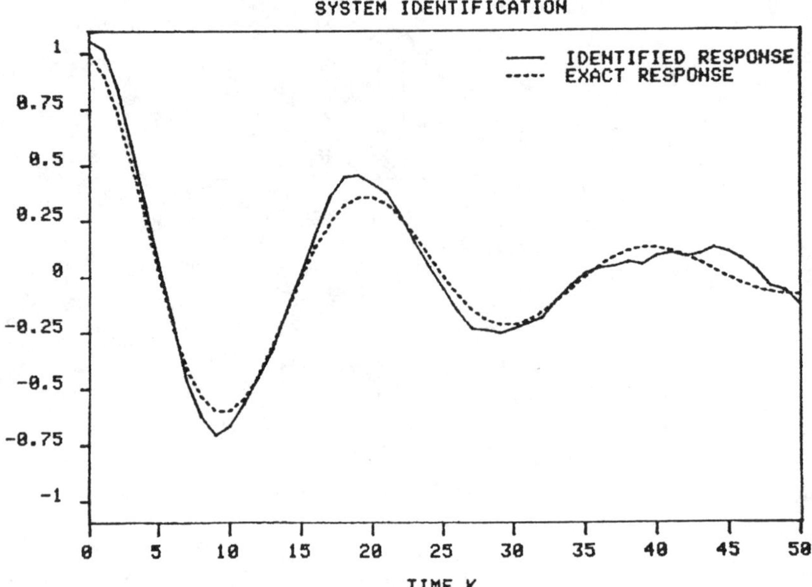


**Figure 2.3**   System Identification by Cross-Correlation

system $H(z)$ was defined by a sinusoidally damped impulse response of length 50, given by

$$h_k = (0.95)^k \cos(0.1\pi k) \qquad \text{for } 0 \le k \le 50 \qquad (2.2.1)$$

Using a random number generator routine, 1500 samples of a unit-variance zero-mean white-noise sequence $x_n$ were generated and filtered through the filter $H$ to obtain the output sequence $y_n$. Then, the first 50 lags of the sample cross-correlation were computed according to

$$\hat{R}_{yx}(k) = \frac{1}{N} \sum_{n=0}^{N-1-k} y_{n+k}\, x_n$$

with $N = 1500$ and $k = 0, 1, \ldots , 50$. Figure 2.3 shows the impulse response identified according to $h_k = \hat{R}_{yx}(k)/\sigma_x^2 = \hat{R}_{yx}(k)$, plotted together with the exact response defined by Eq. (2.2.1).

Other system identification techniques exist that are based on least-squares error criteria. They can be formulated off-line or on-line using adaptive methods [4–6]. Such identification techniques are intimately connected to the analysis procedures of extracting the model parameters of signal models, as we discussed in Section 1.10.

## 2.3 *Noise Reduction and Signal Enhancement Filters*

In signal processing applications for noise removal, or signal enhancement, the ratio $\sigma_y^2/\sigma_x^2$ plays an important role. We have

$$\sigma_y^2 = \int_{-\pi}^{\pi} S_{yy}(\omega)\, d\omega/2\pi = \int_{-\pi}^{\pi} |H(\omega)|^2 S_{xx}(\omega)\, d\omega/2\pi = \sigma_x^2 \int_{-\pi}^{\pi} |H(\omega)|^2 \frac{d\omega}{2\pi}$$

provided $x_n$ is white noise. The ratio $\sigma_y^2/\sigma_x^2$ determines whether the input noise is amplified or attenuated as it is filtered through $H(z)$. It will be referred to as the *noise reduction ratio*. Using Parseval's identity, we find the alternative expressions for it

$$\frac{\sigma_y^2}{\sigma_x^2} = \int_{-\pi}^{\pi} |H(\omega)|^2 \frac{d\omega}{2\pi}$$

$$= \sum_{n=0}^{\infty} h_n^2 \tag{2.3.1}$$

$$= \oint_{u.c.} H(z)\, H(z^{-1}) \frac{dz}{2\pi jz}$$

We may denote any one of these as $\|H\|^2$; that is, the quadratic norm of $H$. Computationally, the most recommended procedure is by the contour integral, and the least recommended is by the frequency integral.

*Example 2.3.1:*  Compute the noise reduction ratio of white noise sent through the first order recursive filter

$$H(z) = \frac{1}{1 - az^{-1}} \qquad \sigma_y^2/\sigma_x^2 = \sum_{n=0}^{\infty} h_n^2 = \sum_{n=0}^{\infty} a^{2n} = \frac{1}{1 - a^2}$$

The alternative derivation using contour integration has already been done.

Consider now the problem of extracting a signal $x_n$ from noisy measurements $y_n$ of the form

$$y_n = x_n + v_n$$

where the measurement noise $v_n$ is typically white noise. We wish to design a filter $H(z)$ to process the available measurements $y_n$ to remove the noise component without affecting the signal component. (Our notation throughout uses the symbol $y_n$ to denote the available noisy measurements, and $x_n$ to denote the desired signal to be extracted.) These two requirements are illustrated in Fig. 2.4.

Often, the separation of signal from noise can be done on the basis of

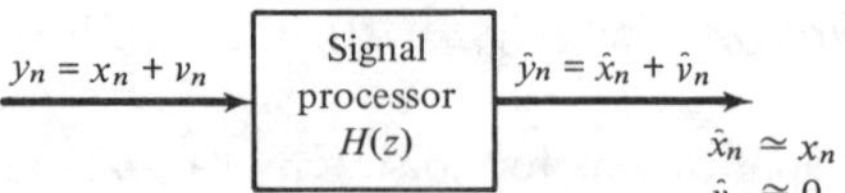

**Figure 2.4**  Signal Processor Requirements

*bandwidth*. If the power spectra of the signal and noise components occupy separate frequency bands, then their separation is easy: Simply design a filter whose frequency response is zero over the entire band over which there is significant noise power, and equal to unity over the band of the signal. An example of this situation arises in Doppler radar, which is designed to detect moving objects; the returned echo signal includes a considerable amount of clutter noise arising from the radar pulses bouncing off stationary objects such as trees, buildings, and the like. The frequency spectrum of the clutter noise is mainly concentrated near DC, whereas the spectrum of the desired signal from moving targets occupies a higher frequency band, not overlapping with the clutter.

On the other hand, if the noise is white, its power spectrum will extend over all frequencies, and therefore it will overlap with the signal band. For example, suppose the signal and noise have power spectra as shown in Fig. 2.5. If we design an ideal bandpass filter $H(\omega)$ whose passband includes the signal band, then after filtering, the output power spectra will look as in Fig. 2.6. A lot of noise energy is removed by the filter, thus tending to reduce the overall output noise variance

$$\sigma_{\hat{v}}^2 = \int_{-\pi}^{\pi} S_{\hat{v}\hat{v}}(\omega)\, d\omega/2\pi$$

At the same time, the signal spectrum is left undistorted.

Some knowledge of the frequency spectra for the desired signal and the interfering noise was required in order to design the filter. The basic idea was to design a filter whose passband coincided with the spectral band of the desired signal, and whose stopband coincided with the spectral band of the noise. Clearly, if noise and signal have highly overlapping spectra, such simple signal processing design techniques will not be successful. Thus, an important question arises: For a given signal spectrum and noise spectrum, what is the best linear filter to separate noise from signal? The answer will lead to the methods of Wiener or optimal filtering to be discussed later on. In the remainder of this section, we present four examples illustrating the above ideas.

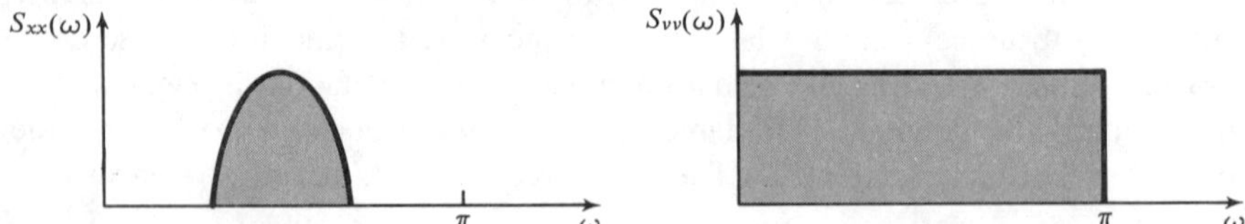

**Figure 2.5**  Signal and Noise Spectra before Processing

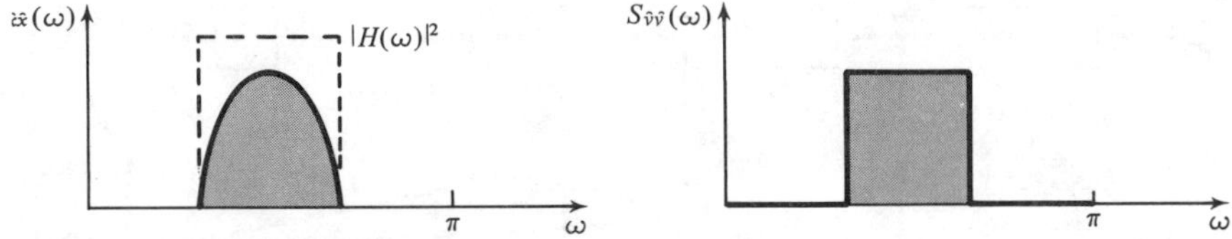

**Figure 2.6**   Signal and Noise Spectra after Processing

*Example 2.3.2   Clutter Rejection Filters in Coherent MTI Radar*

By taking advantage of the Doppler effect, moving target indicator (MTI) radar systems [7] can distinguish between weak echo returns from small moving objects and strong echo returns from stationary objects (clutter), such as trees, buildings, the sea, the weather, and so on. An MTI radar sends out short-duration sinusoidal pulses of some carrier frequency, say $f_0$. The pulses are sent out every $T$ seconds (the pulse repetition interval). A pulse reflected from a target moving with velocity $v$ will suffer a Doppler frequency shift to $f_0 + f$, where $f$ is the Doppler shift given by

$$f = \frac{2v}{\lambda_0} = \frac{2v}{c} f_0$$

The receiver maintains a phase-coherent reference carrier signal, so that the target echo signal and the reference signal can be heterodyned to leave only the relative frequency shift; that is, the Doppler shift. Thus, after the removal of the carrier, the returned echo pulses will have a basic sinusoidal dependence

$$\exp(j2\pi f t)$$

Clutter returns from truly stationary objects ($v = 0$) will correspond to the DC component ($f = 0$) of the returned signal. But, clutter returns from slightly nonstationary objects such as trees or the weather, will not be exactly DC and will be characterized by a small frequency spread about DC. Thus, a typical clutter spectrum will occupy a narrow frequency band about DC as shown:

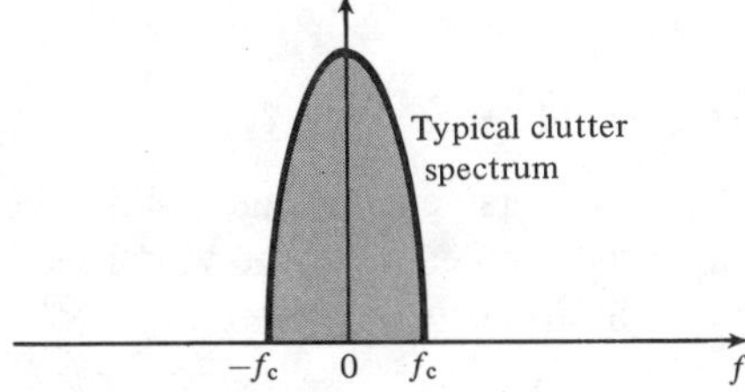


Subsequent processing with a clutter rejection filter can remove the clutter frequency components. According to the previous discussion, such a filter must essentially be an ideal high pass filter with a low frequency stopband

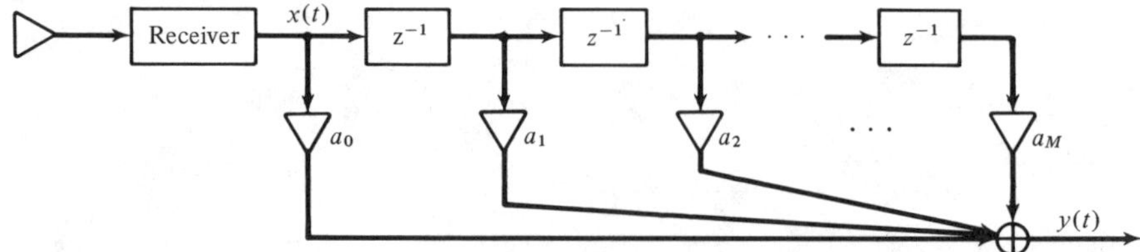

**Figure 2.7**  Tapped Delay Line Clutter Rejection Filter

that coincides with the clutter spectral band. Since the MTI system is a pulsed system with period $T$, such a filter can be designed as simple tapped delay line using delays of $T$ seconds, as shown in Fig. 2.7, where $z^{-1}$ represents a delay by $T$ seconds. The I/O equation of this filter is

$$y(t) = \sum_{m=0}^{M} a_m\, x(t - mT)$$

with transfer function

$$H(z) = a_0 + a_1 z^{-1} + a_2 z^{-2} + \cdots + a_M z^{-M}$$

Its frequency response is obtained by setting $z = \exp(j\omega) = \exp(j2\pi fT)$. Due to the sampled-data nature of this problem, the frequency response of the filter is periodic in $f$ with period $f_s = 1/T$; that is, the pulse repetition frequency. An ideally designed clutter filter would vanish over the clutter passband, as shown in Fig. 2.8. Because of the periodic nature of the frequency response, the filter will also reject the frequency bands around multiples of the sampling frequency $f_s$. If a target is moving at speeds that correspond to such frequencies, that is,

$$n\, f_s = \frac{2v}{c}\, f_0 \qquad n = 1,2,\ldots$$

then such a target cannot be seen; it also gets canceled by the filter. Such speeds are known as "blind speeds." In practice, the single and double delay high-pass filters

$$H(z) = 1 - z^{-1}$$

$$H(z) = (1 - z^{-1})^2 = 1 - 2\,z^{-1} + z^{-2}$$

are commonly used. Nonrecursive tapped delay-line filters are preferred over recursive ones, since the former have short transient response; e.g., $MT$ seconds for a filter with $M$ delays.

*Example 2.3.3:*  Radar measurements of the Earth-Moon distance $D$ are taken of the form

$$y_n = D + v_n$$

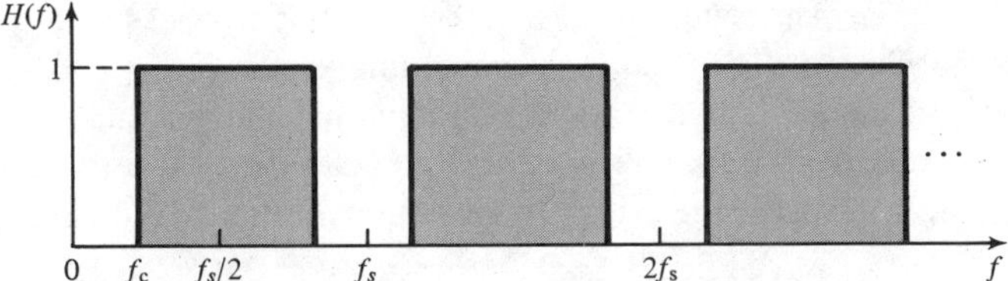

**Figure 2.8**  Ideal Frequency Response of Clutter Rejection Filter

where $v_n$ is zero-mean white noise of variance $\sigma_v^2$, representing measurement errors. Two signal processing schemes are to be compared as to their noise reduction and signal enhancing capability:

a recursive filter  and  a nonrecursive filter

$$\hat{y}_n = a\hat{y}_{n-1} + by_n \qquad\qquad \hat{y}_n = ay_n + by_{n-1}$$

Discuss the selection of the filter parameters so that on the one hand they do not distort the desired signal, and on the other they tend to reduce the noise. Discuss any tradeoffs. Compare the two cases.

The transfer functions of the two filters are

$$H(z) = \frac{b}{1 - az^{-1}} \qquad \text{and} \qquad H(z) = a + b\,z^{-1}$$

The desired signal $x_n = D$ must be able to go through these filters entirely undistorted. Since it is a DC signal, we require the frequency response $H(\omega)$ to be unity at zero frequency $\omega = 0$, or equivalently, at $z = \exp(j\omega) = 1$

$$H(1) = \frac{b}{1 - a} = 1 \qquad \text{and} \qquad H(1) = a + b = 1$$

In both cases, we find the constraint $b = 1 - a$, so that

$$H(z) = \frac{1 - a}{1 - az^{-1}} \qquad \text{and} \qquad H(z) = a + (1 - a)\,z^{-1}$$

Both of these filters will allow the DC constant signal $x_n = D$ to pass through undistorted. There is still freedom in selecting the parameter $a$. The effectiveness of the filter in reducing the noise is decided on the basis of the noise reduction ratio

$$\frac{\sigma_y^2}{\sigma_x^2} = \frac{(1 - a)^2}{1 - a^2} = \frac{1 - a}{1 + a} \qquad \text{and} \qquad \frac{\sigma_y^2}{\sigma_x^2} = a^2 + b^2 = a^2 + (1 - a)^2$$

These are easily derived using either the contour integral formula, or the sum of the impulse responses squared.

Effective noise reduction will be achieved if these ratios are made as small as possible. For the recursive case, stability of the filter requires $a$ to be $-1 < a < 1$. The requirement that the noise ratio be less than one further implies that $0 < a < 1$. And it becomes smaller the closer $a$ is selected to one. For the nonrecursive case, the minimum value of the noise ratio is obtained when $a = 0.5$. The graphical comparison of the noise reduction ratios in the two cases suggests that the recursive filter will do a much better

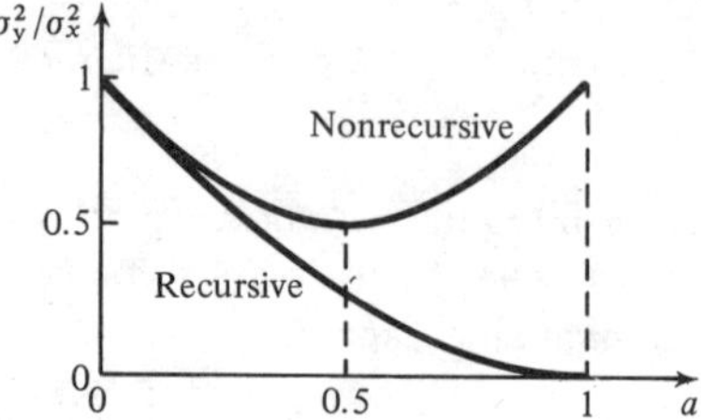


job than the nonrecursive one. But there is a price to be paid for that. The closer $a$ is to unity—that is, the closer the pole is moved to the unit circle—the slower the response of the filter will be, as can be seen by inspecting the impulse response of the filter

$$h_n = ba^n$$

The effectiveness of the recursive filter may also be seen by plotting its magnitude response versus frequency for various values of $a$, as in Fig. 2.9. As the parameter $a$ tends to one, the filter's response becomes more and more narrow around the frequency $\omega = 0$ (this is the signal band in this case). Therefore, the filter is able to remove more power from the noise.

Finally, we should not leave the impression that nonrecursive filters are always bad. For example, if we allow a filter with, say, $M$ taps

$$\hat{y}_n = \frac{1}{M} (y_n + y_{n-1} + \cdots + y_{n-M})$$

the noise ratio is easily found to be

$$\frac{\sigma_y^2}{\sigma_x^2} = \sum_{n=0}^{M} h_n^2 = \frac{1}{M^2} (1 + 1 + \cdots + 1) = \frac{M}{M^2} = \frac{1}{M}$$

which can be made as small as desired by increasing the number of taps $M$.

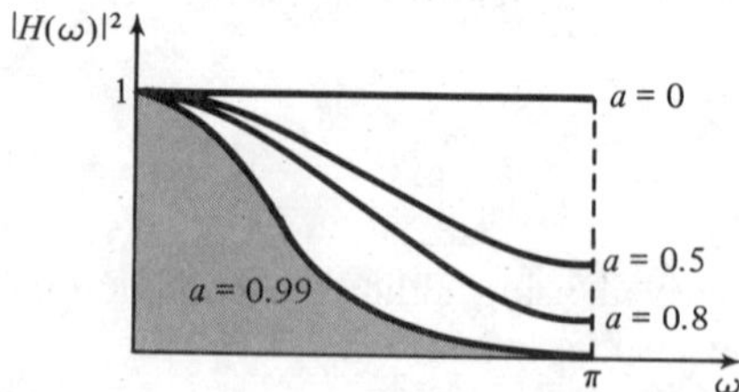


**Figure 2.9**   Magnitude Response for Various Values of $a$

Figures 2.10 through 2.16 demonstrate the various ideas behind this example. Figure 2.10 shows 100 samples of a zero-mean white gaussian noise $v_n$ of variance $\sigma_v^2 = 100$, generated by means of the standard gaussian random number generator subroutine GAUSS. Next, these samples were filtered by the first order recursive filter $\hat{v}_n = a\hat{v}_{n-1} + (1 - a)v_n$, with the parameter $a$ chosen as $a = 0.95$. Figure 2.11 shows the lowpass filtering effect as well as the noise-reducing property of this filter. The output signal $\hat{v}_n$ has been plotted together with the white noise input $v_n$. Figures 2.12 and 2.13 show a comparison of the theoretically expected autocorrelations and the experimentally computed sample autocorrelations from the actual sample values, both for the input and the output signals. The theoretical autocorrelations are

$$R_{vv}(k) = \sigma_v^2 \, \delta(k) \qquad \text{and} \qquad R_{\hat{v}\hat{v}}(k) = R_{\hat{v}\hat{v}}(0) \, a^{|k|}$$

Figures 2.14 through 2.16 show the interplay between noise reduction and speed of response of the filter as the filter parameter $a$ is gradually selected closer and closer to unity. The values $a = 0.8, 0.9,$ and $0.95$ were tried. The input to the filter was the noisy measurement signal $y_n$ and the output $\hat{y}_n$ was computed by iterating the difference equation of the filter starting with zero initial conditions.

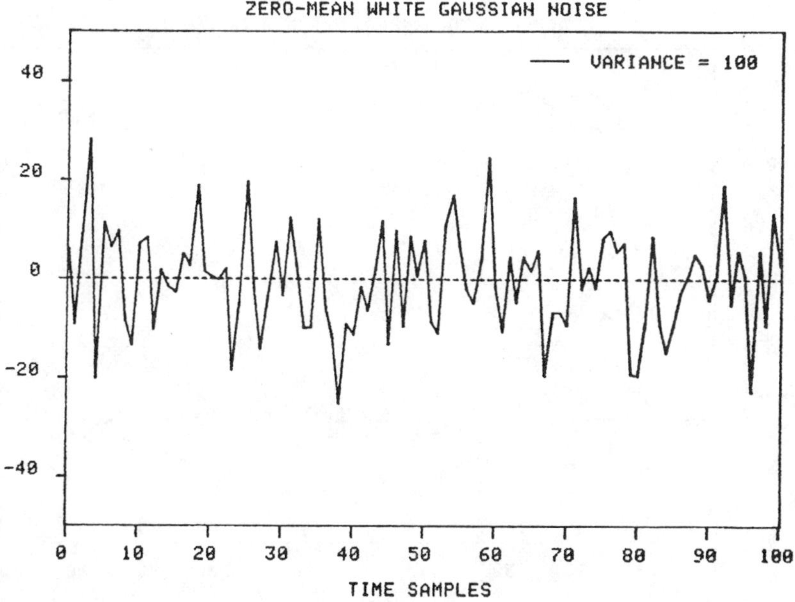


**Figure 2.10** Zero-Mean White Noise

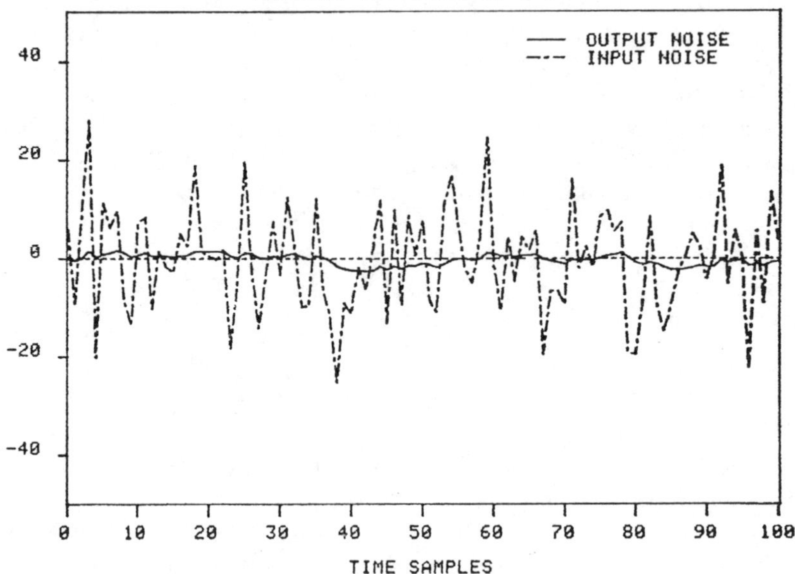



**Figure 2.11**  Filtered White Noise

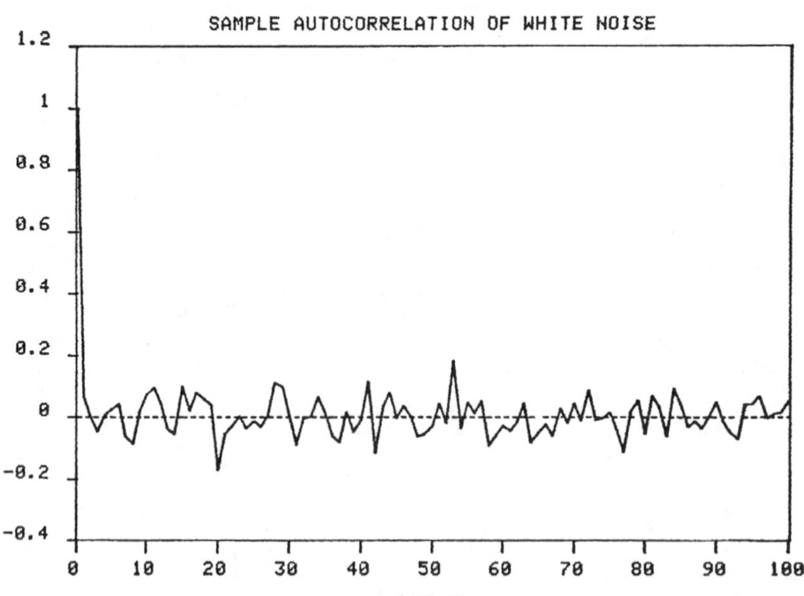



**Figure 2.12**  Sample Autocorrelation of White Noise

68

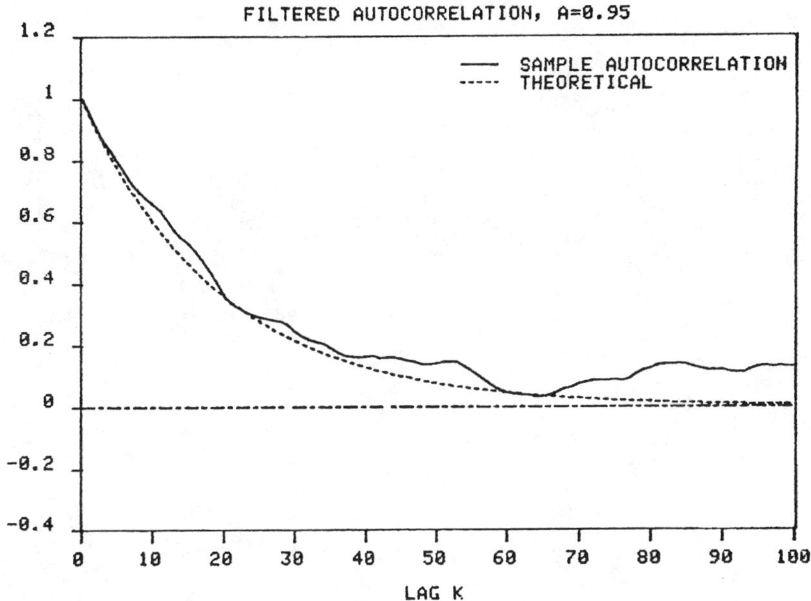

**Figure 2.13**   Sample Autocorrelation of Filtered Noise

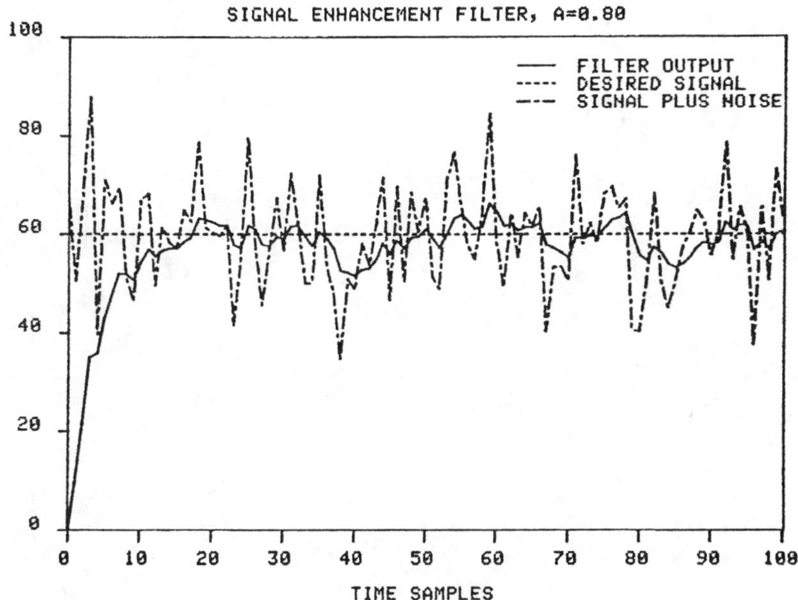

**Figure 2.14**   Filtered Measurements ($a = 0.8$)

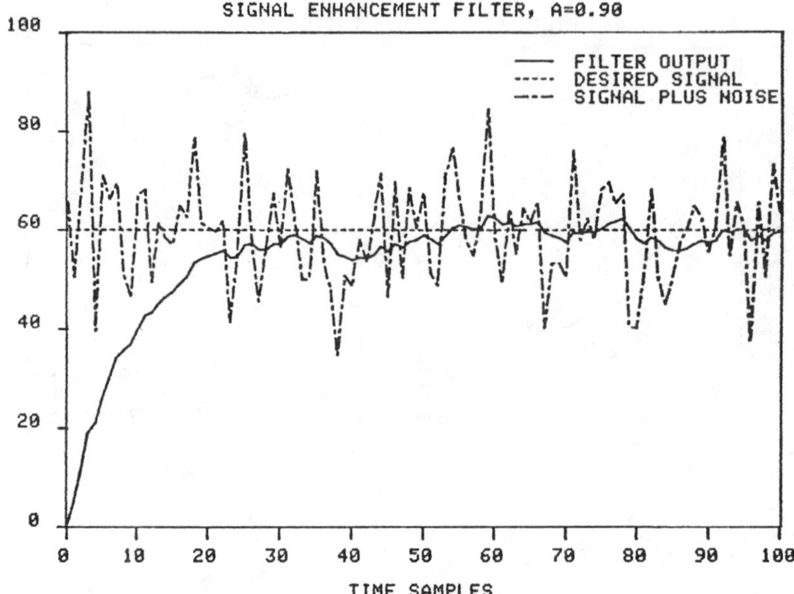


**Figure 2.15**   Filtered Measurements ($a = 0.9$)

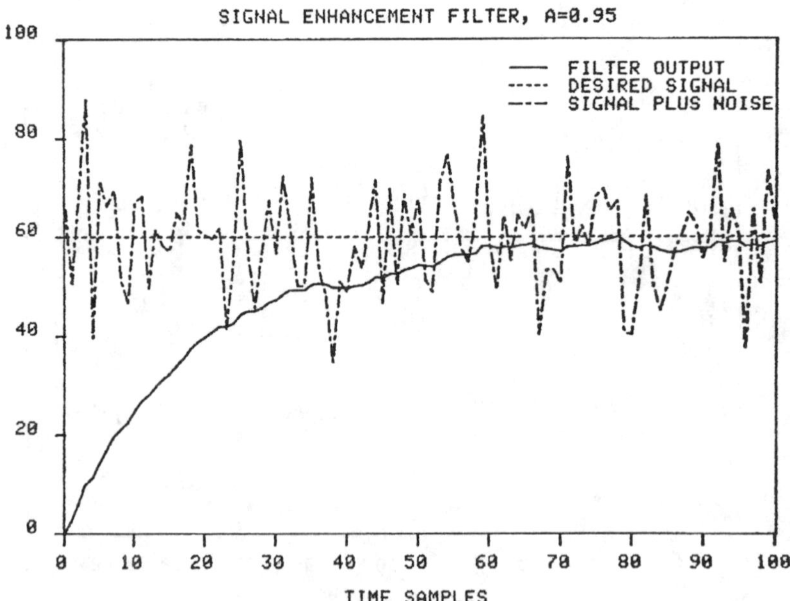


**Figure 2.16**   Filtered Measurements ($a = 0.95$)

*Example 2.3.4:* A digital AM receiver is to lock onto a carrier signal of frequency of 10 kHz. The available signal consists of the carrier signal plus white noise. If the available signal is sampled at a rate of 40 kHz, show that its samples will be of the form

$$y_n = x_n + v_n = \cos(\pi n/2) + v_n$$

where the first term represents the sampled carrier and the second the noise. To separate the signal from the noise, a 2nd order filter is used of the form

$$\hat{y}_n = -a^2\hat{y}_{n-2} + (1 - a^2)\, y_n \qquad (|a|<1)$$

Discuss the noise reduction properties of this filter.

Again, this filter has been chosen so that the desired signal which is a sinusoid of frequency $\omega_0 = \pi/2$ will pass through unchanged. That is, at $z = \exp(j\omega_0) = \exp(j\pi/2) = j$, the frequency response must be unity

$$H(z) = \frac{1 - a^2}{1 + a^2 z^{-2}} \qquad \text{with} \qquad H(z)\Big|_{z=j} = \frac{1 - a^2}{1 - a^2} = 1$$

The noise reduction ratio is most easily computed by the contour integral

$$\frac{\sigma_y^2}{\sigma_x^2} = \oint_{u.c.} H(z)\, H(z^{-1})\, \frac{dz}{2\pi jz} = \oint_{u.c.} \frac{(1 - a^2)^2 z}{(z - aj)(z + aj)(1 + a^2 z^2)}\, \frac{dz}{2\pi j}$$

$$= (\text{residues at } z = \pm ja) = \frac{(1 - a^2)^2}{1 - a^4} = \frac{1 - a^2}{1 + a^2}$$

Selecting the poles $\pm ja$ to be near the unit circle (from inside) will result in a slow but efficient filter in reducing the noise component.

*Example 2.3.5: Signal Enhancement by Digital Averaging.*

Signal averaging computers are routinely used to improve the signal to noise ratio of signals that are corrupted by noise and can be measured repeatedly—for example, in measuring evoked action potentials using scalp electrodes, or in integrating successive returns in pulsed radar. A similar concept is also used in the so-called "beamforming" operation in sonar and radar arrays. The objective is to measure a signal $x(n)$ of duration of $N$ samples, $n = 0,1, \ldots ,N - 1$. The measurement can be performed (evoked) repeatedly. A total of $M$ such measurements are performed and the results are averaged by the signal averaging computer. Let the results of the $m$th measurement, for $m = 1,2, \ldots ,M$, be the samples

$$y_m(n) = x(n) + v_m(n), \qquad \text{for } n = 0,1, \ldots ,N - 1$$

A signal averaging computer averages (integrates) the results of the $M$ measurements

$$\hat{x}(n) = \frac{1}{M} \sum_{m=1}^{M} y_m(n) \qquad \text{for } n = 0,1, \ldots ,N-1$$

by accumulating (integrating) the results of the $M$ measurements, as shown in following diagram

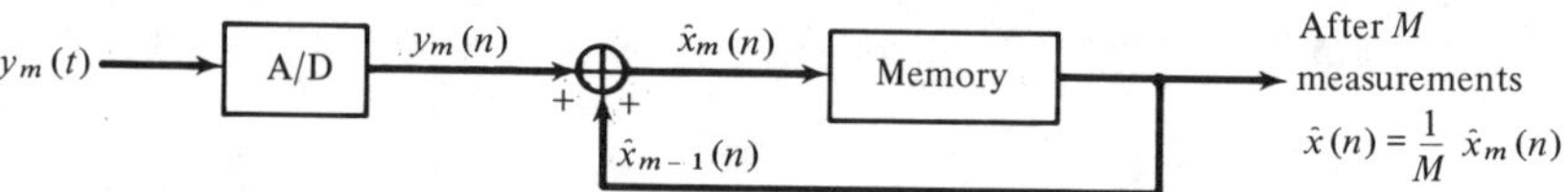


The result of the averaging operation may be expressed as

$$\hat{x}(n) = \frac{1}{M} \sum_{m=1}^{M} y_m(n) = \frac{1}{M} \sum_{m=1}^{M} [x(n) + v_m(n)] = x(n) + \hat{v}(n)$$

where

$$\hat{v}(n) = \frac{1}{M} \sum_{m=1}^{M} v_m(n)$$

Assuming $v_m(n)$ to be mutually uncorrelated; that is, $E[v_m(n)v_l(n)] = \sigma_v^2\,\delta_{ml}$, we compute the variance of the averaged noise $\hat{v}(n)$:

$$\sigma_{\hat{v}}^2 = E[\hat{v}(n)^2] = \frac{1}{M^2} \sum_{m,l=0}^{M} E[v_m(n)v_l(n)] = \frac{1}{M^2} \sum_{m,l=0}^{M} \sigma_v^2\,\delta_{ml}$$

$$= \frac{1}{M^2}\,(\sigma_v^2 + \sigma_v^2 + \cdots + \sigma_v^2) = \frac{1}{M^2}\,M\sigma_v^2 = \frac{1}{M}\,\sigma_v^2$$

Therefore, the signal to noise ratio (SNR) is improved by a factor of $M$.

## 2.4 *Quantization Noise*

In digital filtering operations, such as shown in the following figure, one must

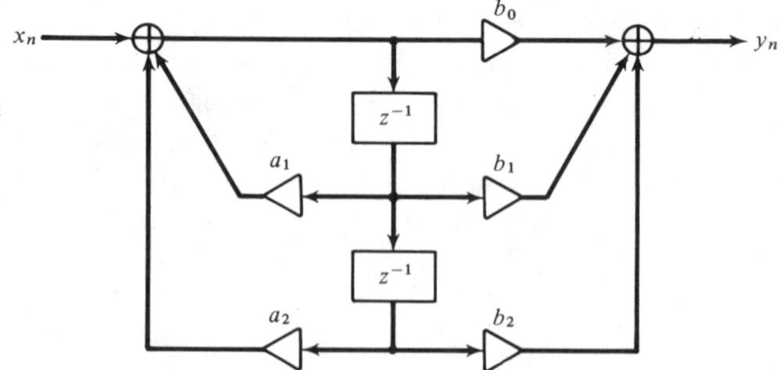

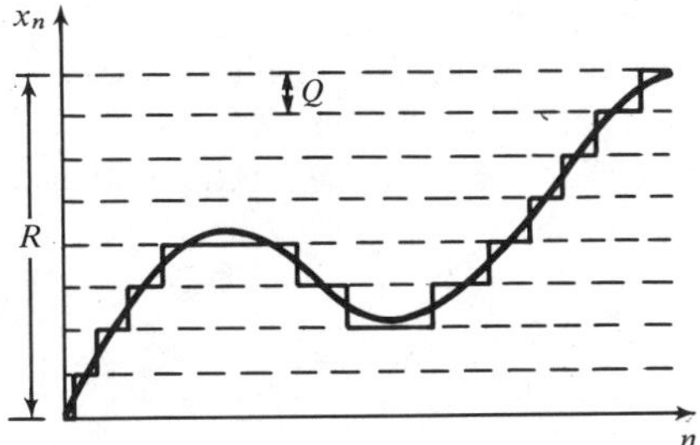

**Figure 2.17**  Uniform Quantizer

deal with the following types of quantization errors [2,3,8]:

1. Quantization of the input samples $x_n$ due to the A/D conversion.
2. Quantization of the filter coefficients $a_i$, $b_i$.
3. Roundoff errors from the internal multiplications.

A typical uniform quantization operation of a sampled signal is shown in Fig. 2.17. The spacing between levels is denoted by $Q$ and the overall range of variation of the signal by $R$. If $b$ bits are assigned to represent each sample value, then the total number of representable signal values is $2^b$, and therefore the number of levels that can fit within the range $R$ is

$$2^b = \frac{R}{Q}$$

which also leads to the so-called "6 dB per bit" rule for the dynamic range of the quantizer

$$\text{dB} = 10 \log_{10} \left(\frac{R}{Q}\right)^2 = b\, 20 \log_{10}(2) = 6b \text{ decibels}$$

The quantization operation may be represented as

$$x_n \longrightarrow \boxed{Q} \longrightarrow [x_n]$$

where $[x_n]$ denotes the quantized value of $x_n$; that is, the nearest level. The quantization error is $\delta_n = [x_n] - x_n$. The quantization operation may be replaced by an equivalent additive source of noise $\delta_n$, as shown in Fig. 2.18. In the case of *large-amplitude wideband* signals; that is, signals that vary rapidly through the entire range $R$, it may be assumed that the quantization error is a *uniformly* distributed *white-noise* signal. It is further assumed that the quantization noise $\delta_n$ is uncorrelated with the input signal $x_n$. In such a case, the quantization noise

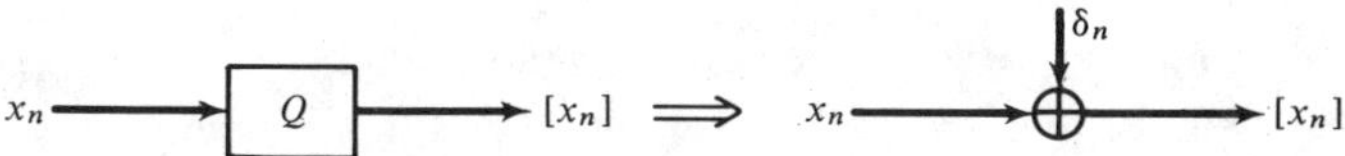

**Figure 2.18**  Equivalent Noise Model for a Quantizer

lends itself to simple statistical treatment. Namely, the quantization operation may be replaced by an equivalent additive white-noise source, acting where the quantization operation is occurring. Since $\delta_n$ is assumed to be uniformly distributed in the range $-Q/2 \leq \delta_n \leq Q/2$, it follows that it has zero mean and variance

$$\sigma_\delta^2 = \frac{Q^2}{12} \tag{2.4.1}$$

## 2.5  *Statistical Treatment of Multiplier Roundoff Error*

Here, we would like to use the results of the previous section in the computation of the roundoff noise arising from the internal multiplications in digital filters. Consider a typical multiplier in a digital filter

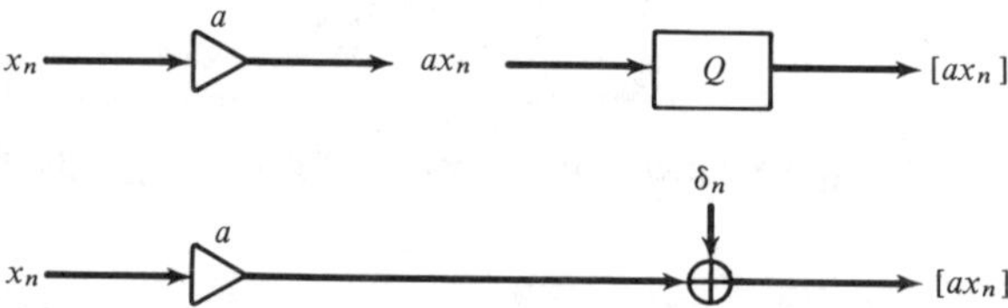

The result of the multiplication requires double precision to be represented fully. If this result is subsequently rounded to single precision, then the overall operation, and its noise-equivalent model, will be of the form

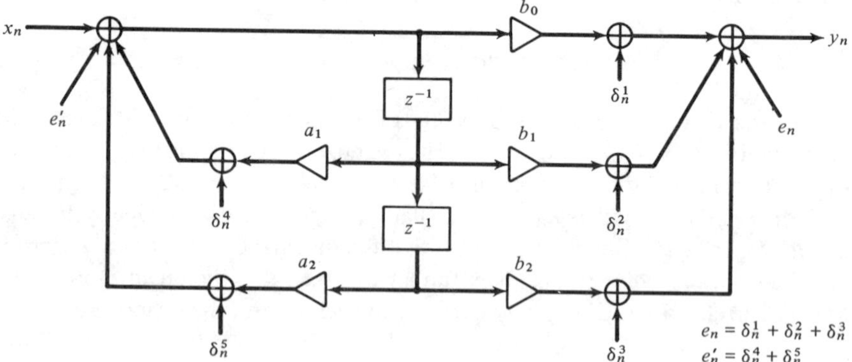

For example, the second order section shown above will be replaced by

with five elementary noise sources acting independently of each other at the locations shown. In placing these noise sources at those locations, we are implicitly assuming that a quantization operation is being performed immediately after each multiplication. This may not always be true, especially in the newer hardware configurations which employ special purpose chips, such as the Bell Lab Digital Signal Processor, or the TRW multiplier-accumulators. Such chips perform the multiplication operation with full double precision. Depending on the specific digital filter implementation, it is possible for such full precision products to accumulate somewhat before the result is finally rounded back to single precision. The methods presented here can easily be extended to such a situation. For illustrative purposes, we shall assume that the quantizing operations occur just after each multiplication.

To find the output noise power resulting from each noise source we must identify the transfer function from each noise source to the output of the filter. For example, the three elementary noises at the forward multipliers may be combined into one acting at the output adder and having combined variance

$$\sigma_e^2 = 3\,\sigma_\delta^2 = 3\,Q^2/12$$

and the two noises at the two feedback multipliers may be replaced by one acting at the input adder and having variance

$$\sigma_{e'}^2 = 2\,\sigma_\delta^2 = 2\,Q^2/12$$

The transfer function from $e_n'$ to the output is $H(z)$ itself, and from $e_n$ to the output it is unity. Adding the output noise variances due to $e_n'$ and $e_n$, we find the total output roundoff noise power

$$\sigma_\epsilon^2 = \sigma_{e'}^2\,\|H\|^2 + \sigma_e^2$$

*Example 2.5.1:* Suppose $H(z) = H_1(z)H_2(z)$,

where $H_1(z) = \dfrac{1}{1 - az^{-1}}$, and $H_2(z) = \dfrac{1}{1 - bz^{-1}}$ with $a > b$.

Determine the output roundoff noise powers when the filter is realized in the following three forms:

1. $H_1(z)$ cascaded by $H_2(z)$
2. $H_2(z)$ cascaded by $H_1(z)$
3. $H(z)$ realized in its canonical form

In case 1, the roundoff noises are as shown. The transfer functions of $e_n$

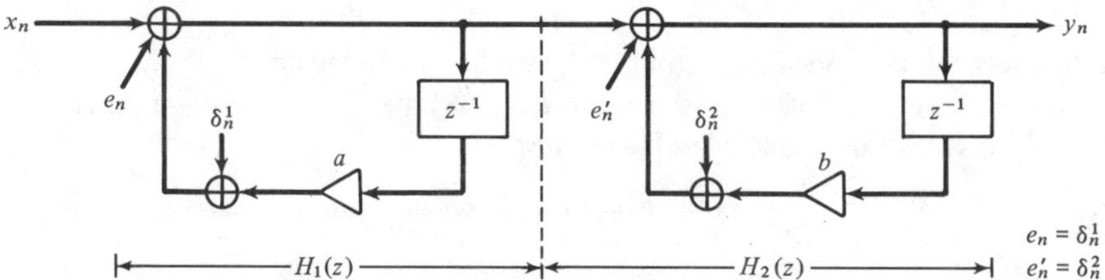

and $e'_n$ to the output are $H(z)$ and $H_2(z)$ respectively. Adding the output noise power from each we find

$$\sigma_\epsilon^2 = \sigma_\delta^2 \|H\|^2 + \sigma_\delta^2 \|H_2\|^2$$

$$= \frac{Q^2}{12} \frac{1 + ab}{(1 - ab)(1 - a^2)(1 - b^2)} + \frac{Q^2}{12} \frac{1}{1 - b^2}$$

Interchanging the role of $H_1$ and $H_2$ we find, for case 2,

$$\sigma_\epsilon^2 = \frac{Q^2}{12} \frac{1 + ab}{(1 - ab)(1 - a^2)(1 - b^2)} + \frac{Q^2}{12} \frac{1}{1 - a^2}$$

And finally in case 3, the canonical realization has two elementary noise

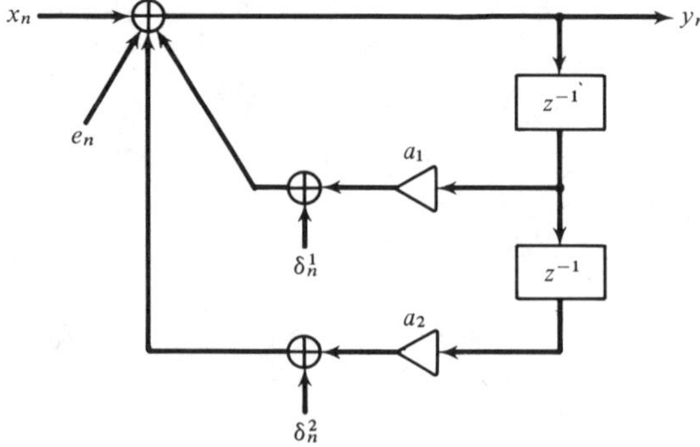


sources as shown. They may be combined into one acting at the input adder. Its variance will be $\sigma_e^2 = 2\,Q^2/12$. The transfer function to the output is $H(z)$ itself; thus,

$$\sigma_\epsilon^2 = \sigma_e^2 \|H\|^2 = 2 \frac{Q^2}{12} \frac{1 + ab}{(1 - ab)(1 - a^2)(1 - b^2)}$$

It can be seen from the above example that the output roundoff power depends on the particular *realization* of the digital filter. A great deal of research has gone into developing realization structures that minimize the roundoff noise [9–16].

## 2.6 *Introduction to Linear Prediction*

In this section, we present a preliminary introduction to the concepts and methods of linear prediction based on the finite past. We have already mentioned how prediction ideas come into play by reinterpreting

$$\hat{y}_n = ay_{n-1} = \text{prediction of } y_n \text{ based on one past sample}$$

$$e_n = y_n - \hat{y}_n = y_n - ay_{n-1} = \text{prediction error}$$

and have indicated how to determine the prediction coefficient $a$ by a least squares minimization criterion. Here, we would like to replace that criterion, based on time averages, with a least squares criterion based on statistical averages:

$$\mathcal{E}(a) = E[e_n^2] = E[(y_n - ay_{n-1})^2] = \min$$

We will no longer assume that $y_n$ is a first order autoregressive Markov process; thus, the prediction error $e_n$ will not quite be white noise. The problem we are posing is to find the best linear predictor based on the previous sample alone, regardless of whether $y_n$ is a first order autoregressive process or not. That is, we seek the projection of $y_n$ on the subspace $Y_{n-1} = \{y_{n-1}\}$ spanned only by previous sample $y_{n-1}$; this projection is the best linear estimate of $y_n$ based on $y_{n-1}$ only. If, accidentally, the signal $y_n$ happened to be first order autoregressive, then $e_n$ would turn out to be white and our methods would determine the proper value for the Markov model parameter $a$.

The best value for the prediction coefficient $a$ is obtained by differentiating $\mathcal{E}$ with respect to $a$, and setting the derivative to zero

$$\frac{\partial \mathcal{E}}{\partial a} = 2E\left[e_n \frac{\partial e_n}{\partial a}\right] = -2E[e_n y_{n-1}] = 0$$

Thus, we obtain the orthogonality equation

$$E[e_n y_{n-1}] = 0$$

which states that the prediction error $e_n$ is decorrelated from $y_{n-1}$. An equivalent way of writing this condition is the normal equation

$$E[e_n y_{n-1}] = E[(y_n - ay_{n-1})y_{n-1}] = E[y_n y_{n-1}] - a E[y_{n-1}^2] = 0$$

or, in terms of the autocorrelations $R_{yy}(k) = E[y_{n+k}y_n]$,

$$R_{yy}(1) = a R_{yy}(0) \qquad \text{or} \qquad a = \frac{R_{yy}(1)}{R_{yy}(0)}$$

The minimum value of the prediction error $\mathcal{E}(a)$ for the above optimal value of $a$, may be written as

$$\begin{aligned}
\min \mathcal{E} &= E[e_n^2] = E[e_n(y_n - ay_{n-1})] = E[e_n y_n] - a E[e_n y_{n-1}] \\
&= E[e_n y_n] = E[(y_n - ay_{n-1})y_n] = E[y_n^2] - a E[y_{n-1} y_n] \\
&= R_{yy}(0) - a R_{yy}(1) = R_{yy}(0) - [R_{yy}(1)^2/R_{yy}(0)] \\
&= (1 - a^2) R_{yy}(0)
\end{aligned}$$

The resulting prediction-error filter has transfer function

$$y_n \longrightarrow \boxed{A'(z)} \longrightarrow e_n$$

A realization of the prediction filter is drawn in Fig. 2.19. The upper output is the prediction error, whose average power has been minimized, and the lower

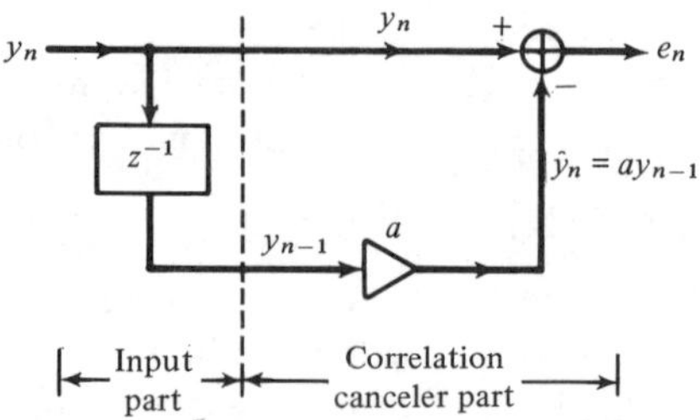



**Figure 2.19**   First Order Linear Predictor

output is the predicted waveform. The original signal may be written as a sum of two terms:

$$y_n = \hat{y}_n + e_n$$

The first term $\hat{y}_n = ay_{n-1}$ is highly correlated with the secondary signal $y_{n-1}$, which in turn is input to the multiplier. The second term $e_n$, by virtue of the orthogonality relations, is completely uncorrelated with $y_{n-1}$. In Fig. 2.19 we have indicated a dividing line between the input part and the correlation canceler part. The latter may be recognized as a special case of the correlation canceler configuration of Fig. 1.1. The input part simply provides the two inputs to the correlation canceler. Since these two inputs are $y_n$ and $y_{n-1}$, the canceler tries to cancel any correlations that may exist between these two signals; in other words, it tries to remove any serial correlations that might be present in $y_n$.

Next, we discuss higher order predictors and find their connection to lower order predictors. First, we change to a more standard notation by replacing the parameter $a$ by $a_1 = -a$. That is, we take

$$\hat{y}_n = -a_1 y_{n-1} = \text{prediction of } y_n \text{ based on } one \text{ past sample}$$

$$e_n = y_n - \hat{y}_n = y_n + a_1 y_{n-1} = \text{prediction error} \qquad (2.6.1)$$

$$\mathcal{E}(a_1) = E[e_n^2] = E[(y_n + a_1 y_{n-1})^2] = \text{minimum}$$

It will prove instructive to discuss, in parallel, the second order case of predicting $y_n$ on the basis of two past samples $y_{n-1}$ and $y_{n-2}$

$$\hat{y}_n' = -[a_1' y_{n-1} + a_2' y_{n-2}]$$

$$= \text{prediction of } y_n \text{ based on } two \text{ past samples}$$

$$e_n' = y_n - \hat{y}_n' = y_n + a_1' y_{n-1} + a_2' y_{n-2} = \text{prediction error}$$

$$\mathcal{E}(a_1', a_2') = E[e_n'^2] = E[(y_n + a_1' y_{n-1} + a_2' y_{n-2})^2] = \text{minimum}$$

The second order predictor $\hat{y}_n'$ of $y_n$ is the orthogonal projection of $y_n$ onto the subspace spanned by the past two samples $Y_{n-1}' = \{y_{n-1}, y_{n-2}\}$. A realization of the second order predictor is shown in Fig. 2.20. Again, it may be recognized

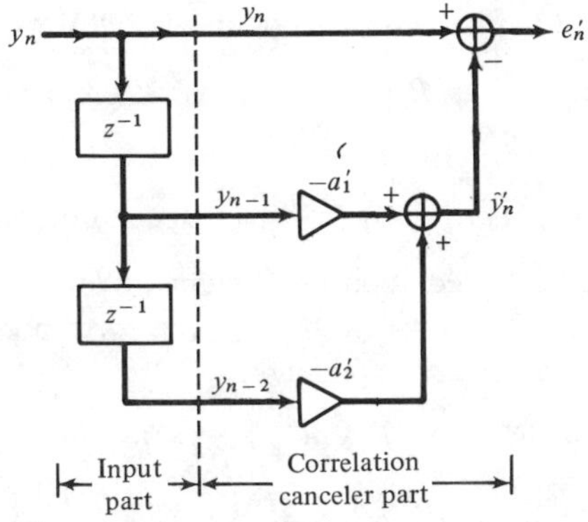



**Figure 2.20** Second Order Linear Predictor

as a special case of the correlation canceler. The input part provides the necessary inputs to the canceler. The main input to the canceler is $y_n$ and the secondary input is the 2-vector $\begin{bmatrix} y_{n-1} \\ y_{n-2} \end{bmatrix}$. The canceler tries to remove any correlations between $y_n$ and $y_{n-1}$ as well as $y_{n-2}$. That is, it tries to remove even more sequential correlations than the first order predictor did.

The corresponding prediction-error filters are for the two cases

$$A(z) = 1 + a_1 z^{-1} \quad \text{and} \quad A'(z) = 1 + a'_1 z^{-1} + a'_2 z^{-2}$$

Our objective is to determine the best choice of the prediction-error filters $(1, a_1)$ and $(1, a'_1, a'_2)$ such that the corresponding mean-squared prediction errors are minimized. The minimization conditions in the two cases become

$$\frac{\partial \mathcal{E}}{\partial a_1} = 2E\left[ e_n \frac{\partial e_n}{\partial a_1} \right] = 2E[e_n y_{n-1}] = 0$$

$$\frac{\partial \mathcal{E}'}{\partial a'_1} = 2E\left[ e'_n \frac{\partial e'_n}{\partial a'_1} \right] = 2E[e'_n y_{n-1}] = 0,$$

$$\frac{\partial \mathcal{E}'}{\partial a'_2} = 2E[e'_n y_{n-2}] = 0$$

Inserting

$$e_n = \sum_{m=0}^{1} a_m y_{n-m} \qquad \text{(we set } a_0 = 1\text{)}$$

$$e'_n = \sum_{m=0}^{2} a'_m y_{n-m} \qquad \text{(again, } a'_0 = 1\text{)}$$

into these orthogonality equations, we obtain the two sets of normal equations

$$R(1) + a_1 R(0) = 0 \qquad \text{(first order predictor)} \qquad (2.6.2)$$

$$R(1) + a_1' R(0) + a_2' R(1) = 0 \qquad (2.6.3)$$

$$R(2) + a_1' R(1) + a_2' R(0) = 0 \qquad \text{(second order predictor)}$$

which determine the best prediction coefficients. We have also simplified our previous notation and set $R(k) = E[y_{n+k} y_n]$. The corresponding minimal values for the mean-squared errors are expressed as

$$\mathcal{E} = E[e_n^2] = E[e_n y_n] = R(0) + a_1 R(1) \qquad (2.6.4)$$

$$\mathcal{E}' = E[e_n'^2] = E[e_n' y_n] = R(0) + a_1' R(1) + a_2' R(2) \qquad (2.6.5)$$

We have already shown the first of these. The second is derived by a similar procedure

$$\mathcal{E}' = E[e_n'^2] = E[e_n'(y_n + a_1' y_{n-1} + a_2' y_{n-2})]$$

$$= E[e_n' y_n] + a_1' E[e_n' y_{n-1}] + a_2' E[e_n' y_{n-2}] = E[e_n' y_n]$$

$$= E[(y_n + a_1' y_{n-1} + a_2' y_{n-2})y_n], \text{ etc.}$$

The orthogonality equations, together with the equations for the prediction errors, can be put into a matrix form as follows

$$\begin{bmatrix} R(0) & R(1) \\ R(1) & R(0) \end{bmatrix} \begin{bmatrix} 1 \\ a_1 \end{bmatrix} = \begin{bmatrix} \mathcal{E} \\ 0 \end{bmatrix}$$

$$\begin{bmatrix} R(0) & R(1) & R(2) \\ R(1) & R(0) & R(1) \\ R(2) & R(1) & R(0) \end{bmatrix} \begin{bmatrix} 1 \\ a_1' \\ a_2' \end{bmatrix} = \begin{bmatrix} \mathcal{E}' \\ 0 \\ 0 \end{bmatrix} \qquad (2.6.6)$$

*Example 2.6.1:* Rederive the results (2.6.3) and (2.6.5) for the second order predictor using the correlation canceler formulation of Section 1.4. In the notation of Section 1.4, the primary input to the canceler is the 1-vector $\mathbf{x} = [y_n]$ and the secondary input is the 2-vector $\mathbf{y} = \begin{bmatrix} y_{n-1} \\ y_{n-2} \end{bmatrix}$. Then,

$$R_{xy} = E[y_n \, \overparen{y_{n-1}, y_{n-2}}] = [E[y_n y_{n-1}], E[y_n y_{n-2}]] = [R(1), R(2)]$$

$$R_{yy} = E\left[ \begin{pmatrix} y_{n-1} \\ y_{n-2} \end{pmatrix} \overparen{y_{n-1}, y_{n-2}} \right] = \begin{bmatrix} R(0) & R(1) \\ R(1) & R(0) \end{bmatrix}$$

Therefore,

$$H = R_{xy} R_{yy}^{-1} = [R(1), R(2)] \begin{bmatrix} R(0) & R(1) \\ R(1) & R(0) \end{bmatrix}^{-1}$$

If we denote this row vector by $H = -[a_1', a_2']$, we find

$$-[a_1', a_2'] = [R(1), R(2)] \begin{bmatrix} R(0) & R(1) \\ R(1) & R(0) \end{bmatrix}^{-1}$$

which is the solution of Eq. (2.6.3). The corresponding estimate $\hat{\mathbf{x}} = H\mathbf{y}$ is then

$$\hat{y}_n' = -[a_1', a_2'] \begin{bmatrix} y_{n-1} \\ y_{n-2} \end{bmatrix} = -[a_1'\, y_{n-1} + a_2'\, y_{n-2}]$$

and the minimum value of the mean-squared estimation error is

$$\mathcal{E}' = E[e_n'^2] = R_{xx} - H\,R_{yx} = E[y_n^2] + [a_1', a_2'] \begin{bmatrix} R(1) \\ R(2) \end{bmatrix}$$

$$= R(0) + a_1'\, R(1) + a_2'\, R(2)$$

which agrees with Eq. (2.6.5).

## 2.7 *Gapped Functions and the Levinson Recursion*

Instead of solving the matrix equations (2.6.6) directly, or independently, of each other, we would like to develop an iterative procedure for constructing the solution $(1, a_1', a_2')$ in terms of the solution $(1, a_1)$. The procedure is known as Levinson's algorithm. To this end, it proves convenient to work with the elegant concept of the "gapped" functions, first introduced into this context by Robinson and Treitel [17]. The gapped functions for the first and second order predictors are defined by

$$g(k) = E[e_n\, y_{n-k}] \qquad \text{(for first order predictor)}$$

$$g'(k) = E[e_n'\, y_{n-k}] \qquad \text{(for second order predictor)}$$

They are the cross-correlations between the prediction-error sequences and the sequence $y_n$. These definitions are motivated by the orthogonality equations, which are the determining equations for the prediction coefficients. That is, if the best coefficients $(1, a_1)$ and $(1, a_1', a_2')$ are used, then the gapped functions must vanish at lags $k = 1$ for the first order case, and $k = 1, 2$ for the second order case; that is,

$$g(1) = E[e_n\, y_{n-1}] = 0$$

$$g'(1) = E[e_n'\, y_{n-1}] = 0, \qquad g'(2) = E[e_n'\, y_{n-2}] = 0$$

Thus, the functions $g(k)$ and $g'(k)$ develop gaps of lengths one and two, respectively, as seen in Fig. 2.21. A special role is played by the value of the

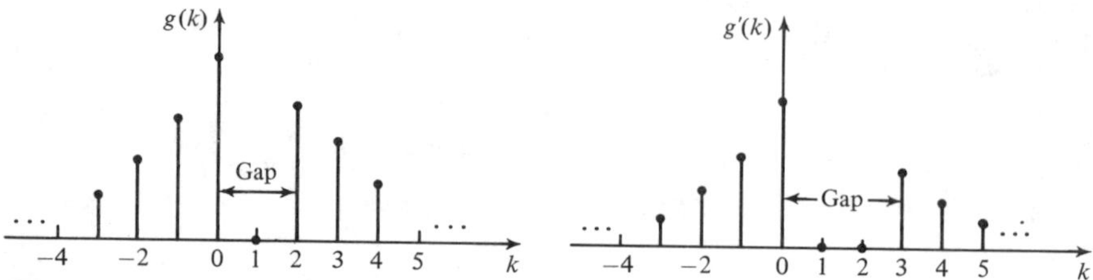

**Figure 2.21**  Gapped Functions of Orders One and Two

gapped functions at $k = 0$. It follows from the expressions (2.6.4) for the minimized prediction errors that

$$\mathcal{E} = g(0) = E[e_n\, y_n] \qquad \text{and} \qquad \mathcal{E}' = g'(0) = E[e'_n\, y_n] \qquad (2.7.1)$$

The gapped functions may also be expressed as the *convolution* of the prediction-error filters $(1,a_1)$ and $(1,a'_1,a'_2)$ with the autocorrelation function $R(k) = E[y_{n+k}\, y_n]$, as can be seen from the definition

$$g(k) = E[e_n\, y_{n-k}] = E\left[\left(\sum_{m=0}^{1} a_m\, y_{n-m}\right) y_{n-k}\right] \qquad (2.7.2)$$

$$= \sum_{m=0}^{1} a_m E[y_{n-m}\, y_{n-k}] = \sum_{m=0}^{1} a_m\, R(k - m)$$

and similarly

$$g'(k) = E[e'_n\, y_{n-k}] = \sum_{m=0}^{2} a'_m\, R(k - m) \qquad (2.7.3)$$

The *Levinson recursion*, which iteratively constructs the best linear predictor of order two from the best predictor of order one, can be derived with the help of the gapped functions. The basic idea is to use the gapped function of order one, which already has a gap of length one, and construct from it a new gapped function with gap of length two.

Starting with $g(k)$, first *reflect* it about the origin, then *delay* it sufficiently until the gap of the reflected function is aligned with the gap of $g(k)$. In the present case, the required delay is only two units, as can be seen from Fig. 2.22. Any linear combination of the two gapped functions $g(k)$ and $g(2 - k)$ will have gap of at least length one. Now, select the coefficients in the linear combination so that the gap becomes of length two

$$g'(k) = g(k) - \gamma_2\, g(2 - k) \qquad (2.7.4)$$

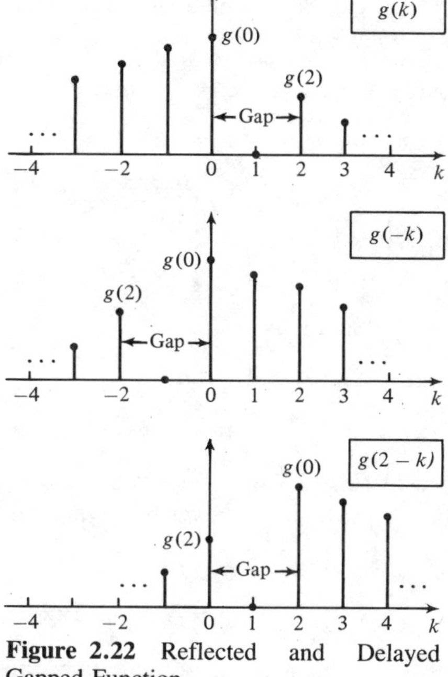



**Figure 2.22** Reflected and Delayed Gapped Function

with the extra gap condition $g'(2) = 0$

$$g'(2) = g(2) - \gamma_2 \, g(0) = 0$$

which determines the coefficient $\gamma_2$ as

$$\gamma_2 = \frac{g(2)}{g(0)} = \frac{R(2) + a_1 R(1)}{R(0) + a_1 R(1)} \tag{2.7.5}$$

The coefficient $\gamma_2$ is called the *reflection coefficient* or PARCOR coefficient and if selected as above, it will ensure that the new gapped function $g'(k)$ has a gap of length two. To find the new prediction-error filter, we write Eq. (2.7.4) in the $z$-domain, noting that the $z$-transform of $g(-k)$ is $G(z^{-1})$, and that of $g(2 - k)$ is $z^{-2}G(z^{-1})$

$$G'(z) = G(z) - \gamma_2 \, z^{-2} \, G(z^{-1})$$

Using the convolutional equations (2.7.2) and (2.7.3), expressed in the $z$-domain, we find

$$A'(z) \, S_{yy}(z) = A(z) \, S_{yy}(z) - \gamma_2 \, z^{-2} \, A(z^{-1}) \, S_{yy}(z^{-1})$$

Since $S_{yy}(z^{-1}) = S_{yy}(z)$, it can be canceled from both sides, giving the desired relationship between the new and the old prediction-error filters

$$A'(z) = A(z) - \gamma_2 \, z^{-2} \, A(z^{-1}) \qquad \text{(Levinson recursion)} \qquad (2.7.6)$$

and equating coefficients

$$\begin{bmatrix} 1 \\ a_1' \\ a_2' \end{bmatrix} = \begin{bmatrix} 1 \\ a_1 \\ 0 \end{bmatrix} - \gamma_2 \begin{bmatrix} 0 \\ a_1 \\ 1 \end{bmatrix} \qquad \begin{aligned} a_1' &= a_1 - \gamma_2 \, a_1 \\[4pt] a_2' &= -\gamma_2 \end{aligned}$$

Introducing the reverse polynomials

$$A^R(z) = z^{-1} \, A(z^{-1}) = a_1 + z^{-1}$$

$$A'^R(z) = z^{-2} \, A'(z^{-1}) = a_2' + a_1' \, z^{-1} + z^{-2}$$

and taking the reverse of Eq. (2.7.6), we obtain a more convenient recursion that involves both the forward and the reverse polynomials:

$$A'(z) = A(z) - \gamma_2 \, z^{-1} \, A^R(z)$$
$$A'^R(z) = z^{-1} \, A^R(z) - \gamma_2 \, A(z) \qquad\qquad (2.7.7)$$

It is of interest also to express the new prediction error in terms of the old one. Using $\mathcal{E}' = g'(0)$ and the above recursions, we find

$$\mathcal{E}' = g'(0) = g(0) - \gamma_2 \, g(2) = g(0) - \gamma_2^2 \, g(0)$$

$$= (1 - \gamma_2^2)\mathcal{E}$$

or

$$\mathcal{E}' = (1 - \gamma_2^2)\mathcal{E} \qquad\qquad (2.7.8)$$

Since both $\mathcal{E}$ and $\mathcal{E}'$ are positive quantities, it follows that $\gamma_2$ must have magnitude less than one.

Using Eq. (2.6.4), we also obtain

$$\mathcal{E} = E[e_n^2] = g(0) = R(0) + a_1 R(1)$$
$$= (1 - \gamma_1^2) \, R(0) = (1 - \gamma_1^2) \, \sigma_y^2 \qquad\qquad (2.7.9)$$

where, by convention, the reflection coefficient $\gamma_1$ for the first order predictor was defined as $\gamma_1 = -a_1$. Equation (2.7.9) implies that $\gamma_1$ also has magnitude less than one. Combining Eqs. (2.7.8) and (2.7.9), we find

$$\mathcal{E}' = (1 - \gamma_2^2)\mathcal{E} = (1 - \gamma_2^2)(1 - \gamma_1^2) \, \sigma_y^2 \qquad\qquad (2.7.10)$$

The Levinson recursion (2.7.7) leads directly to the so-called *lattice filters* of linear prediction. Instead of realizing just the filter $A'(z)$, the lattice realizations *simultaneously* realize both $A'(z)$ and its reverse $A'^R(z)$. The input to both filters is the sequence $y_n$ being predicted. Since $A'(z)$ is related to $A(z)$, first a lattice realization of $A(z)$ will be constructed. Writing $A(z) = 1 + a_1 z^{-1} = 1 - \gamma_1 z^{-1}$

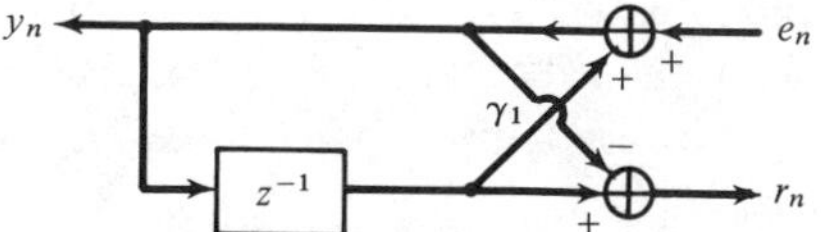

**Figure 2.23**  Lattice Realization of First
Order Prediction-Error Filter

and $A^R(z) = -\gamma_1 + z^{-1}$, a simultaneous realization of both, with a common input $y_n$, is shown in Fig. 2.23, where a common multiplier $\gamma_1$ is indicated for both branches. The transfer functions from $y_n$ to $e_n$ and $r_n$ are $A(z)$ and $A^R(z)$, respectively. Using Eq. (2.7.7), it should be evident that a simultaneous realization of both $A'(z)$ and $A'^R(z)$ can be obtained by simply adding one more lattice section involving the coefficient $\gamma_2$, as shown in Fig. 2.24. Again, the transfer functions from the common input $y_n$ to $e_n'$ and $r_n'$ are $A'(z)$ and $A'^R(z)$, respectively.

Such lattice realizations are alternatives to the direct form realizations given in Figs. 2.19 and 2.20. One of the nicest properties of lattice realizations is that higher order predictors can be easily obtained by simply *adding* more lattice sections. Another important property is that lattice filters are better behaved (less sensitive) under quantization of the multiplier coefficients than the direct form realizations.

The linear prediction problem solved here was to find the best predictor based on just *one* or *two* past samples. It must be contrasted with the full linear prediction problem mentioned in Sections 1.10 and 1.15, which was based on the *entire* past of $y_n$. It is, of course, the latter that whitens the error $e_n$ and provides the signal model of $y_n$. However, in practice, the full prediction problem is difficult to solve because it requires the determination of an infinite number of prediction coefficients $\{a_1, a_2, \ldots\}$. Thus, the problem of linear prediction based on the finite past is of more practical interest. A more complete discussion of linear prediction, Levinson's algorithm, and lattice filters, will be presented in Chapter 5.

## 2.8  *Introduction to Data Compression and DPCM*

In this section, we discuss the application of linear prediction ideas to the problem of data compression by *differential PCM* (DPCM) methods. For illustrative purposes, we work with a second order predictor. The predictor is to be used

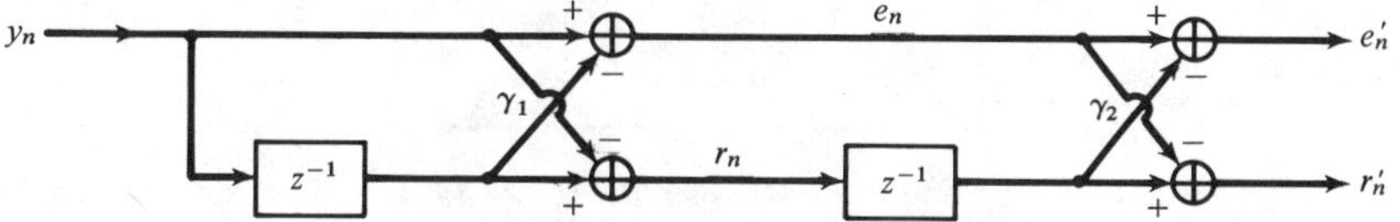

**Figure 2.24**  Lattice Realization of Second Order Prediction-Error Filter

to compress the dynamic range of a signal $y_n$ so that it may be transmitted more efficiently. Suppose we have already found the best predictor coefficients $(1,a_1,a_2)$ that minimize the prediction error by solving Eq. (2.6.3) (for simplicity, the primes have been dropped):

$$\mathcal{E} = E[e_n^2] = \min$$

$$e_n = y_n - \hat{y}_n = y_n + a_1 y_{n-1} + a_2 y_{n-2}$$

The basic idea in data compression is that if the predictor is good, then the prediction error $e_n$ will be small, or rather it will have a compressed dynamic range compared to the original signal. If we were to code and transmit $e_n$ rather than $y_n$, we would need fewer bits to represent each sample $e_n$ than we would need for $y_n$. At the receiving end, the original waveform $y_n$ can be reconstructed by processing $e_n$ through the inverse of the prediction error filter. The overall system is represented as follows

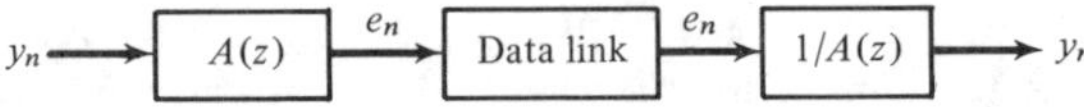


For meaningful reconstruction, it is necessary that the inverse filter $1/A(z)$ be stable (and causal). This requires that the zeros of the prediction-error filter $A(z)$ lie inside the unit circle in the $z$-plane. Two proofs of this fact will be presented later on, in Sections 3.7 and 5.6. The gain in the dynamic ratio that we expect to achieve with this method of data compression is given by the ratio

$$G = \frac{\sigma_y^2}{\sigma_e^2} = \frac{\sigma_y^2}{\mathcal{E}} = \frac{1}{(1 - \gamma_2^2)(1 - \gamma_1^2)}$$

where we used Eq. (2.7.10). This is always greater than one, since both $\gamma_1$ and $\gamma_2$ have magnitude less than one. Even without this result, we could have concluded that the above ratio is greater than one. Indeed, the quantity $\sigma_y^2 = R_{yy}(0)$ is the prediction error for the trivial choice of the prediction-error coefficients $\mathbf{a} = (1,a_1,a_2) = (1,0,0)$, whereas $\mathcal{E} = \sigma_e^2$ corresponds to the choice that minimizes the prediction error; therefore, $\sigma_y^2 > \sigma_e^2$.

Next, we discuss the question of *quantizing* the prediction-error sequence $e_n$ for the purpose of transmission or storage. First, we note that *any* prediction-error filter $A(z)$ may be realized as

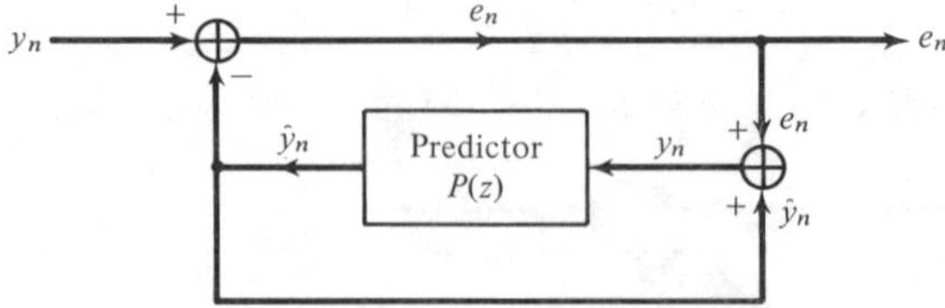


where $P(z) = 1 - A(z)$ is the corresponding predictor filter; for example, $P(z) = -[a_1 z^{-1} + a_2 z^{-2}]$ for the second order case. The conventional differ-

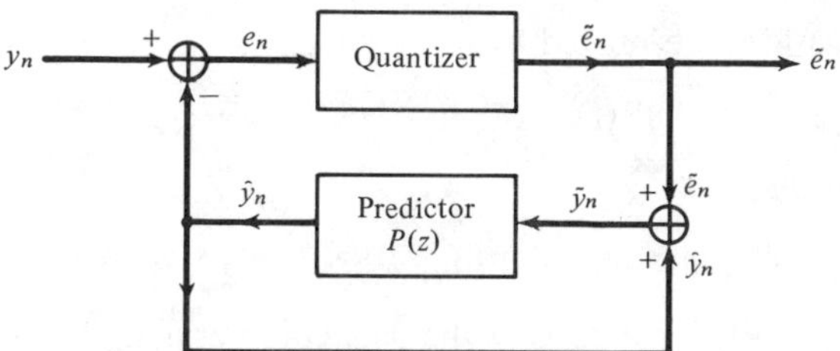

**Figure 2.25** DPCM Encoder

ential PCM encoder is a predictor realized in this manner with a quantizer inserted as shown in Fig. 2.25. The presence of the quantizer introduces a quantization error $\delta_n$ such that

$$\tilde{e}_n = e_n + \delta_n \tag{2.8.1}$$

where $\delta_n$ may be assumed to be zero-mean uniform white noise of variance $\sigma_\delta^2 = Q^2/12$, where $Q$ is the step size of the quantizer. This particular realization ensures that, at the reconstructing end, the quantization errors *do not accumulate*. This follows from the property that

$$\tilde{y}_n - y_n = (\tilde{e}_n + \hat{y}_n) - y_n = \tilde{e}_n - e_n = \delta_n \tag{2.8.2}$$

which states that $\tilde{y}_n$ differs from $y_n$ only by the quantization error $\delta_n$ suffered by the current input $e_n$ to the quantizer. The complete DPCM system is shown in Fig. 2.26. It is evident from this figure that $\tilde{y}_n - y_n$ given by Eq. (2.8.2) is the *reconstruction error*, resulting only from the (irreversible) quantization error $\delta_n$. The data compression gain afforded by such a system is conveniently expressed in terms of the following SNRs:

$$\text{SNR(DPCM)} = \frac{\sigma_e^2}{\sigma_\delta^2} = \text{signal-to-quantization noise of the DPCM signal } e_n$$

$$\text{SNR(PCM)} = \frac{\sigma_y^2}{\sigma_\delta^2} = \text{signal-to-quantization noise of the PCM signal } y_n$$

$$G = \frac{\sigma_y^2}{\sigma_e^2} = \text{gain of the predictor system}$$

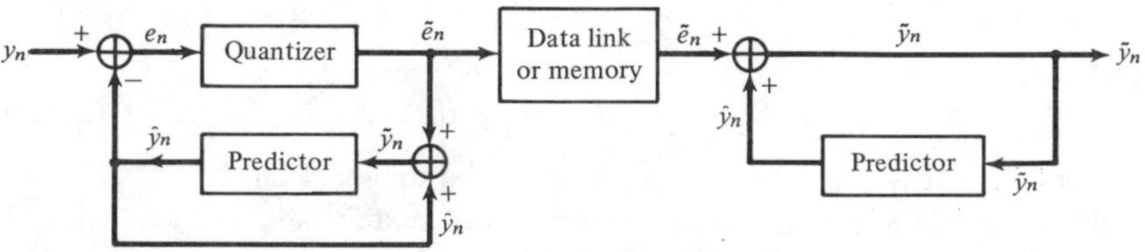

**Figure 2.26** DPCM System for Digital Data Transmission, or Storage

These three quantities are related by

$$\text{SNR(DPCM)} = \text{SNR(PCM)}/G$$

or, expressed in dB,

$$10 \log_{10} \text{SNR(DPCM)} = 10 \log_{10} \text{SNR(PCM)} - 10 \log_{10} G \quad (2.8.3)$$

Therefore, the quantity $10 \log_{10} G$ is the data compression gain afforded by the DPCM system *over* PCM. The best DPCM coder is thus the one maximizing the predictor gain $G$, or equivalently, minimizing the mean-squared prediction error

$$\mathcal{E} = \sigma_e^2 = E[e_n^2] = \min \quad (2.8.4)$$

The presence of the quantizer makes this minimization problem somewhat different from the ordinary prediction problem. However, it can be handled easily using the standard assumptions regarding the quantization noise $\delta_n$; namely, that $\delta_n$ is white noise and that it is uncorrelated with the input sequence $y_n$. First, we note that minimizing $\mathcal{E}$ is equivalent to minimizing

$$\tilde{\mathcal{E}} = E[\tilde{e}_n^2] = E[(e_n + \delta_n)^2] = E[e_n^2] + E[\delta_n^2] = \mathcal{E} + \sigma_\delta^2,$$

or

$$\tilde{\mathcal{E}} = E[\tilde{e}_n^2] = \min \quad (2.8.5)$$

Replacing the quantizer by the equivalent noise source (2.8.1), we may redraw the DPCM coder with $\delta_n$ acting at the input adder:

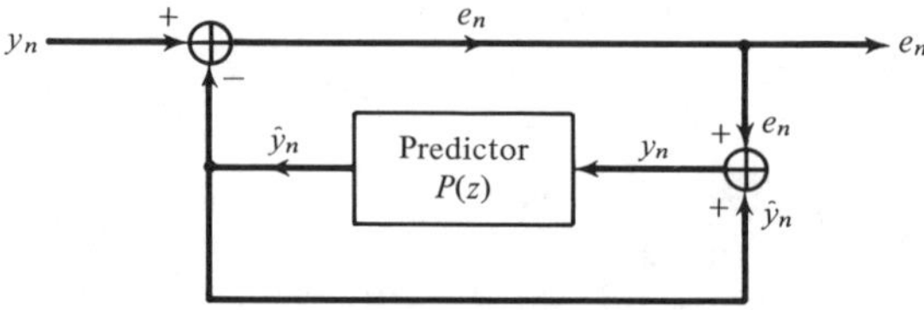


It is evident from this figure that the minimization problem (2.8.5) is equivalent to an effective linear prediction problem of predicting the noisy sequence $\tilde{y}_n = y_n + \delta_n$. Since $y_n$ and $\delta_n$ are mutually uncorrelated, the autocorrelation function of $\tilde{y}_n$; $\tilde{R}(k) = E[\tilde{y}_{n+k} \tilde{y}_n]$ is expressible as the sum of the individual autocorrelations of $y_n$ and $\delta_n$; that is,

$$E[\tilde{y}_{n+k} \tilde{y}_n] = E[(y_{n+k} + \delta_{n+k})(y_n + \delta_n)] = E[y_{n+k} y_n] + E[\delta_{n+k} \delta_n]$$

or

$$\tilde{R}(k) = R(k) + \sigma_\delta^2 \delta(k) \quad (2.8.6)$$

where we used $E[\delta_{n+k} \delta_n] = \sigma_\delta^2 \delta(k)$. Only the diagonal entries of the autocorrelation matrix $\tilde{R}$ are different from those of $R$, and are shifted by an amount

$$\tilde{R}(0) = R(0) + \sigma_\delta^2 = R(0)\,[1 + \epsilon] \quad (2.8.7)$$

where $\epsilon = \sigma_\delta^2/\sigma_y^2 = 1/\text{SNR(PCM)}$. The optimal prediction coefficients are obtained by solving the corresponding normal equations (2.6.3) or (2.6.6), but with $R(0)$ replaced by $\tilde{R}(0)$. Typically, SNR(PCM) is fairly large, and therefore $\epsilon$ is only a small correction which may be ignored without degrading much the performance of the system.

We also point out that a similar change in the autocorrelation matrix, given by (2.8.7), occurs in a different context in the least-squares design of wave-shaping filters from noisy data. In that context, $\epsilon$ is referred to as the Backus-Gilbert parameter. This will be discussed in Chapter 5.

DPCM encoding methods have been applied successfully to speech and image data compression [18,19]. In speech, a compression gain of 4 to 12 dB over PCM can be obtained. The gain can be increased even further using adaptive quantizers and/or adaptive predictors. For images, using 3d order predictors, the gain is typically 20 dB over PCM.

Finally, we remark that DPCM transmission systems are susceptible to channel errors. Unlike quantization errors which do not accumulate at the reconstructing end, channel errors do accumulate and get amplified. This may be seen as follows: Let the received sequence $\tilde{e}_n$ be corrupted by white gaussian channel noise $v_n$, as shown in Fig. 2.27. Then, both $\tilde{e}_n$ and the channel noise $v_n$ get filtered through the reconstructing inverse filter $B(z) = 1/A(z)$. This filter is designed to decompress $\tilde{e}_n$ back to $\tilde{y}_n$ and thus, it has a gain which is greater than one. This will also cause the channel noise $v_n$ to be amplified as it goes through $B(z)$. The output to input noise power is given by the quadratic norm $\|B\|^2$:

$$\frac{\sigma_{\hat{v}}^2}{\sigma_v^2} = \|B\|^2 = \left\|\frac{1}{A}\right\|^2 = \oint_{u.c.} \frac{1}{A(z)A(z^{-1})} \frac{dz}{2\pi j z}$$

which is always greater than one. For example, for the second order predictor it can be shown that

$$\|B\|^2 = \left\|\frac{1}{A}\right\|^2 = \frac{1}{(1 - \gamma_1^2)(1 - \gamma_2^2)}$$

where $\gamma_1$ and $\gamma_2$ are the reflection coefficients of the prediction problem (2.8.5).

To combat against channel errors, some channel encoding with error-protection must be done prior to transmission [20,21].

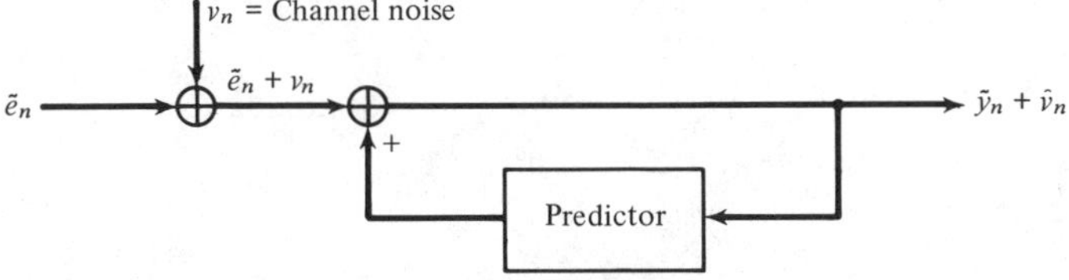


**Figure 2.27**   Channel Noise in a DPCM Receiver

## Problems

**Problem 2.1:**
Let $x(n)$ be a zero-mean white-noise sequence of unit variance. For each of the following filters compute the output autocorrelation $R_{yy}(k)$ for all $k$, using $z$-transforms:

1. $y(n) = x(n) - x(n - 1)$
2. $y(n) = x(n) - 2 x(n - 1) + x(n - 2)$
3. $y(n) = -0.5 y(n - 1) + x(n)$
4. $y(n) = 0.25 y(n - 2) + x(n)$

Also, sketch the output power spectrum $S_{yy}(\omega)$ versus frequency $\omega$.

**Problem 2.2:**
Let $y_n$ be the output of a (stable and causal) filter $H(z)$ driven by the signal $x_n$, and let $w_n$ be another unrelated signal. Assume all signals are stationary random signals. Show the following relationships between power spectral densities:

(a) $S_{yw}(z) = H(z) S_{xw}(z)$
(b) $S_{wy}(z) = S_{wx}(z) H(z^{-1})$

**Problem 2.3:**
A stationary random signal $y_n$ is sent through a finite filter $A(z) = a_0 + a_1 z^{-1} + \cdots + a_M z^{-M}$ to obtain the output signal $e_n$:

$$y_n \longrightarrow \boxed{A(z)} \longrightarrow e_n \qquad e_n = \sum_{m=0}^{M} a_m\, y_{n-m}$$

Show that the average power of the output signal $e_n$ can be expressed in the two alternative forms:

$$E[e_n^2] = \int_{-\pi}^{\pi} S_{yy}(\omega)\, |A(\omega)|^2\, \frac{d\omega}{2\pi} = \mathbf{a}^T R_{yy}\, \mathbf{a}$$

where $\mathbf{a} = [a_0, a_1, \ldots, a_M]^T$ and $R_{yy}$ is the $(M + 1) \times (M + 1)$ autocorrelation matrix of $y_n$ having matrix elements $R_{yy}(i,j) = E[y_i\, y_j] = R_{yy}(i - j)$.

**Problems 2.4:**
Consider two autoregressive random signals $y_n$ and $y_n'$ generated by the signal models:

$$\epsilon_n \longrightarrow \boxed{1/A(z)} \longrightarrow y_n \quad \text{and} \quad \epsilon_n' \longrightarrow \boxed{1/A'(z)} \longrightarrow y_n'$$

$$A(z) = 1 + a_1 z^{-1} + \cdots + a_M z^{-M}$$

and

$$A'(z) = 1 + a_1' z^{-1} + \cdots + a_M' z^{-M}$$

(a) Suppose $y_n$ is filtered through the analysis filter $A'(z)$ of $y_n'$ producing the output signal $e_n$; that is,

$$y_n \longrightarrow \boxed{A(z)} \longrightarrow e_n \qquad e_n = \sum_{m=0}^{M} a_m' y_{n-m}$$

If $y_n$ were to be filtered through its own analysis filter $A(z)$, it would produce the innovations sequence $\epsilon_n$. Show that the average power of $e_n$, compared to the average power of $\epsilon_n$, is given by

$$\frac{\sigma_e^2}{\sigma_\epsilon^2} = \frac{\mathbf{a}'^T R_{yy} \mathbf{a}'}{\mathbf{a}^T R_{yy} \mathbf{a}} = \int_{-\pi}^{\pi} \left| \frac{A'(\omega)}{A(\omega)} \right|^2 \frac{d\omega}{2\pi} = \left\| \frac{A'}{A} \right\|^2$$

where $\mathbf{a}$, $\mathbf{a}'$, and $R_{yy}$ have the same meaning as in Problem 2.3. This ratio can be taken as a measure of *similarity* between the two signal models. The log of this ratio is Itakura's *LPC distance measure* used in speech recognition.

(b) Alternatively, show that if $y_n'$ were to be filtered through $y_n$'s analysis filter $A(z)$ resulting in $e_n' = \sum_{m=0}^{M} a_m y_{n-m}'$, then

$$\frac{\sigma_{e'}^2}{\sigma_{\epsilon'}^2} = \frac{\mathbf{a}^T R_{yy}' \mathbf{a}}{\mathbf{a}'^T R_{yy}' \mathbf{a}'} = \int_{-\pi}^{\pi} \left| \frac{A(\omega)}{A'(\omega)} \right|^2 \frac{d\omega}{2\pi} = \left\| \frac{A}{A'} \right\|^2$$

**Problem 2.5:**
The autocorrelation function of complex-valued signals is defined by

$$R_{yy}(k) = E[y_{n+k} y_n^*]$$

(a) Show that stationarity implies $R_{yy}(-k) = R_{yy}(k)^*$.

(b) If $y_n$ is filtered through a (possibly complex-valued) filter $A(z) = a_0 + a_1 z^{-1} + \cdots + a_M z^{-M}$, show that the average power of the output signal $e_n$ can be expressed as

$$\mathcal{E} = E[e_n^* e_n] = \mathbf{a}^\dagger R_{yy} \mathbf{a}$$

where $\mathbf{a}^\dagger$ denotes the hermitian conjugate of $\mathbf{a}$, and $R_{yy}$ has matrix elements

$$R_{yy}(i,j) = R_{yy}(i - j).$$

**Problem 2.6:**
(a) Let $y_n = A_1 \exp[j(\omega_1 n + \phi_1)]$ be a complex sinusoid of amplitude $A_1$ and frequency $\omega_1$. The randomness of $y_n$ arises only from the phase $\phi_1$ which

is assumed to be a random variable uniformly distributed over the interval $0 \leq \phi_1 \leq 2\pi$. Show that the autocorrelation function of $y_n$ is

$$R_{yy}(k) = |A_1|^2 \exp(j\omega_1 k)$$

(b) Let $y_n$ be the sum of two complex sinusoids

$$y_n = A_1 \exp[j(\omega_1 n + \phi_1)] + A_2 \exp[j(\omega_2 n + \phi_2)]$$

with uniformly distributed random phases $\phi_1$ and $\phi_2$ which are also assumed to be independent of each other. Show that the autocorrelation of $y_n$ is

$$R_{yy}(k) = |A_1|^2 \exp(j\omega_1 k) + |A_2|^2 \exp(j\omega_2 k)$$

*Problem 2.7:* Sinusoids in Noise

Suppose $y_n$ is the sum of $L$ complex sinusoids with random phases, in the presence of uncorrelated noise:

$$y_n = v_n + \sum_{i=1}^{L} A_i \exp[j(\omega_i n + \phi_i)]$$

where $\phi_i$; $i = 1, 2, \ldots L$, are uniformly distributed random phases which are assumed to be mutually independent, and $v_n$ is zero-mean white noise of variance $\sigma_v^2$. Also, assume that $v_n$ is independent of $\phi_i$.

(a) Show that $E[e^{j\phi_i} e^{-j\phi_k}] = \delta_{ik}$; $i, k = 1, 2, \ldots, L$

(b) Show that the autocorrelation of $y_n$ is

$$R_{yy}(k) = \sigma_v^2 \delta(k) + \sum_{i=1}^{L} |A_i|^2 \exp(j\omega_i k)$$

(c) Suppose $y_n$ is filtered through a filter $A(z) = a_0 + a_1 z^{-1} + \cdots + a_M z^{-M}$ of order $M$, producing the output signal $e_n$. Show that the average output power is expressible as

$$\mathcal{E} = E[e_n^* e_n] = \mathbf{a}^\dagger R_{yy} \mathbf{a} = \sigma_v^2 \, \mathbf{a}^\dagger \mathbf{a} + \sum_{i=1}^{L} |A_i|^2 |A(\omega_i)|^2$$

where $\mathbf{a}$, $\mathbf{a}^\dagger$, $R_{yy}$ have the same meaning as Problem 2.5, and $A(\omega_i)$ is the frequency response of the filter evaluated at the sinusoid frequencies $\omega_i$; $i = 1, 2, \ldots, L$; that is,

$$A(\omega_i) = \sum_{m=0}^{M} a_m e^{-j\omega_i m}$$

(d) If the noise $v_n$ is correlated with autocorrelation $Q(k)$, so that $E[v_{n+k} v_n^*] = Q(k)$, show that in this case

$$\mathcal{E} = E[e_n^* e_n] = \mathbf{a}^\dagger R_{yy} \mathbf{a} = \mathbf{a}^\dagger Q \mathbf{a} + \sum_{i=1}^{L} |A_i|^2 |A(\omega_i)|^2$$

where $Q$ is the noise covariance matrix, $Q(i,j) = Q(i - j)$.

*Problem 2.8:* Computer Experiment

Consider the linear system defined by Eq. (2.2.1). Generate 1500 samples of a unit-variance, zero-mean, white-noise sequence $x_n$; $n = 0,1, \ldots ,1499$ and filter them through the filter $H$ to obtain the output sequence $y_n$. Compute the sample cross-correlation $\hat{R}_{yx}(k)$ for $k = 0,1, \ldots ,50$ to obtain estimates of the impulse response $h_k$. Plot the estimated impulse response versus time together with the simulated response (2.2.1) on the same graph. Repeat, using a different realization of $x_n$.

*Problem 2.9:*

A filter is defined by $y(n) = -0.64\, y(n - 2) + 0.36\, x(n)$.

(a) Suppose the input is zero-mean, unit-variance, white noise. Compute the output spectral density $S_{yy}(z)$ and power spectrum $S_{yy}(\omega)$ and plot it roughly versus frequency.

(b) Compute the output autocorrelation $R_{yy}(k)$ for all lags $k$.

(c) Compute the noise reduction ratio of this filter.

(d) What signal $s(n)$ can pass through this filter and remain entirely unaffected (at least in the steady-state regime)?

(e) How can the filter coefficients be changed so that (i) the noise reduction capability of the filter is improved, while at the same time (ii) the above signal $s(n)$ still goes through unchanged. Explain any tradeoffs.

*Problem 2.10:* Computer Experiment

(a) Generate 1000 samples of a zero-mean, unit-variance, white gaussian noise sequence $x(n)$; $n = 0,1, \ldots ,999$, and filter them through the filter defined by the difference equation

$$y(n) = a\, y(n - 1) + (1 - a)\, x(n)$$

with $a = 0.95$. To avoid the transient effects introduced by the filter, discard the first 900 output samples and save the last 100 samples of $y(n)$. Compute the sample autocorrelation of $y(n)$ from this length-100 block of samples.

(b) Determine the theoretical autocorrelation $R_{yy}(k)$, and on the same graph plot the theoretical and sample autocorrelations versus $k$. Do they agree?

**Problem 2.11:**

Following the procedure of Example (2.6.1), rederive the results of Eqs. (2.6.2) and (2.6.4) for the first order predictor using the correlation canceling formulation of Section 1.4.

**Problem 2.12:**

Let $y(n) = (1,1,1,1)$ for $n = 0,1,2,3$. We want to "predict" the fifth sample in this sequence.

(a) Compute the sample autocorrelation of this sequence.

(b) Using the Yule-Walker method, determine the best first order predictor of the form

$$\hat{y}(n) = -a_1 y(n-1)$$

What is the predicted value of the fifth sample? What is the mean squared prediction error?

(c) Since we only have sample autocorrelations to work with, let us define the gapped function $g(k)$ as the *convolution* of the prediction-error filter $(1,a_1)$ with the sample autocorrelation $\hat{R}_{yy}(k)$, in accordance with Eq. (2.7.2). Verify that $g(k)$ has a gap of length one.

(d) It is desired next, to determine the best second order predictor

$$\hat{y}'(n) = -[a_1' y(n-1) + a_2' y(n-2)]$$

Using the gapped function $g(k)$, construct a new gapped function $g'(k)$ having a gap of length two. Determine the prediction-error filter $(1,a_1'a_2')$.

(e) Compute the predicted value of the fifth sample in this case, and the mean-squared prediction error. Is the predicted fifth value what you expected? Is the value predicted by the second order predictor "better" than that predicted by the first order predictor?

(f) Determine the zeros of the prediction filter $(1,a_1',a_2')$ and verify that they lie inside the unit circle in the $z$-plane.

**Problem 2.13:**

(a) Repeat parts (a) and (b) of Problem 2.12 for the sequence $y_n = (-1, 1, -1, 1)$.

(b) Repeat for $y_n = (1,2,3,4)$.

**Problem 2.14:**

Show that the inverse lattice filter of Fig. 2.23 is realized as

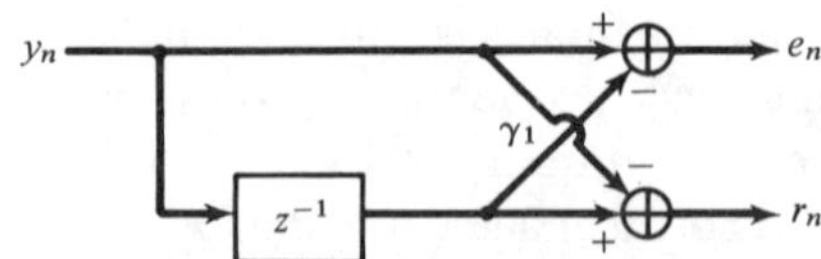

Show that the transfer function from $e_n$ to $y_n$ is the synthesis filter $1/A(z)$. (Note the different sign conventions at the upper adder.)

### Problem 2.15:

The second order synthesis lattice filter is realized as follows:

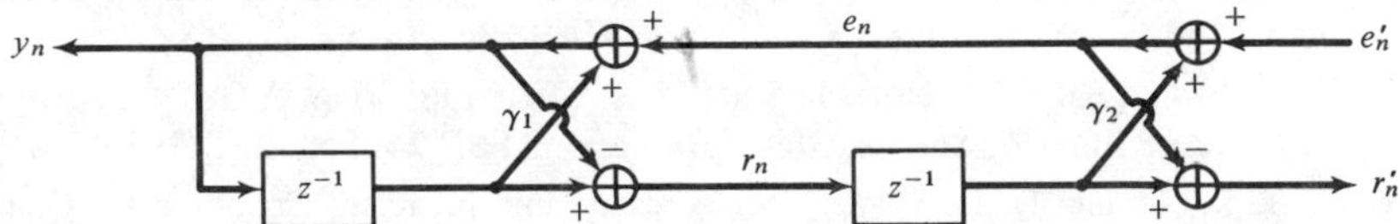

Show that the transfer function from $e_n'$ to $y_n$ is the synthesis filter $1/A'(z)$.

### Problem 2.16:

Consider the second order prediction-error filter $A'(z)$ given in Eq. (2.7.6). Show that the quadratic norm of the synthesis filter $1/A'(z)$ is given by

$$\left\|\frac{1}{A'}\right\|^2 = \oint_{u.c.} \frac{1}{A'(z)A'(z^{-1})} \frac{dz}{2\pi jz} = \frac{1}{(1 - \gamma_1^2)(1 - \gamma_2^2)}$$

where $\gamma_1$ and $\gamma_2$ are the corresponding reflection coefficients. (*Hint:* factor $A'(z)$ into its zeros, which are both inside the unit circle, perform the indicated contour integration, and rewrite the result in terms of $\gamma_1$ and $\gamma_2$.)

The above result was used in Section 2.8 in the discussion of the channel errors in DPCM systems.

## References


1. A. Papoulis, *Probability, Random Variables, and Stochastic Processes*, New York, McGraw-Hill, 1965.

2. A. V. Oppenheim and R. W. Schafer, *Digital Signal Processing*, Englewood Cliffs, Prentice-Hall, 1975.

3. S. Tretter, *Introduction to Discrete-Time Signal Processing*, New York, Wiley, 1974.

4. K. J. Aström and P. Eykhoff, System Identification—A Survey, *Automatica*, **7**, 123–162 (1971).

5. P. Eykhoff, *System Identification: Parameter and State Estimation*, New York, Wiley, 1974.

6. G. C. Goodwin and R. L. Payne, *Dynamic System Identification, Experimental Design and Data Analysis*, New York, Academic, 1977.

7. M. I. Skolnik, *Introduction to Radar Systems*, New York, McGraw-Hill, 1980.

8. L. Rabiner and B. Gold, *Theory and Application of Digital Signal Processing*, Englewood-Cliffs, Prentice-Hall, 1975.

9. L. B. Jackson, Roundoff Noise Analysis for Fixed-Point Digital Filters Realized in Cascade or Parallel Form, *IEEE Trans. Audio Electroacoust.*, **AU-18**, 107–122 (1970).

10. B. Liu, Effect of Finite Word Length on the Accuracy of Digital Filters—A Review, *IEEE Trans. Circuit Th.*, **CT-18**, 670–677 (1971).

11. C. T. Mullis and R. A. Roberts, Synthesis of Minimum Roundoff Noise Fixed Point Digital Filters, *IEEE Trans. Circuits Syst.*, **CAS-23**, 551–562 (1976).

12. S. Y. Hwang, Roundoff Noise in State-Space Digital Filtering: A General Analysis, *IEEE Trans. Acoust., Speech, Signal Process.*, **ASSP-24**, 256–262 (1976).

13. A. B. Sripad and D. L. Snyder, A Necessary and Sufficient Condition for Quantization Errors to be Uniform and White, *IEEE Trans. Acoust., Speech, Signal Process.*, **ASSP-25**, 442–448 (1977).

14. T. L. Chang and S. A. White, An Error Cancellation Digital Filter Structure and its Distributed Arithmetic Implementation, *IEEE Trans. Circuits Syst.*, **CAS-28**, 339–342 (1981).

15. W. E. Higgins and D. C. Munson, Jr., Noise Reduction Strategies for Digital Filters: Error Spectrum Shaping versus the Optimal Linear State-Space Formulation, *IEEE Trans. Acoust., Speech, Signal Process.*, **ASSP-30**, 963–973 (1982).

16. C. T. Mullis and R. A. Roberts, An Interpretation of Error Spectrum Shaping in Digital Filters, *IEEE Trans. Acoust., Speech, Signal Process.*, **ASSP-30**, 1013–1015 (1982).

17. E. Robinson and S. Treitel, Maximum Entropy and the Relationship of the Partial Autocorrelation to the Reflection Coefficients of a Layered System, *IEEE Trans. Acoust., Speech, Signal Process.*, **ASSP-28**, 22 (1980).

18. J. L. Flanagan, et al., Speech Coding, *IEEE Trans. Commun.*, **COM-27**, 710–736, (1979).

19. A. K. Jain, Image Data Compression: A Review, *Proc. IEEE*, **69**, 349–389 (1981).

20. J. W. Modestino and D. G. Daut, Combined Source-Channel Coding of Images, *IEEE Trans. Commun.*, **COM-27**, 1644–1659 (1979).

21. D. G. Daut and J. W. Modestino, Two-Dimensional DPCM Image Transmission over Fading Channels, *IEEE Trans. Commun.*, **COM-31**, 315–328 (1983).

# 3

---

# Spectral Factorization

In this chapter we discuss the concept of minimal-phase signals and filters, state the spectral factorization theorem and demonstrate its importance in making signal models, and present a proof of the minimal-phase property of the prediction-error filter of linear prediction.

## 3.1 *Minimal-Phase Signals and Filters* [1–4]

A *minimal-phase* sequence $\mathbf{a} = (a_0, a_1, \ldots, a_M)$ has a $z$-transform with all its zeros inside the unit circle in the complex $z$-plane

$$A(z) = a_0 + a_1 z^{-1} + \cdots + a_M z^{-M} \qquad (3.1.1)$$

$$= a_0 (1 - z_1 z^{-1}) (1 - z_2 z^{-1}) \cdots (1 - z_M z^{-1})$$

with $|z_i| < 1$; $i = 1, 2, \ldots, M$. Such a polynomial is also called a *minimal-delay* polynomial. Define the following related polynomials

$$A^*(z) = a_0^* + a_1^* z^{-1} + \cdots + a_M^* z^{-M} = \text{complex-conjugated coefficients}$$

$$\bar{A}(z) = a_0^* + a_1^* z + \cdots + a_M^* z^M = \text{conjugated and reflected}$$

$$A^R(z) = a_M^* + a_{M-1}^* z^{-1} + \cdots + a_0^* z^{-M} = \text{reversed and conjugated}$$

We note the relationships

$$\bar{A}(z) = A^*(z^{-1}) \qquad \text{and} \qquad A^R(z) = z^{-M} \bar{A}(z) = z^{-M} A^*(z^{-1}) \qquad (3.1.2)$$

We also note that when we set $z = \exp(j\omega)$ to obtain the corresponding frequency responses, $\bar{A}(\omega)$ becomes the complex conjugate of $A(\omega)$

$$\bar{A}(\omega) = A(\omega)^* \tag{3.1.3}$$

It is easily verified that all these polynomials have the *same* magnitude spectrum. That is,

$$|A(\omega)|^2 = |\bar{A}(\omega)|^2 = |A^*(\omega)|^2 = |A^R(\omega)|^2 \tag{3.1.4}$$

For example, in the case of a doublet $\mathbf{a} = (a_0, a_1)$ and its reverse $\mathbf{a}^R = (a_1^*, a_0^*)$ we verify explicitly

$$\begin{aligned}
|A(\omega)|^2 = A(\omega)\, A(\omega)^* &= (a_0 + a_1 e^{-j\omega})\,(a_0^* + a_1^* e^{j\omega}) \\
&= a_0^* a_0 + a_1^* a_1 + 2\,Re(a_0^* a_1 e^{-j\omega}) \\
&= (a_1^* + a_0^* e^{-j\omega})\,(a_1 + a_0 e^{j\omega}) \\
&= A^R(\omega)\,(A^R(\omega))^* = |A^R(\omega)|^2
\end{aligned}$$

Thus, on the basis of the magnitude spectrum, one cannot distinguish the doublet $\mathbf{a} = (a_0, a_1)$ from its reverse $\mathbf{a}^R = (a_1^*, a_0^*)$.

In the more general case of a polynomial of degree $M$, factored into doublets as in Eq. (3.1.1), we note that each doublet can be replaced by its reverse

$$(1, -z_i) \rightarrow (-z_i^*, 1), \qquad \text{or} \qquad (1 - z_i z^{-1}) \rightarrow (-z_i^* + z^{-1})$$

without affecting the overall magnitude spectrum $|A(\omega)|^2$. Since there are $M$ such factors, there will be a total of $2^M$ different $M$th degree polynomials, or equivalently $2^M$ different length-$(M + 1)$ sequences, all having the *same magnitude spectrum*. Every time a factor $(1 - z_i z^{-1})$ is reversed to become $(-z_i^* + z^{-1})$, the corresponding zero changes from $z = z_i$ to $z = 1/z_i^*$. If $z_i$ is inside the unit circle, then $1/z_i^*$ is outside, as shown

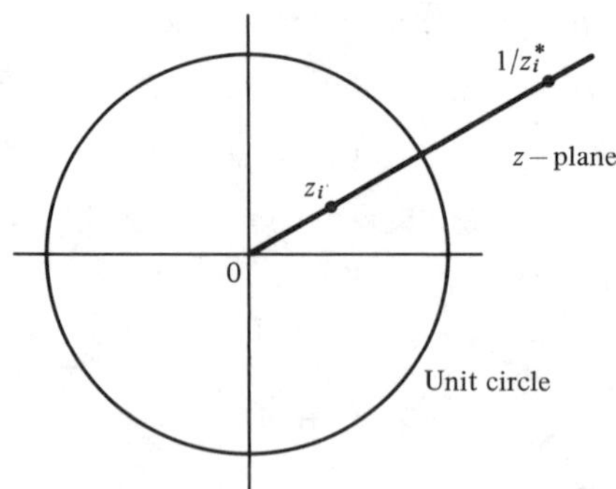


To enumerate all these sequences, start by taking all zeros $z_i$ to be inside the unit circle and successively keep reversing each factor until all $2^M$ possibilities have been exhausted. At the last step, all the factors will have been flipped, corresponding to all the zeros being outside the unit circle. The resulting polynomial and sequence are referred to as *maximal-phase*, or *maximal-delay*.

As an example consider the two doublets

$$\mathbf{a} = (2,1) \quad \text{and} \quad \mathbf{b} = (3,2)$$

and form the four different sequences, where $*$ denotes convolution

$$\mathbf{c}_0 = \mathbf{a} * \mathbf{b} = (2,1) * (3,2) = (6,7,2), \qquad C_0(z) = A(z)\,B(z)$$

$$\mathbf{c}_1 = \mathbf{a}^R * \mathbf{b} = (1,2) * (3,2) = (3,8,4), \qquad C_1(z) = A^R(z)\,B(z)$$

$$\mathbf{c}_2 = \mathbf{a} * \mathbf{b}^R = (2,1) * (2,3) = (4,8,3), \qquad C_2(z) = A(z)\,B^R(z)$$

$$\mathbf{c}_3 = \mathbf{a}^R * \mathbf{b}^R = (1,2) * (2,3) = (2,7,6), \qquad C_3(z) = A^R(z)\,B^R(z)$$

All four sequences have the same magnitude spectrum.

## 3.2  *Partial Energy and Minimal Delay*

Since the total energy of a sequence $\mathbf{a} = (a_0, a_1, \ldots, a_M)$ is given by Parseval's equality as

$$\sum_{m=0}^{M} |a_m|^2 = \int_{-\pi}^{\pi} |A(\omega)|^2 \frac{d\omega}{2\pi}$$

it follows that all of the above $2^M$ sequences, having the same magnitude spectrum, will also have the same *total energy*. However, the *distribution* of the total energy over time may be different. And this will allow an alternative characterization of the minimal phase sequences, first given by Robinson.

Define the *partial* energy by

$$P_a(n) = \sum_{m=0}^{n} |a_m|^2 = |a_0|^2 + |a_1|^2 + \cdots + |a_n|^2; \qquad n = 0,1, \ldots, M$$

then, for the above example, the partial energies for the four different sequences are given in the table

|          | $\mathbf{c}_0$ | $\mathbf{c}_1$ | $\mathbf{c}_2$ | $\mathbf{c}_3$ |
| -------- | -------------- | -------------- | -------------- | -------------- |
| $P(0)$   | 36             | 9              | 16             | 4              |
| $P(1)$   | 85             | 73             | 80             | 53             |
| $P(2)$   | 89             | 89             | 89             | 89             |

We note that $\mathbf{c}_0$ which has both its zeros inside the unit circle (i.e., minimal phase) is also the sequence that has most of its energy concentrated at the *earlier* times; that is, it makes its impact as early as possible, *with minimal delay*. In contrast, the maximal-phase sequence $\mathbf{c}_3$ has most of its energy concentrated at its tail; thus, making most of its impact at the end, *with maximal delay*.

## 3.3  *Invariance of the Autocorrelation Function*

This section will present yet another characterization of the above class of sequences. It will be important in proving the minimal-phase property of the linear prediction filters.

The *sample autocorrelation* of a (possibly complex-valued) sequence $\mathbf{a} = (a_0, a_1, \ldots, a_M)$ is defined by

$$R_{aa}(k) = \sum_{n=0}^{M-k} a_{n+k}\, a_n^*, \qquad \text{for } 0 \le k \le M$$

$$R_{aa}(k) = R_{aa}(-k)^*, \qquad \text{for } -M \le k \le -1$$

(3.3.1)

It is easily verified that the corresponding power spectral density is factored as

$$S_{aa}(z) = \sum_{k=-M}^{M} R_{aa}(k)\, z^{-k} = A(z)\, \bar{A}(z) \tag{3.3.2}$$

The magnitude spectrum is obtained by setting $z = e^{j\omega}$

$$S_{aa}(\omega) = |A(\omega)|^2 \tag{3.3.3}$$

with an inversion formula

$$R_{aa}(k) = \int_{-\pi}^{\pi} |A(\omega)|^2 e^{j\omega k}\, \frac{d\omega}{2\pi} \tag{3.3.4}$$

It follows from Eq. (3.3.4) that the above $2^M$ different sequences having the same magnitude spectrum, also have the *same* sample autocorrelation. They cannot be distinguished on the basis of their autocorrelation.

Therefore, there are $2^M$ different spectral factorizations of $S_{aa}(z)$ of the form

$$S_{aa}(z) = A(z)\, \bar{A}(z) \tag{3.3.5}$$

but there is only one with minimal-phase factors. The procedure for obtaining it is straightforward: Find the zeros of $S_{aa}(z)$, which come in pairs $z_i$ and $1/z_i^*$. Thus there are $2M$ such zeros. Group those that lie inside the unit circle into a common factor. This defines $A(z)$ as a minimal-phase polynomial.

## 3.4  *Minimal-Delay Property*

Here, we discuss the effect of flipping a zero from the inside to the outside of the unit circle, on the minimal-delay and the minimal-phase properties of the signal. Suppose $A(z)$ is of degree $M$ and has a zero $z_1$ inside the unit circle. Let

$B(z)$ be the polynomial that results by flipping this zero to the outside; that is, $z_1 \rightarrow 1/z_1^*$

$$A(z) = (1 - z_1 z^{-1}) \, F(z)$$

$$\tag{3.4.1}$$

$$B(z) = (- z_1^* + z^{-1}) \, F(z)$$

where $F(z)$ is a polynomial of degree $M - 1$. Both $A(z)$ and $B(z)$ have the same magnitude spectrum. We may think of this operation as sending $A(z)$ through an *all-pass* filter

$$B(z) = \frac{-z_1^* + z^{-1}}{1 - z_1 z^{-1}} \, A(z)$$

In terms of the polynomial coefficients, Eq. (3.4.1) becomes

$$a_n = f_n - z_1 f_{n-1}$$

$$\tag{3.4.2}$$

$$b_n = -z_1^* f_n + f_{n-1}$$

for $n = 0, 1, \ldots, M$, from which we obtain

$$|a_n|^2 - |b_n|^2 = (1 - |z_1|^2)(|f_n|^2 - |f_{n-1}|^2)$$

$$\tag{3.4.3}$$

Summing to get the partial energies $P_a(n) = \sum_{m=0}^{n} |a_m|^2$, we find

$$P_a(n) - P_b(n) = (1 - |z_1|^2) \, |f_n|^2$$

$$\tag{3.4.4}$$

for $0 \leq n \leq M$. Thus, the partial energy of the sequence **a** remains greater than that of **b**, for *all* times $n$; that is, $A(z)$ is of earlier delay than $B(z)$. The total energy is, of course, the same as follows from the fact that $F(z)$ is of degree $M - 1$, thus, missing the $M$th term or $f_M = 0$. We have then

$$P_a(n) \geq P_b(n)$$

for $0 \leq n \leq M$, and in particular, we have

$$P_a(M) = P_b(M) \qquad \text{and} \qquad P_a(0) \geq P_b(0)$$

The last inequality can also be stated as $|a_0| \geq |b_0|$, and will be important in our proof of the minimal-phase property of the prediction-error filter of linear prediction.

## 3.5 *Minimal-Phase Property*

The effect of reversing the zero $z_1$ on the phase responses of $A(z)$ and $B(z)$ of Eq. (3.4.1), can be seen as follows. For $z = \exp(j\omega)$, define the *phase lag* as the negative of the *phase response*

$$A(\omega) = |A(\omega)| \, e^{j \mathrm{Arg} A(\omega)}$$

$$\theta_A(\omega) = -\mathrm{Arg} A(\omega) = \text{phase-lag response}$$

and similarly for $B(z)$. Since $A(\omega)$ and $B(\omega)$ have the same magnitude, they will differ only by a phase

$$\frac{A(\omega)}{B(\omega)} = \exp[j(\theta_B - \theta_A)] = \frac{1 - z_1 e^{-j\omega}}{-z_1^* + e^{-j\omega}} = \frac{e^{j\omega} - z_1}{e^{-j\omega} - z_1^*} e^{-j\omega} = e^{2j\psi} e^{-j\omega}$$

With reference to the figure below let $\psi = \mathrm{Arg}(e^{j\omega} - z_1)$ then we easily find

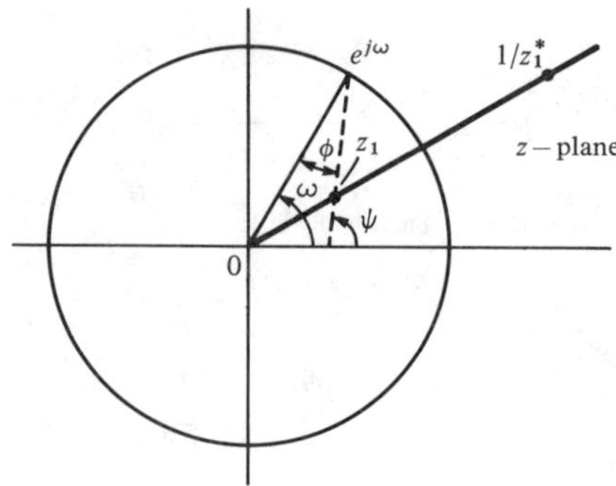


the angle relationships

$$\theta_B(\omega) - \theta_A(\omega) = 2\psi - \omega = 2(\omega + \phi) - \omega = \omega + 2\phi \geq 0$$

which show that $\theta_B(\omega) \geq \theta_A(\omega)$ for all $\omega$ in $[0,\pi]$. Thus, the phase lag of $A(z)$ remains *smaller* than that of $B(z)$. The phase-lag curve for the case when $A(z)$ has all its zeros inside the unit circle will remain below all the other phase-lag curves. The term *minimal-phase* strictly speaking means *minimal phase-lag*.

## 3.6 *Spectral Factorization Theorem*

We finish our digression on minimal-phase sequences by quoting the spectral factorization theorem [5]:

Any *rational* power spectral density $S_{yy}(z)$ of a stationary signal $y_n$, can be factored in a minimal-phase form

$$S_{yy}(z) = \sigma_\epsilon^2 B(z) B(z^{-1}) \tag{3.6.1}$$

where

$$B(z) = \frac{N(z)}{D(z)} \tag{3.6.2}$$

with *both* $D(z)$ and $N(z)$ minimal-phase polynomials; that is, both having all their zeros inside the unit circle. By adjusting the overall constant $\sigma_\epsilon^2$, both $D(z)$ and $N(z)$ may be taken to be *monic* polynomials. Then, they are *unique*.

This theorem guarantees the existence of a *causal and stable* random signal generator filter for the signal $y_n$ of the type discussed in Section 1.10:

$$\epsilon_n \longrightarrow \boxed{B(z)} \longrightarrow y_n \quad \text{(signal model for } y_n\text{)}$$

with $\epsilon_n$ white noise of variance $\sigma_\epsilon^2$. The minimal-phase property of $B(z)$ also guarantees the *stability* of the (causal) inverse filter $1/B(z)$; that is, the whitening filter

$$y_n \longrightarrow \boxed{1/B(z)} \longrightarrow \epsilon_n = \text{white noise}$$

The proof of the spectral factorization theorem is straightforward. Since $S_{yy}(z)$ is the power spectral density of a (real-valued) stationary process $y_n$, it will satisfy the symmetry conditions $S_{yy}(z) = S_{yy}(z^{-1})$. Therefore, if $z_i$ is zero then $1/z_i$ is also a zero, and if $z_i$ is truly complex then the reality of $R_{yy}(k)$ implies that $z_i^*$ will also be a zero. Thus, both $z_i$ and $1/z_i^*$ are zeros. Therefore, the numerator polynomial of $S_{yy}(z)$ is of the type of Eq. (3.3.5) and can be factored into its minimal phase polynomials $N(z)N(z^{-1})$. This is also true of the denominator of $S_{yy}(z)$.

All sequential correlations in the original signal $y_n$ arise from the *filtering* action of $B(z)$ on the white noise input $\epsilon_n$. This follows from Eq. (2.1.14):

$$R_{yy}(k) = \sigma_\epsilon^2 \sum_n b_{n+k}\, b_n \qquad B(z) = \sum_{n=0}^{\infty} b_n\, z^{-n} \tag{3.6.3}$$

Effectively, we have modeled the statistical autocorrelation $R_{yy}(k)$ by the sample autocorrelation of the impulse response of the synthesis filter $B(z)$. Since $B(z)$ is causal, such factorization, corresponds to a LU, or Cholesky, factorization of the autocorrelation matrix.

This matrix representation can be seen as follows: Let $B$ be the lower triangular Toeplitz matrix defined exactly as in Eq. (1.10.2)

$$b_{ni} = b_{n-i}$$

and let the autocorrelation matrix of $y_n$ be

$$R_{yy}(i,j) = R_{yy}(i-j)$$

Then, the transpose matrix $B^T$ will have matrix elements

$$(B^T)_{ni} = b_{i-n}$$

and Eq. (3.6.3) can be written in the form

$$R_{yy}(i,j) = R_{yy}(i-j) = \sigma_\epsilon^2 \sum_n b_{n+i-j}\, b_n$$

$$= \sigma_\epsilon^2 \sum_k b_{i-k}\, b_{j-k} = \sigma_\epsilon^2 \sum_k (B)_{ik}\, (B^T)_{kj} = \sigma_\epsilon^2\, (B\, B^T)_{ij}$$

Thus, in matrix notation

$$R_{yy} = \sigma_\epsilon^2 \, B \, B^T \tag{3.6.4}$$

This equation is a special case of the more general LU factorization of the Gram-Schmidt construction given by Eq. (1.5.17). Indeed, the assumption of stationarity implies that the quantity

$$\sigma_\epsilon^2 = E[\epsilon_n^2]$$

is independent of the time $n$, and therefore the diagonal matrix $R_{\epsilon\epsilon}$ of Eq. (1.5.17) becomes a multiple of the identity matrix.

## 3.7  *Minimal-Phase Property of the Prediction-Error Filter*

As mentioned in Section 2.8, the minimal-phase property of the prediction-error filter $A(z)$ of linear prediction is an important property because it guarantees the stability of the causal inverse synthesis filter $1/A(z)$. There are many proofs of this property in the literature [6–9]. Here, we would like to present a simple proof [10] which is based directly on the fact that the optimal prediction coefficients minimize the mean-squared prediction error. Although we have only discussed first and second order linear predictors, for the purposes of this proof we will work with the more general case of an $M$th order predictor defined by

$$\hat{y}_n = - \, [a_1 y_{n-1} + a_2 y_{n-2} + \cdots + a_M y_{n-M}]$$

which is taken to represent the best prediction of $y_n$ based on the *past $M$* samples $Y_n = \{y_{n-1}, y_{n-2}, \ldots, y_{n-M}\}$. The corresponding prediction error is

$$e_n = y_n + a_1 y_{n-1} + a_2 y_{n-2} + \cdots + a_M \, y_{n-M}$$

The best set of prediction coefficients $\{a_1, a_2, \ldots, a_M\}$ is found by minimizing the mean-squared prediction error

$$\mathcal{E}(a_1, a_2, \ldots, a_M) = E[e_n^* e_n] = \sum_{m,k=0}^{M} a_m^* \, E[y_{n-m} y_{n-k}] \, a_k \tag{3.7.1}$$

$$= \sum_{m,k=0}^{M} a_m^* \, R_{yy}(m - k) \, a_k$$

where we set $a_0 = 1$. For the proof of the minimal phase property, we do not need the explicit solution of this minimization problem; we only use the fact that the optimal coefficients minimize Eq. (3.7.1). The key to the proof is based on the observation that Eq. (3.7.1) can be written in the alternative form

$$\mathcal{E}(\mathbf{a}) = \sum_{k=-M}^{M} R_{yy}(k) \, R_{aa}(k) \tag{3.7.2}$$

where $R_{aa}(k)$ is the sample autocorrelation of the prediction-error filter sequence $\mathbf{a} = [1,a_1,a_2, \ldots ,a_M]^T$, as defined in Eq. (3.3.1). The equivalence of Eqs. (3.7.1) and (3.7.2) can be seen easily, either by rearranging the summation indices of Eq. (3.7.1), or by using the results of Problems 2.3 and 2.5.

*Example 3.7.1:* We demonstrate this explicitly for the $M = 2$ case. Using the definition (3.3.1) we have

$$R_{aa}(0) = |a_0|^2 + |a_1|^2 + |a_2|^2 = 1 + |a_1|^2 + |a_2|^2$$

$$R_{aa}(1) = R_{aa}(-1)^* = a_1 a_0^* + a_2 a_1^* = a_1 + a_2 a_1^*$$

$$R_{aa}(2) = R_{aa}(-2)^* = a_2 a_0^* = a_2$$

Since $y_n$ is real-valued stationary, we have $R_{yy}(k) = R_{yy}(-k)$. Then, Eq. (3.7.1) becomes explicitly

$$\mathcal{E}(\mathbf{a}) = \sum_{m,k=0}^{M} a_m^* R_{yy}(m - k) a_k = [1,a_1^*,a_2^*] \begin{bmatrix} R_{yy}(0) & R_{yy}(1) & R_{yy}(2) \\ R_{yy}(1) & R_{yy}(0) & R_{yy}(1) \\ R_{yy}(2) & R_{yy}(1) & R_{yy}(0) \end{bmatrix} \begin{bmatrix} 1 \\ a_1 \\ a_2 \end{bmatrix}$$

$$= R_{yy}(0) [1 + a_1^* a_1 + a_2^* a_2] + R_{yy}(1) [(a_1 + a_2 a_1^*) + (a_1^* + a_2^* a_1)]$$
$$+ R_{yy}(2) [a_2 + a_2^*] = R_{yy}(0)R_{aa}(0) + R_{yy}(1) [R_{aa}(1) + R_{aa}(-1)]$$
$$+ R_{yy}(2) [R_{aa}(2) + R_{aa}(-2)]$$

Let $\mathbf{a} = [1,a_1,a_2, \ldots ,a_M]^T$ be the optimal set of coefficients which minimize $\mathcal{E}(\mathbf{a})$, and let $z_i;\ i = 1,2, \ldots ,M$ be the zeros of the corresponding prediction-error filter; that is,

$$1 + a_1 z^{-1} + a_2 z^{-2} + \cdots + a_M z^{-M}$$
$$= (1 - z_1 z^{-1}) (1 - z_2 z^{-1}) \cdots (1 - z_M z^{-1}) \quad (3.7.3)$$

Reversing any one of the zero factors in this equation; that is, replacing $(1 - z_i z^{-1})$ by its reverse $(-z_i^* + z^{-1})$, results in a sequence that has the same sample autocorrelation as $\mathbf{a}$. As we have seen, there are $2^M$ such sequences, all with the same sample autocorrelation. We would like to show that among these, $\mathbf{a}$ is the one having the minimal phase property.

To this end, let $\mathbf{b} = [b_0,b_1, \ldots ,b_M]^T$ be any one of the above $2^M$ sequences, and define the normalized sequence

$$\mathbf{c} = \mathbf{b}/b_0 = [1,b_1/b_0,b_2/b_0, \ldots ,b_M/b_0]^T \quad (3.7.4)$$

Using the fact that $\mathbf{b}$ has the same sample autocorrelation as $\mathbf{a}$, we find for the sample autocorrelation of $\mathbf{c}$:

$$R_{cc}(k) = R_{bb}(k)/|b_0|^2 = R_{aa}(k)/|b_0|^2 \quad (3.7.5)$$

The performance index (3.7.2) evaluated at $\mathbf{c}$ is then

$$\mathcal{E}(\mathbf{c}) = \sum_{k=-M}^{M} R_{yy}(k) R_{cc}(k) = \sum_{k=-M}^{M} R_{yy}(k) R_{aa}(k)/|b_0|^2 \qquad (3.7.6)$$

or

$$\mathcal{E}(\mathbf{c}) = \mathcal{E}(\mathbf{a})/|b_0|^2 \qquad (3.7.7)$$

Since $\mathbf{a}$ minimizes $\mathcal{E}$, it follows that $\mathcal{E}(\mathbf{c}) \geqslant \mathcal{E}(\mathbf{a})$. Therefore, Eq. (3.7.7) implies that

$$|b_0| \leqslant 1 \qquad (3.7.8)$$

This must be true of all $\mathbf{b}$s in the above class. Eq. (3.7.8) then, immediately implies the minimal-phase property of $\mathbf{a}$. Indeed, choosing $\mathbf{b}$ to be that sequence which is obtained from Eq. (3.7.3) by reversing only the $i$th zero factor $(1 - z_i z^{-1})$ and not the other zero factors, it follows that in this case

$$b_0 = -z_i^*$$

and therefore Eq. (3.7.8) implies that

$$|z_i| \leqslant 1 \qquad (3.7.9)$$

which shows that all the zeros of $A(z)$ are inside the unit circle and thus, $A(z)$ is minimal-phase. An alternative proof based on the Levinson recursion and Rouche's theorem of complex analysis will be presented in Chapter 5.


## Problems

**Problem 3.1:**
Prove Eq. (3.3.2).


**Problem 3.2:**
Using Eq. (3.4.1), show Eqs. (3.4.3) and (3.4.4)


**Problem 3.3:**
A random signal $y_n$ has autocorrelation function

$$R_{yy}(k) = (0.5)^{|k|} \qquad \text{for all } k$$

Find a random signal generator model for $y_n$.


**Problem 3.4:**
Repeat Problem 3.3 when

$$R_{yy}(k) = (0.5)^{|k|} + (-0.5)^{|k|} \qquad \text{for all } k$$

**Problem 3.5:**

The autocorrelation of a stationary random signal $y(n)$ is

$$R_{yy}(k) = \frac{1 - R^2}{1 + R^2} R^{|k|} \cos(\pi k/2) \qquad \text{for all } k$$

where $R$ is a parameter such that $0 < R < 1$.

(a) Compute the power spectrum of $y(n)$, and sketch it roughly versus frequency for various values of $R$. What happens as $R$ tends to 1?

(b) Find the signal generator filter for $y(n)$ and determine its difference equation and its poles and zeros.

**Problem 3.6:**

A stationary random signal $y_n$ has a rational spectral density given by

$$S_{yy}(z) = \frac{2.18 - 0.6\,(z + z^{-1})}{1.25 - 0.5\,(z + z^{-1})}$$

Determine the signal model filter $B(z)$ and the parameter $\sigma_\epsilon^2$. Write the difference equation generating $y_n$.

**Problem 3.7:**

Let $y_n = c\,x_n + v_n$. It is given that

$$S_{xx}(z) = \frac{Q}{(1 - az^{-1})(1 - az)}, \qquad S_{vv}(z) = R, \qquad S_{xv}(z) = 0$$

where $a,c,Q,R$ are given constants (assume $|a| < 1$ for stability of $x_n$).

(a) Show that the filter model for $y_n$ is of the form

$$B(z) = \frac{1 - fz^{-1}}{1 - az^{-1}}$$

where $f$ has magnitude *less* than one and is the solution of the algebraic quadratic equation

$$a\,R\,(1 + f^2) = [c^2\,Q + R\,(1 + a^2)]\,f$$

and show that the other solution has magnitude greater than one.

(b) Show that $f$ can alternatively be expressed as

$$f = \frac{Ra}{R + c^2 P}$$

where $P$ is the *positive* solution of the quadratic equation

$$Q = P - \frac{PRa^2}{R + c^2 P}$$

known as the *algebraic Riccati* equation. Show that the other solution is negative.

Show that positivity of $P$ is essential to guarantee that $f$ has magnitude less than one.

(c) Show that the scale factor $\sigma_\epsilon^2$ that appears in the spectral factorization (3.6.1) can also be expressed in terms of $P$ as

$$\sigma_\epsilon^2 = R + c^2 P$$

This method of solution of the spectral factorization problem by reducing it to the solution of an algebraic Riccati equation is quite general, and can be extended to the multichannel case.

**Problem 3.8:**

Consider a stable (but not necessarily causal) sequence $\{b_n; -\infty < n < \infty\}$ with a $z$-transform $B(z)$

$$B(z) = \sum_{n=-\infty}^{\infty} b_n z^{-n}$$

Define an infinite Toeplitz matrix $B$ by

$$B_{ni} = b_{n-i}, \qquad \text{for } -\infty < n,i < \infty$$

This establishes a correspondence between stable $z$-transforms or stable sequences and infinite Toeplitz matrices.

(a) Show that if the sequence $b_n$ is causal, then $B$ is lower triangular, as shown here:

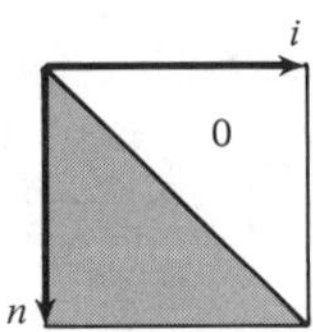


In the literature of integral operators and kernels, such matrices are rotated by 90 degrees as shown:

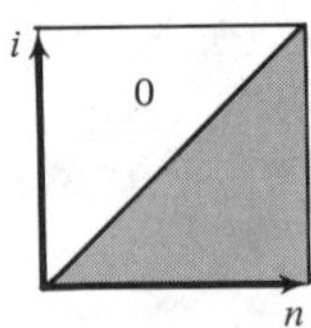


so that the $n$-axis is the horizontal-axis. For this reason, in that context they are called ''right Volterra kernels,'' or ''causal kernels.''

(b) Show that the transpose $B^T$ corresponds to the reflected (about the origin) sequence $b_{-n}$ and to the $z$-transform $B(z^{-1})$.

(c) Show that the convolution of two sequences $a_n$ and $b_n$

$$c_n = a_n * b_n \quad \text{or} \quad C(z) = A(z)\,B(z)$$

corresponds to the commutative matrix product

$$C = A\,B = B\,A$$

**Problem 3.9:**
Prove Eq. (3.7.2) for any $M$.

## References


1. E. Robinson and S. Treitel, *Geophysical Signal Analysis,* Englewood Cliffs, Prentice-Hall, 1980.
2. E. A. Robinson, *Statistical Communication and Detection*, New York, Hafner, 1967.
3. E. A. Robinson, *Multichannel Time-Series Analysis with Digital Computer Programs*, San Francisco, Holden-Day, 1967.
4. A. V. Oppenheim and R. W. Schafer, *Digital Signal Processing*, Englewood Cliffs, Prentice-Hall, 1975.
5. P. Whittle, *Prediction and Regulation*, New York, Van Nostrand Reinhold, 1963.
6. J. D. Markel and A. H. Gray, Jr., *Linear Prediction and Speech*, New York, Springer-Verlag, 1976.
7. E. A. Robinson and S. Treitel, Digital Signal Processing in Geophysics, in A. V. Oppenheim, Ed., *Applications of Digital Signal Processing*, Englewood Cliffs, Prentice-Hall, 1978.
8. S. Lang and J. McClellan, A Simple Proof of Stability for All-Pole Linear Prediction Models, *Proc. IEEE*, **67**, 860–861 (1979).
9. S. Kay and L. Pakula, Simple Proofs of the Minimum Phase Property of the Prediction Error Filter, *IEEE Trans. Acoust., Speech, Signal Process.*, **ASSP-31**, 501 (1983).
10. S. J. Orfanidis, A Proof of the Minimal Phase Property of the Prediction Error Filter, *Proc. IEEE*, **71**, 905 (1983).

# 4
___

# Linear Estimation
# of Signals

THE PROBLEM OF ESTIMATING one signal from another is one of the most important in signal processing. In many applications, the desired signal is not available or observable directly. Instead, the observable signal is a degraded or distorted version of the original signal. The signal estimation problem is to recover, in the best way possible, the desired signal from its degraded replica. We mention some typical examples: (1) The desired signal may be corrupted by strong additive noise, such as weak evoked brain potentials measured against the strong background of ongoing EEGs; or weak radar returns from a target in the presence of strong clutter. (2) An antenna array designed to be sensitive towards a particular ''look'' direction may be vulnerable to strong jammers from other directions due to sidelobe leakage; the signal processing task here is to null the jammers while at the same time maintaining the sensitivity of the array towards the desired look direction. (3) A signal transmitted over a communications channel can suffer phase and amplitude distortions and can be subject to additive channel noise; the problem is to recover the transmitted signal from the distorted received signal. (4) A Doppler radar processor tracking a moving target must take into account dynamical noise—such as small purely random accelerations—affecting the dynamics of the target, as well as measurement errors. (5) An image recorded by an imaging system is subject to distortions such as blurring due to motion or to the finite aperture of the system, or other geometric distortions; the problem here is to undo the distortions introduced by the imaging system and restore the original image. A related problem, of interest in medical image processing, is that of reconstructing an image from its projections. (6) In remote sensing and inverse scattering applications, the basic problem is, again, to infer one signal from another; for example, to infer the temperature profile of the atmosphere from measurements of the spectral distribution of infrared energy; or to deduce the structure of a dielectric medium, such as the ionosphere, by studying its response to electromagnetic wave scattering; or, in oil exploration, to infer the layered structure of the earth by measuring its response to an impulsive input near its surface.

In this chapter, we pose the signal estimation problem and discuss some of the criteria used in the design of signal estimation algorithms.

We do not present a complete discussion of all methods of signal recovery and estimation that have been invented for applications as diverse as those mentioned above. Our emphasis is on traditional linear least-squares estimation methods, not only because they are widely used, but also because they have served as the motivating force for the development of other estimation techniques and as the yardstick for evaluating them. We develop the theoretical solution of the Wiener filter both in the stationary and nonstationary cases, and discuss its connection to the orthogonal projection, Gram-Schmidt constructions, and correlation canceling ideas of Chapter 1. By means of an example, we introduce Kalman filter concepts and discuss their connection to Wiener filtering and to signal modeling. Practical implementations of the Wiener filter are discussed in Chapters 5 and 6. Other signal recovery methods for deconvolution applications that are based on alternative design criteria are briefly discussed in Chapter 5, where we also discuss some interesting connections between Wiener filtering/ linear prediction methods and inverse scattering methods.

## 4.1  *Linear and Nonlinear Estimation of Signals*

The signal estimation problem can be stated as follows: We wish to estimate a random signal $x_n$ on the basis of available observations of a related signal $y_n$. The available signal $y_n$ is to be processed by an optimal processor that produces the best possible estimate of $x_n$:

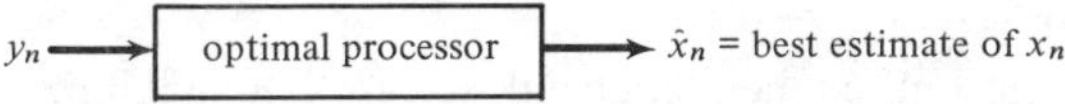


The resulting estimate $\hat{x}_n$ will be a functional of the observations $y_n$. If the optimal processor is linear, such as a linear filter, then the estimate $\hat{x}_n$ will be a linear function of the observations. We are going to concentrate mainly on linear processors. However, we would like to point out that, depending on the estimation criterion, there are cases where the estimate $\hat{x}_n$ may turn out to be a nonlinear function of the $y_n$s.

We discuss briefly four major estimation criteria for designing such optimal processors. They are:

(1) The maximum a posteriori (MAP) criterion
(2) The maximum likelihood (ML) criterion
(3) The mean squared (MS) criterion
(4) The linear mean squared (LMS) criterion

The LMS criterion is a special case of the MS criterion. It requires, *a priori*, that the estimate $\hat{x}_n$ be a *linear* function of the $y_n$s. Note that the acronym LMS is also used, especially in the context of adaptive filtering, for *least mean squared*. The main advantage of the LMS processor is that it requires only knowledge of second order statistics for its design, whereas the other, nonlinear, processors require more detailed knowledge of probability densities.

To explain the various estimation criteria, let us assume that the desired signal $x_n$ is to be estimated over a finite time interval $n_a \leq n \leq n_b$. Without loss of generality, we may assume that the observed signal $y_n$ is also available over the same interval. Define the vectors

$$\mathbf{x} = \begin{bmatrix} x_{n_a} \\ x_{n_a+1} \\ \vdots \\ x_{n_b} \end{bmatrix} \qquad \mathbf{y} = \begin{bmatrix} y_{n_a} \\ y_{n_a+1} \\ \vdots \\ y_{n_b} \end{bmatrix}$$

For each value of $n$, we seek the functional dependence

$$\hat{x}_n = \hat{x}_n(\mathbf{y})$$

of $\hat{x}_n$ on the given observation vector $\mathbf{y}$ that provides the *best estimate* of the $n$th sample $x_n$.

1. The criterion for the MAP estimate is to maximize the a posteriori conditional density of $x_n$ given that $\mathbf{y}$ already occurred; namely,

$$p(x_n/\mathbf{y}) = \text{maximum} \tag{4.1.1}$$

in other words, the optimal estimate $\hat{x}_n$ is that $x_n$ which maximizes this quantity for the given vector $\mathbf{y}$; $\hat{x}_n$ is therefore the most probable choice resulting from the given set of observations $\mathbf{y}$.

2. The ML criterion, on the other hand, selects $\hat{x}_n$ to maximize the conditional density of $\mathbf{y}$ given $x_n$; that is,

$$p(\mathbf{y}/x_n) = \text{maximum} \tag{4.1.2}$$

This criterion selects $\hat{x}_n$ as though the already collected observations $\mathbf{y}$ were the most likely ones to occur.

3. The MS criterion minimizes the mean-squared estimation error

$$\mathcal{E}_n = E[e_n^2] = \text{min}, \qquad \text{where } e_n = x_n - \hat{x}_n(\mathbf{y}) \tag{4.1.3}$$

That is, the best choice of functional dependence $\hat{x}_n = \hat{x}_n(\mathbf{y})$ is sought that minimizes this expression. We know from our results in Section 1.4 that the required choice is the corresponding *conditional mean*

$$\hat{x}_n(\mathbf{y}) = E[x_n/\mathbf{y}] = \text{MS estimate} \qquad (4.1.4)$$

computed with respect to the conditional density $p(x_n/\mathbf{y})$.

4. Finally, the LMS criterion requires the estimate to be a linear function of the observations

$$\hat{x}_n = \sum_{i=n_a}^{n_b} h(n,i)y_i \qquad (4.1.5)$$

For each $n$, the weights $h(n,i)$, $n_a \leq i \leq n_b$ are selected so as to *minimize* the mean-squared estimation error

$$\mathcal{E}_n = E[e_n^2] = E[(x_n - \hat{x}_n)^2] = \text{minimum} \qquad (4.1.6)$$

With the exception of the LMS estimate, all the other estimates $\hat{x}_n(\mathbf{y})$ are, in general, nonlinear in $\mathbf{y}$.

*Example 4.1.1:*  If both $x_n$ and $\mathbf{y}$ are zero-mean and jointly gaussian, then Examples 1.4.1 and 1.4.2 imply that the MS and the LMS estimates of $x_n$ are the *same*. Furthermore, since $p(x_n/\mathbf{y})$ is gaussian it will be a curve symmetric about its maximum which occurs at its mean; that is, at $E[x_n/\mathbf{y}]$. Therefore, the MAP estimate of $x_n$ is equal to the MS estimate. In conclusion, for *zero-mean jointly gaussian* $x_n$ and $\mathbf{y}$, the three estimates MAP, MS, and LMS *coincide*.

*Example 4.1.2:*  To see the nonlinear character and the differences among the various estimates, consider the following example: A discrete-amplitude, constant-in-time signal $x$ can take on the three values

$$x = -1 \qquad x = 0 \qquad x = +1$$

each with probability of $1/3$. This signal is placed on a known carrier waveform $c_n$ and transmitted over a noisy channel. The received samples are of the form

$$y_n = c_n x + v_n$$

$n = 1, 2, \ldots, M$, where $v_n$ are zero-mean white gaussian noise samples of variance $\sigma_v^2$, assumed to be independent of $x$. The above set of measurements can be written in an obvious vector notation

$$\mathbf{y} = \mathbf{c}x + \mathbf{v}$$

(a) Determine the conditional densities $p(y/x)$ and $p(x/y)$.

(b) Determine and compare the four alternative estimates MAP, ML, MS, and LMS of $x$.

To compute $p(y/x)$, note that if $x$ is given, then the only randomness left in $\mathbf{y}$ arises from the noise term $\mathbf{v}$. Since $v_n$ are uncorrelated and gaussian, they will be independent; therefore,

$$p(\mathbf{y}/x) = p(\mathbf{v}) = \prod_{n=1}^{M} p(v_n)$$

$$= (2\pi\sigma_v^2)^{-M/2} \exp\left[ -\frac{1}{2\sigma_v^2} \sum_{n=1}^{M} v_n^2 \right]$$

$$= (2\pi\sigma_v^2)^{-M/2} \exp\left[ -\frac{1}{2\sigma_v^2} \mathbf{v}^2 \right]$$

$$= (2\pi\sigma_v^2)^{-M/2} \exp\left[ -\frac{1}{2\sigma_v^2} (\mathbf{y} - \mathbf{c}x)^2 \right]$$

Using Bayes' rule we find $p(x/\mathbf{y}) = p(\mathbf{y}/x)p(x)/p(\mathbf{y})$. Since

$$p(x) = [\delta(x - 1) + \delta(x) + \delta(x + 1)]/3$$

we find

$$p(x/\mathbf{y}) = [p(\mathbf{y}/1)\delta(x - 1) + p(\mathbf{y}/0)\delta(x) + p(\mathbf{y}/-1)\delta(x + 1)]/A$$

where the constant $A$ is

$$A = 3p(\mathbf{y}) = 3 \int p(\mathbf{y}/x)p(x)\,dx = p(\mathbf{y}/1) + p(\mathbf{y}/0) + p(\mathbf{y}/-1)$$

To find the MAP estimate of $x$, the quantity $p(x/\mathbf{y})$ must be maximized with respect to $x$. Since the expression for $p(x/\mathbf{y})$ forces $x$ to be one of the three values $+1, 0, -1$, it follows that the maximum among the three coefficients $p(\mathbf{y}/1)$, $p(\mathbf{y}/0)$, $p(\mathbf{y}/-1)$ determines the value of $x$. Thus, for a given $\mathbf{y}$ we select that $x$ such that

$$p(\mathbf{y}/x) = \text{maximum of } \{p(\mathbf{y}/1), p(\mathbf{y}/0), p(\mathbf{y}/-1)\}$$

Using the gaussian nature of $p(\mathbf{y}/x)$, we find equivalently

$$(\mathbf{y} - \mathbf{c}x)^2 = \text{minimum of } \{(\mathbf{y} - \mathbf{c})^2, \mathbf{y}^2, (\mathbf{y} + \mathbf{c})^2\}$$

Subtracting $\mathbf{y}^2$ from both sides, dividing by $\mathbf{c}^T\mathbf{c}$, and denoting

$$\bar{y} = \frac{\mathbf{c}^T\mathbf{y}}{\mathbf{c}^T\mathbf{c}}$$

we find the equivalent equation

$$x^2 - 2x\bar{y} = \min\{1 - 2\bar{y}, 0, 1 + 2\bar{y}\}$$

and in particular, applying these for $x = +1, 0, -1$, we find

$$
\hat{x}_{\text{MAP}} = \begin{cases} 1 \text{ if } \bar{y} > \dfrac{1}{2} \\[2ex] 0 \text{ if } -\dfrac{1}{2} < \bar{y} < \dfrac{1}{2} \\[2ex] -1 \text{ if } \bar{y} < -\dfrac{1}{2} \end{cases}
$$

To determine the ML estimate, we must maximize $p(\mathbf{y}/x)$ with respect to $x$. The ML estimate does not require knowledge of the probability density $p(x)$ of $x$. Therefore, differentiating $p(\mathbf{y}/x)$ with respect to $x$ and setting the derivative to zero gives

$$
\frac{\partial}{\partial x} p(\mathbf{y}/x) = 0 \qquad \text{or} \qquad \frac{\partial}{\partial x} \ln p(\mathbf{y}/x) = 0 \qquad \text{or} \qquad \frac{\partial}{\partial x} (\mathbf{y} - \mathbf{c}x)^2 = 0
$$

which gives

$$
\hat{x}_{\text{ML}} = \frac{\mathbf{c}^T \mathbf{y}}{\mathbf{c}^T \mathbf{c}} = \bar{y}
$$

The MS estimate is obtained by computing the conditional mean

$$
E[x/\mathbf{y}] = \int xp(x/\mathbf{y}) \, dx
$$

$$
= \int x[p(\mathbf{y}/1)\delta(x - 1) + p(\mathbf{y}/0)\delta(x) + p(\mathbf{y}/-1)\delta(x + 1)] \, dx/A
$$

$$
= [p(\mathbf{y}/1) - p(\mathbf{y}/-1)]/A
$$

or

$$
\hat{x}_{\text{MS}} = \frac{p(\mathbf{y}/+1) - p(\mathbf{y}/-1)}{p(\mathbf{y}/+1) + p(\mathbf{y}/0) + p(\mathbf{y}/-1)}
$$

Canceling some common factors from numerator and denominator, we find a simpler expression

$$
\hat{x}_{\text{MS}} = \frac{2\text{sh}(2a\bar{y})}{e^a + 2\text{ch}(2a\bar{y})} \qquad \text{where} \qquad a = \frac{\mathbf{c}^T \mathbf{c}}{2\sigma_v^2}
$$

Finally, the LMS estimate can be computed as in Example 1.4.3. We find

$$
\hat{x}_{\text{LMS}} = \frac{1}{\dfrac{\sigma_v^2}{\sigma_x^2} + \mathbf{c}^T \mathbf{c}} \mathbf{c}^T \mathbf{y} = \frac{\mathbf{c}^T \mathbf{c}}{\dfrac{\sigma_v^2}{\sigma_x^2} + \mathbf{c}^T \mathbf{c}} \bar{y}
$$

All four estimates have been expressed in terms of $\bar{y}$. Note that the ML estimate is linear but has a different slope than the LMS estimate. The nonlinearity of the various estimates is best seen in the following figure:

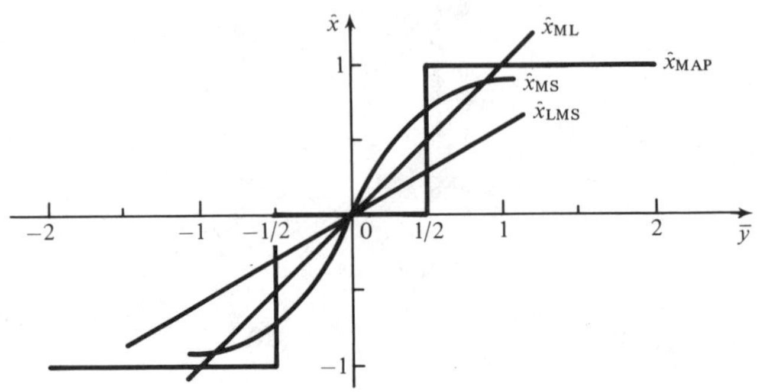


## 4.2   *Orthogonality and Normal Equations*

From now on, we will concentrate on the optimal linear estimate defined by Eqs. (4.1.5) and (4.1.6). For each time instant $n$, at which an estimate $\hat{x}_n$ is sought, the optimal weights $h(n,i)$; $n_a \leq i \leq n_b$ must be determined that minimize the error criterion (4.1.6). In general, a new set of optimal weights must be computed for each time instant $n$. In the special case when the processes $x_n$ and $y_n$ are stationary and the observations are available for a long time; that is, $n_a = -\infty$, the weights become time-invariant in the sense that $h(n,i) = h(n - i)$, and the linear processor becomes an ordinary *time-invariant linear filter*. We will discuss the solution for $h(n,i)$ both for the time-invariant and the more general cases. The problem of determining the optimal weights $h(n,i)$ according to the mean-squared minimization criterion (4.1.6) is in general referred to as the *Wiener filtering problem* [1–11]. An interesting historical account of the development of this problem and its ramifications is given in the review article by Kailath [12].

Wiener filtering problems are conventionally divided into three types:

(1) The optimal *smoothing* problem
(2) The optimal *filtering* problem, and
(3) The optimal *prediction* problem.

In all cases, the optimal estimate of $x_n$ at a given time instant $n$ is given by an expression of the form (4.1.5), as a linear combination of the available observations $y_n$ in the interval $n_a \leq n \leq n_b$. The division into three types of problems depends on *which* of the available observations in that interval are taken into account *in making up* the linear combination (4.1.5).

1. In the smoothing problem, *all* the observations in the interval $[n_a, n_b]$ are taken into account. The shaded part in the following figure denotes the range of observations that are used in the summation of Eq. (4.1.5):

$$\hat{x}_n = \sum_{i=n_a}^{n_b} h(n,i)y_i$$

Since some of the observations are to the future of $x_n$, the linear operation is not causal. This does not present a problem if the sequence $y_n$ is already available and stored in memory.

2. The optimal filtering problem, on the other hand, requires the linear operation (4.1.5) to be *causal*; that is, only those observations that are in the *present* and *past* of the current sample $x_n$ must be used in making up the estimate $\hat{x}_n$. This requires that the matrix of optimal weights $h(n,i)$ be *lower* triangular; that is,

$$h(n,i) = 0, \qquad \text{for } n < i$$

Thus, in reference to the figure below, only the shaded portion of the observation interval is used at the current time instant:

$$\hat{x}_n = \sum_{i=n_a}^{n} h(n,i)y_i$$

The estimate $\hat{x}_n$ depends on the present and all the past observations, from the fixed starting point $n_a$ to the current time instant $n$. As $n$ increases, more and more observations are taken into account in making up the estimate $\hat{x}_n$, and the actual computation of $\hat{x}_n$ becomes less and less efficient. It is desirable, then, to be able to recast the expression for $\hat{x}_n$ in a time-recursive form. This is what is done in Kalman filtering. But, there is another way to make the Wiener filter computationally manageable. Instead of allowing a growing number of observations, only the current and the *past M* observations $y_i$; $i = n, n-1, \ldots, n - M$ are taken into account. In this case, only $(M+1)$ filter weights are to be computed at each time instant $n$. This is depicted below:

$$\hat{x}_n = \sum_{i=n-M}^{n} h(n,i)y_i = \sum_{m=0}^{M} h(n, n-m)y_{n-m}$$

This is referred to as the *finite impulse response* (FIR) Wiener filter. Because of its simple implementation, the FIR Wiener filter has enjoyed widespread popularity. Depending on the particular application, the practical implementation of the filter may vary. In Section 4.3 we present the theoretical formulation that applies to the stationary case; in Chapter 5 we reconsider it as a waveshaping and spiking filter and discuss a number of deconvolution applications. In Chapter 6 we consider its adaptive implementation using the Widrow-Hoff LMS algorithm and discuss a number of applications such as channel equalization and echo cancellation; we also discuss two alternative adaptive implementations—the so-called "gradient lattice," and the "recursive least-squares".

3. Finally, the linear prediction problem is a special case of the optimal filtering problem with the additional stipulation that observations only up to time instant $n - D$ must be used in obtaining the current estimate $\hat{x}_n$; this is equivalent to the problem of predicting $D$ units of time into the future. The range of observations used in this case is shown below:

$$\hat{x}_n = \sum_{i=n_a}^{n-D} h(n,i)y_i$$

Of special interest to us will be the case of *one-step* prediction, corresponding to the choice $D = 1$. This is depicted below:

$$\hat{x}_n = \sum_{i=n_a}^{n-1} h(n,i)y_i$$

If we demand that the prediction be based only on the past $M$ samples (from the current sample), we obtain the FIR version of the prediction problem, referred to as *linear prediction based on the past M samples*, which is depicted below:

$$\hat{x}_n = \sum_{i=n-M}^{n-1} h(n,i)y_i = \sum_{m=1}^{M} h(n,n-m)y_{n-m}$$

Next, we set up the orthogonality and normal equations for the optimal weights. We begin with the smoothing problem. The estimation error is in this case

$$e_n = x_n - \hat{x}_n = x_n - \sum_{i=n_a}^{n_b} h(n,i)y_i \tag{4.2.1}$$

Differentiating the mean-squared estimation error (4.1.6) with respect to *each* weight $h(n,i)$, $n_a \leq i \leq n_b$, and setting the derivative to zero, we obtain the *orthogonality equations* that are enough to determine the weights:

$$\frac{\partial \mathcal{E}_n}{\partial h(n,i)} = 2E\left[ e_n \frac{\partial e_n}{\partial h(n,i)} \right] = -2E[e_n y_i] = 0, \qquad \text{for } n_a \leq i \leq n_b$$

or

$$R_{ey}(n,i) = E[e_n y_i] = 0 \qquad \text{(orthogonality equations)} \qquad (4.2.2)$$

for $n_a \leq i \leq n_b$. Thus, the estimation error $e_n$ is orthogonal (uncorrelated) to *each* observation $y_i$ used in *making up* the estimate $\hat{x}_n$. The orthogonality equations provide exactly as many equations as there are unknown weights.

Inserting Eq. (4.2.1) for $e_n$, the orthogonality equations may be written in an equivalent form, known as the *normal equations*

$$E\left[ \left( x_n - \sum_{k=n_a}^{n_b} h(n,k) y_k \right) y_i \right] = 0$$

or

$$E[x_n y_i] = \sum_{k=n_a}^{n_b} h(n,k) E[y_k y_i] \qquad \text{(normal equations)} \qquad (4.2.3)$$

for $n_a \leq i \leq n_b$. These determine the optimal weights at the current time instant $n$. In the vector notation of Section 4.1, we write Eq. (4.2.3) as

$$E[\mathbf{x}\mathbf{y}^T] = H E[\mathbf{y}\mathbf{y}^T]$$

where $H$ is the *matrix of weights* $h(n,i)$. The optimal $H$ and the estimate are then

$$\hat{\mathbf{x}} = H\mathbf{y} = E[\mathbf{x}\mathbf{y}^T] E[\mathbf{y}\mathbf{y}^T]^{-1} \mathbf{y}$$

This is identical to the correlation canceler of Section 1.4. The orthogonality equations (4.2.2) are precisely the correlation cancellation conditions. Extracting the $n$th row of this matrix equation, we find an explicit expression for the $n$th estimate $\hat{x}_n$

$$\hat{x}_n = E[x_n \mathbf{y}^T] E[\mathbf{y}\mathbf{y}^T]^{-1} \mathbf{y}$$

which is recognized as the projection of the random variable $x_n$ onto the subspace spanned by the *available* observations; namely, $Y = \{y_{n_a}, y_{n_a+1}, \ldots, y_{n_b}\}$. This is a general result: The minimum mean-squared linear estimate $\hat{x}_n$ is the projection of $x_n$ onto the subspace spanned by *all* the observations that are used *to make up* that estimate. This result is a direct consequence of the quadratic minimization criterion (4.1.6) and the orthogonal projection theorem discussed in Section 1.5.

Using the methods of Section 1.4, the minimized estimation error at time instant $n$ is easily computed by

$$\mathcal{E}_n = E[e_n e_n] = E[e_n x_n] = E\left[\left(x_n - \sum_{i=n_a}^{n_b} h(n,i)y_i\right)x_n\right]$$

$$= E[x_n^2] - \sum_{i=n_a}^{n_b} h(n,i)E[y_i x_n]$$

$$= E[x_n^2] - E[x_n \mathbf{y}^T]E[\mathbf{y}\mathbf{y}^T]^{-1}E[\mathbf{y}x_n]$$

which corresponds to the diagonal entries of the covariance matrix of the estimation error $\mathbf{e}$:

$$R_{ee} = E[\mathbf{e}\mathbf{e}^T] = E[\mathbf{x}\mathbf{x}^T] - E[\mathbf{x}\mathbf{y}^T]E[\mathbf{y}\mathbf{y}^T]^{-1}E[\mathbf{y}\mathbf{x}^T]$$

The optimal filtering problem is somewhat more complicated because of the causality condition. In this case, the estimate at time $n$ is given by

$$\hat{x}_n = \sum_{i=n_a}^{n} h(n,i)y_i \tag{4.2.4}$$

Inserting this into the minimization criterion (4.1.6) and differentiating with respect to $h(n,i)$ for $n_a \le i \le n$, we find again the orthogonality conditions

$$R_{ey}(n,i) = E[e_n y_i] = 0 \qquad \text{for } n_a \le i \le n \tag{4.2.5}$$

where the most important difference from Eq. (4.2.2) is the restriction on the *range* of $i$'s; that is, $e_n$ is decorrelated only from the present and past values of $y_i$. Again, the estimation error $e_n$ is orthogonal to each observation $y_i$ that is being used in making up the estimate. The orthogonality equations can be converted into the normal equations as follows:

$$E[e_n y_i] = E\left[\left(x_n - \sum_{k=n_a}^{n} h(n,k)y_k\right)y_i\right] = 0$$

or

$$E[x_n y_i] = \sum_{k=n_a}^{n} h(n,k)E[y_k y_i] \qquad \text{for } n_a \le i \le n \tag{4.2.6}$$

or

$$R_{xy}(n,i) = \sum_{k=n_a}^{n} h(n,k)R_{yy}(k,i) \qquad \text{for } n_a \le i \le n \tag{4.2.7}$$

Such equations are generally known as *Wiener-Hopf equations*. Introducing the vector of observations *up to* the current time *n*; namely,

$$\mathbf{y}_n = [y_{n_a}, y_{n_a+1}, \ldots, y_n]^T$$

we may write Eq. (4.2.6) in vector form as

$$E[x_n \mathbf{y}_n^T] = [h(n,n_a), h(n,n_a+1), \ldots, h(n,n)] E[\mathbf{y}_n \mathbf{y}_n^T]$$

which can be solved for the vector of weights

$$[h(n,n_a), h(n,n_a+1), \ldots, h(n,n)] = E[x_n \mathbf{y}_n^T] E[\mathbf{y}_n \mathbf{y}_n^T]^{-1}$$

and for the estimate $\hat{x}_n$:

$$\hat{x}_n = E[x_n \mathbf{y}_n^T] E[\mathbf{y}_n \mathbf{y}_n^T]^{-1} \mathbf{y}_n \qquad (4.2.8)$$

Again, $\hat{x}_n$ is recognized as the *projection* of $x_n$ onto the space spanned by the observations that are used in making up the estimate; namely, $Y_n = \{y_{n_a}, y_{n_a+1}, \ldots, y_n\}$. This solution of Eqs. (4.2.6) and (4.2.7) will be discussed in more detail in Section 4.8, using covariance factorization methods.

## 4.3   *Stationary Wiener Filter*

In this section, we make two assumptions that simplify the structure of Eqs. (4.2.6) and (4.2.7). The first is to assume *stationarity* for all signals so that the cross-correlation and autocorrelation appearing in Eq. (4.2.7) become functions of the *differences* of their arguments. The second assumption is to take the initial time $n_a$ to be the *infinite past*; $n_a = -\infty$, that is, the observation interval is $Y_n = \{y_i; -\infty < i \leq n\}$.

The assumption of stationarity can be used as follows: Suppose we have the solution of $h(n,i)$ of Eq. (4.2.7) for the best weights to estimate $x_n$, and wish to determine the best weights $h(n + \Delta, i)$, $n_a \leq i \leq n + \Delta$ for estimating the sample $x_{n+\Delta}$ at the future time $n + \Delta$. Then, the new weights will satisfy the same equations as (4.2.7) with the changes

$$R_{xy}(n + \Delta, i) = \sum_{k=n_a}^{n+\Delta} h(n + \Delta, k) R_{yy}(k,i) \qquad \text{for } n_a \leq i \leq n + \Delta \qquad (4.3.1)$$

Making a change of variables $i \to i + \Delta$ and $k \to k + \Delta$, we rewrite Eq. (4.3.1) as

$$R_{xy}(n + \Delta, i + \Delta) = \sum_{k=n_a-\Delta}^{n} h(n + \Delta, k + \Delta) R_{yy}(k + \Delta, i + \Delta)$$

$$\text{for } n_a - \Delta \leq i \leq n \qquad (4.3.2)$$

Now, if we assume stationarity, Eqs. (4.2.7) and (4.3.2) become

$$R_{xy}(n - i) = \sum_{k=n_a}^{n} h(n,k)R_{yy}(k - i) \quad \text{for } n_a \leq i \leq n$$

$$R_{xy}(n - i) = \sum_{k=n_a-\Delta}^{n} h(n + \Delta, k + \Delta)R_{yy}(k - i)$$

$$\text{for } n_a - \Delta \leq i \leq n$$

(4.3.3)

If it were not for the differences in the ranges of $i$ and $k$, these two equations would be the same. But this is exactly what happens when we make the second assumption that $n_a = -\infty$. Therefore, by uniqueness of the solution, we find in this case

$$h(n + \Delta, k + \Delta) = h(n,k)$$

and since $\Delta$ is arbitrary, it follows that $h(n,k)$ must be a function of the difference of its arguments, that is,

$$h(n,k) = h(n - k) \tag{4.3.4}$$

Thus, the optimal linear processor becomes a *shift-invariant causal linear filter*; the estimate is given by

$$\hat{x}_n = \sum_{i=-\infty}^{n} h(n - i)y_i = \sum_{i=0}^{\infty} h(i)y_{n-i} \tag{4.3.5}$$

and Eq. (4.3.3) becomes in this case

$$R_{xy}(n - i) = \sum_{k=-\infty}^{n} h(n - k)R_{yy}(k - i) \quad \text{for } -\infty < i \leq n$$

With the change of variables $n - i \rightarrow n$, and $n - k \rightarrow k$, we find

$$R_{xy}(n) = \sum_{k=0}^{\infty} R_{yy}(n - k)h(k) \quad \text{for } n \geq 0 \tag{4.3.6}$$

and written in matrix form

$$\begin{bmatrix} R_{yy}(0) & R_{yy}(1) & R_{yy}(2) & R_{yy}(3) & \cdots \\ R_{yy}(1) & R_{yy}(0) & R_{yy}(1) & R_{yy}(2) & \cdots \\ R_{yy}(2) & R_{yy}(1) & R_{yy}(0) & R_{yy}(1) & \cdots \\ R_{yy}(3) & R_{yy}(2) & R_{yy}(1) & R_{yy}(0) & \cdots \\ \vdots & \vdots & \vdots & \vdots & \end{bmatrix} \begin{bmatrix} h(0) \\ h(1) \\ h(2) \\ h(3) \\ \vdots \end{bmatrix} = \begin{bmatrix} R_{xy}(0) \\ R_{xy}(1) \\ R_{xy}(2) \\ R_{xy}(3) \\ \vdots \end{bmatrix} \tag{4.3.7}$$

These are the *discrete-time Wiener-Hopf equations*. Were it not for the restriction $n \geq 0$ (which reflects the requirement of causality), they could be solved easily by $z$-transform methods. As written above, they require methods of *spectral factorization* for their solution.

Before we discuss such methods, we mention in passing the continuous-time version of the Wiener-Hopf equation:

$$R_{xy}(t) = \int_0^\infty R_{yy}(t - t')h(t') \, dt' \qquad \text{for } t \geq 0$$

We also consider the FIR Wiener filtering problem in the stationary case. The observation interval in this case is $Y_n = \{y_i;\ n - M \leq i \leq n\}$. Using the same arguments as above we have $h(n,i) = h(n - i)$, and the estimate $\hat{x}_n$ is obtained by an ordinary FIR linear filter

$$\hat{x}_n = \sum_{i=n-M}^{n} h(n - i)y_i = h(0)y_n + h(1)y_{n-1} + \cdots + h(M)y_{n-M} \qquad (4.3.8)$$

where the $M + 1$ optimal filter weights $h(0), h(1), \ldots, h(M)$ are obtained by the $(M + 1) \times (M + 1)$ matrix version of the Wiener-Hopf, normal, equations

$$\begin{bmatrix} R_{yy}(0) & R_{yy}(1) & R_{yy}(2) & \cdots & R_{yy}(M) \\ R_{yy}(1) & R_{yy}(0) & R_{yy}(1) & \cdots & R_{yy}(M - 1) \\ R_{yy}(2) & R_{yy}(1) & R_{yy}(0) & \cdots & R_{yy}(M - 2) \\ \vdots & \vdots & \vdots & & \vdots \\ R_{yy}(M) & R_{yy}(M - 1) & R_{yy}(M - 2) & \cdots & R_{yy}(0) \end{bmatrix} \begin{bmatrix} h(0) \\ h(1) \\ h(2) \\ \vdots \\ h(M) \end{bmatrix}$$

$$= \begin{bmatrix} R_{xy}(0) \\ R_{xy}(1) \\ R_{xy}(2) \\ \vdots \\ R_{xy}(M) \end{bmatrix} \qquad (4.3.9)$$

Exploiting the Toeplitz property of the matrix $R_{yy}$, the above matrix equation can be solved efficiently using Levinson's algorithm. This will be discussed in Chapter 5. In Chapter 6, we will consider adaptive implementations of the FIR Wiener filter which produce the optimal filter weights *adaptively* without requiring prior knowledge of the autocorrelation and cross-correlation matrices $R_{yy}$ and $R_{xy}$ and without requiring any matrix inversion.

We summarize our results on the stationary Wiener filter in Fig. 4.1. The optimal filter weights $h(n);\ n = 0,1,2, \ldots$, are computed from Eq. (4.3.7)

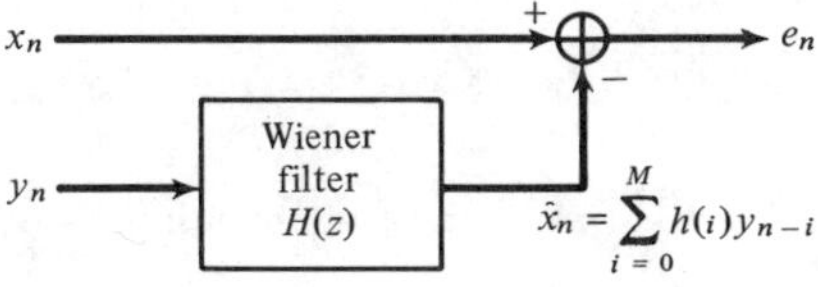

or Eq. (4.3.9). The action of the filter is precisely that of the *correlation canceler*: The filter processes the observation signal $y_n$ *causally* to produce the *best possible estimate* $\hat{x}_n$ of $x_n$, and then it proceeds to cancel it from the output $e_n$. As a result, the output $e_n$ is no longer correlated with any of the present and past values of $y_n$; that is, $E[e_n y_{n-i}] = 0$, for all $i = 0, 1, 2, \ldots$. As we remarked in Section 1.4, it is better to think of $\hat{x}_n$ as the optimal estimate of *that part* of the primary signal $x_n$ which happens to be correlated with the secondary signal $y_n$; this follows from the property that if $x_n = x_1(n) + x_2(n)$ with $R_{x_2 y} = 0$, then $R_{xy} = R_{x_1 y}$. Therefore, the solution of Eq. (4.3.7) for the best weights to estimate $x_n$ is also the solution for the best weights to estimate $x_1(n)$. The filter may also be thought of as the *optimal signal separator* of the two signal components $x_1(n)$ and $x_2(n)$.

## 4.4 Construction of the Wiener Filter by Prewhitening

The normal equations (4.3.6) would have a trivial solution if the sequence $y_n$ were a white-noise sequence with delta-function autocorrelation. Thus, the solution procedure is first to whiten the sequence $y_n$ and then solve the normal equations. To this end, let $y_n$ have a signal model, as guaranteed by the spectral factorization theorem

$$S_{yy}(z) = \sigma_\epsilon^2 B(z) B(z^{-1}) \qquad \epsilon_n \longrightarrow \boxed{B(z)} \longrightarrow y_n \qquad (4.4.1)$$

where $\epsilon_n$ is the driving white noise, and $B(z)$ a minimal-phase filter.

The problem of estimating $x_n$ in terms of the sequence $y_n$ becomes equivalent to the problem of estimating $x_n$ in terms of the white-noise sequence $\epsilon_n$:

$$\epsilon_n \longrightarrow \boxed{B(z)} \longrightarrow y_n \longrightarrow \boxed{H(z)} \longrightarrow \hat{x}_n$$

If we could determine the combined filter

$$F(z) = B(z)H(z) \qquad \epsilon_n \longrightarrow \boxed{F(z)} \longrightarrow \hat{x}_n$$

we would then solve for the desired Wiener filter $H(z)$

$$H(z) = \frac{F(z)}{B(z)} \qquad (4.4.2)$$

Since $B(z)$ is minimal-phase, the indicated inverse $1/B(z)$ is guaranteed to be *stable and causal*. Let $f_n$ be the impulse response of $F(z)$. Then, it satisfies the normal equations of the type of Eq. (4.3.6):

$$R_{x\epsilon}(n) = \sum_{i=0}^{\infty} f_i R_{\epsilon\epsilon}(n - i), \qquad n \geq 0 \qquad (4.4.3)$$

Since $R_{\epsilon\epsilon}(n - i) = \sigma_\epsilon^2 \delta(n - i)$, Eq. (4.4.3) collapses to

$$R_{x\epsilon}(n) = \sigma_\epsilon^2 f_n$$

or

$$f_n = R_{x\epsilon}(n)/\sigma_\epsilon^2, \qquad \text{for } n \geq 0 \tag{4.4.4}$$

Next, compute the corresponding $z$-transform $F(z)$

$$F(z) = \sum_{n=0}^{\infty} f_n z^{-n} = \sum_{n=0}^{\infty} R_{x\epsilon}(n) z^{-n}/\sigma_\epsilon^2 = [S_{x\epsilon}(z)]_+/\sigma_\epsilon^2 \tag{4.4.5}$$

where $[S_{x\epsilon}(z)]_+$ denotes the *causal part* of the double-sided $z$-transform $S_{x\epsilon}(z)$. Generally, the causal part of a $z$-transform

$$G(z) = \sum_{n=-\infty}^{\infty} g_n z^{-n}$$

is defined as

$$[G(z)]_+ = \sum_{n=0}^{\infty} g_n z^{-n}$$

The causal instruction in Eq. (4.4.5) was necessary since the above solution for $f_n$ was valid only for $n \geq 0$. Since $y_n$ is the output of the filter $B(z)$ driven by $\epsilon_n$, it follows that

$$S_{xy}(z) = S_{x\epsilon}(z)B(z^{-1}) \qquad \text{or} \qquad S_{x\epsilon}(z) = S_{xy}(z)/B(z^{-1})$$

Combining Eqs. (4.4.2) and (4.4.5), we finally find

$$H(z) = \frac{1}{\sigma_\epsilon^2 B(z)} \left[\frac{S_{xy}(z)}{B(z^{-1})}\right]_+ \qquad \text{(Wiener filter)} \tag{4.4.6}$$

Thus, the construction of the optimal filter first requires spectral factorization of $S_{yy}(z)$ to obtain $B(z)$, and then use of the above formula. This is the optimal *realizable* Wiener filter based on the *infinite past*. If the causal instruction is ignored, one obtains the optimal *unrealizable* Wiener filter

$$H_{\text{unreal}}(z) = \frac{S_{xy}(z)}{\sigma_\epsilon^2 B(z)B(z^{-1})} = \frac{S_{xy}(z)}{S_{yy}(z)} \tag{4.4.7}$$

The *minimum value* of the *mean-squared estimation error* can be conveniently expressed by a contour integral, as follows

$$\mathcal{E} = E[e_n^2] = E[e_n(x_n - \hat{x}_n)] = E[e_n x_n] - E[e_n \hat{x}_n] = E[e_n x_n] = R_{ex}(0)$$

$$= \oint_{\text{u.c.}} S_{ex}(z)\, dz/2\pi jz = \oint_{\text{u.c.}} [S_{xx}(z) - S_{\hat{x}x}(z)]\, dz/2\pi jz$$

or

$$\mathcal{E} = \oint_{\text{u.c.}} [S_{xx}(z) - H(z)S_{yx}(z)]\, \frac{dz}{2\pi jz} \tag{4.4.8}$$

## 4.5 *Wiener Filter Example*

This example, in addition to illustrating the above ideas, will also serve as a short introduction to *Kalman filtering*. It is desired to estimate the signal $x_n$ on the basis of noisy observations

$$y_n = x_n + v_n,$$

where $v_n$ is white noise of unit variance, $\sigma_v^2 = 1$, uncorrelated with $x_n$. The signal $x_n$ is a first order Markov process, having a signal model

$$x_{n+1} = 0.6\, x_n + w_n$$

where $w_n$ is white noise of variance $\sigma_w^2 = 0.82$.

Enough information is given above to compute the required power spectral densities $S_{xy}(z)$ and $S_{yy}(z)$. First we note that the signal generator transfer function for $x_n$ is

$$w_n \longrightarrow \boxed{M(z)} \longrightarrow x_n, \qquad M(z) = \frac{1}{z - 0.6}$$

so that

$$S_{xx}(z) = \sigma_w^2 M(z)M(z^{-1}) = \frac{0.82}{(z - 0.6)(z^{-1} - 0.6)} = \frac{0.82}{(1 - 0.6z^{-1})(1 - 0.6z)}$$

Then we find

$$S_{xy}(z) = S_{x(x+v)} = S_{xx} + S_{xv} = S_{xx}(z) = \frac{0.82}{(1 - 0.6z^{-1})(1 - 0.6z)}$$

$$S_{yy}(z) = S_{(x+v)(x+v)} = S_{xx} + S_{xv} + S_{vx} + S_{vv} = S_{xx} + S_{vv}$$

$$= \frac{0.82}{(1 - 0.6z^{-1})(1 - 0.6z)} + 1 = \frac{0.82 + (1 - 0.6z^{-1})(1 - 0.6z)}{(1 - 0.6z^{-1})(1 - 0.6z)}$$

$$= \frac{2(1 - 0.3z^{-1})(1 - 0.3z)}{(1 - 0.6z^{-1})(1 - 0.6z)} = 2 \cdot \frac{1 - 0.3z^{-1}}{1 - 0.6z^{-1}} \cdot \frac{1 - 0.3z}{1 - 0.6z}$$

$$= \sigma_\epsilon^2 B(z)B(z^{-1})$$

Then according to Eq. (4.4.6), we must compute the causal part of

$$G(z) = \frac{S_{xy}(z)}{B(z^{-1})} = \frac{\dfrac{0.82}{(1 - 0.6z^{-1})(1 - 0.6z)}}{\dfrac{(1 - 0.3z)}{(1 - 0.6z)}} = \frac{0.82}{(1 - 0.6z^{-1})(1 - 0.3z)}$$

This may be done by partial fraction expansion, but the fastest way is to use the contour inversion formula to compute $g_k$, for $k \geq 0$, and then resum this series:

$$g_k = \oint_{\text{u.c.}} G(z)z^k \, dz/2\pi jz = \oint_{\text{u.c.}} \frac{0.82z^k}{(1 - 0.3z)(z - 0.6)} \frac{dz}{2\pi j}$$

$$= (\text{residue at } z = 0.6) = \frac{0.82(0.6)^k}{(1 - 0.3 \times 0.6)} = (0.6)^k, \qquad k \geq 0$$

Resumming, we find the required causal part

$$[G(z)]_+ = \frac{1}{1 - 0.6z^{-1}}$$

Finally, the optimal Wiener estimation filter is

$$H(z) = \frac{1}{\sigma_\epsilon^2 B(z)} \left[ \frac{S_{xy}(z)}{B(z^{-1})} \right]_+ = \frac{[G(z)]_+}{\sigma_\epsilon^2 B(z)} = \frac{0.5}{1 - 0.3z^{-1}} \qquad (4.5.1)$$

which can be realized as the difference equation

$$\hat{x}_n = 0.3\hat{x}_{n-1} + 0.5y_n, \qquad y_n \longrightarrow \boxed{H(z)} \longrightarrow \hat{x}_n \qquad (4.5.2)$$

The estimation error is also easily computed using the contour formula of Eq. (4.4.8):

$$\mathcal{E} = E[e_n^2] = \oint_{\text{u.c.}} [S_{xx}(z) - H(z)S_{yx}(z)] \, dz/2\pi jz = 0.5$$

To appreciate the improvement afforded by filtering, this error must be compared with the error in case no processing is made and $y_n$ is itself taken to represent a noisy estimate of $x_n$. The estimation error in the latter case is $y_n - x_n = v_n$, so that $\sigma_v^2 = 1$. Thus, the gain afforded by processing is

$$\frac{\sigma_e^2}{\sigma_v^2} = 0.5, \qquad \text{or} \qquad 3 \text{ dB}$$

## 4.6  *Wiener Filter as Kalman Filter*

We would like to cast this example in a Kalman filter form. The difference equation (4.5.2) for the Wiener filter seems to have the ''wrong'' state transition matrix; namely, 0.3 instead of 0.6, which is the state matrix for the state model of $x_n$. However, it is not accidental that the Wiener filter difference equation may rewritten in the alternative form

$$\hat{x}_n = 0.6\hat{x}_{n-1} + 0.5(y_n - 0.6\hat{x}_{n-1})$$

The quantity $\hat{x}_n$ is the best estimate of $x_n$, at time $n$, based on all the observations up to that time; that is, $Y_n = \{y_i; -\infty < i \le n\}$. To simplify the subsequent notation, we denote it by $\hat{x}_{n/n}$. It is the projection of $x_n$ on the space $Y_n$. Similarly, $\hat{x}_{n-1}$ denotes the best estimate of $x_{n-1}$ based on the observations up to time $n-1$; that is, $Y_{n-1} = \{y_i; i \le n-1\}$. The above filtering equation is written in this notation as

$$\hat{x}_{n/n} = 0.6\hat{x}_{n-1/n-1} + 0.5(y_n - 0.6\hat{x}_{n-1/n-1}) \qquad (4.6.1)$$

It allows the computation of the *current best* estimate $\hat{x}_{n/n}$ in terms of the *previous best* estimate $\hat{x}_{n-1/n-1}$ and the new observation $y_n$ that becomes available at the current time instant $n$.

The various terms of Eq. (4.6.1) have nice interpretations: Suppose that the best estimate $\hat{x}_{n-1/n-1}$ of the previous sample $x_{n-1}$ is available. Even before the next observation $y_n$ comes in, we may use this estimate to make a reasonable prediction as to what the next best estimate ought to be. Since we know the system dynamics of $x_n$, we may try to "boost" $\hat{x}_{n-1/n-1}$ to the next time instant $n$ according to the system dynamics; that is, we take

$$\hat{x}_{n/n-1} = 0.6\hat{x}_{n-1/n-1} = \text{prediction of } x_n \text{ on the basis of } Y_{n-1} \qquad (4.6.2)$$

Since $y_n = x_n + v_n$, we may use this prediction of $x_n$ to make a prediction of the next measurement $y_n$; that is, we take

$$\hat{y}_{n/n-1} = \hat{x}_{n/n-1} = \text{prediction of } y_n \text{ on the basis of } Y_{n-1} \qquad (4.6.3)$$

If this prediction were perfect, and if the next observation $y_n$ were noise free, then this would be the value that we would observe. Since we actually observe $y_n$, the observation or innovations residual will be

$$\alpha_n = y_n - \hat{y}_{n/n-1} = \text{innovations residual} \qquad (4.6.4)$$

This quantity represents that part of $y_n$ that *cannot* be predicted on the basis of the previous observations $Y_{n-1}$. It represents the truly new information contained in the observation $y_n$. Actually, if we are making the best prediction possible, then the most we can expect of our prediction is to make the innovations residual a white-noise (uncorrelated) signal; that is, if we make the best possible prediction, then what remains should be unpredictable. According to the general discussion of the relationship between signal models and linear prediction given in Section 1.15, it follows that if $\hat{y}_{n/n-1}$ is the best predictor of $y_n$ then $\alpha_n$ must be the whitening sequence that drives the signal model of $y_n$. We shall verify this fact shortly. This establishes an intimate connection between the *Wiener/Kalman filtering* problem and the *signal modeling* problem. If we overestimate the observation $y_n$, the innovation residual will be negative; and if we underestimate it, the residual will be positive. In either case, we would like to correct our tentative estimate in the right direction. This may be accomplished by

$$\hat{x}_{n/n} = \hat{x}_{n/n-1} + G(y_n - \hat{y}_{n/n-1}) \tag{4.6.5}$$

$$= 0.6\hat{x}_{n-1/n-1} + G(y_n - 0.6\hat{x}_{n-1/n-1})$$

where the gain $G$, known as the *Kalman gain*, should be a positive quantity. The *prediction/correction* procedure defined by Eqs. (4.6.2) through (4.6.5) is known as the *Kalman filter*. It should be clear that any value for the gain $G$ will provide an estimate, even if suboptimal, of $x_n$. Our solution for the Wiener filter has precisely the above structure of a Kalman filter; but it has a gain $G = 0.5$. This value for the gain is the "best" possible, or optimal, value. It is a very instructive exercise to show this in two ways: First, with $G$ arbitrary, the estimation filter of Eq. (4.6.5) has transfer function

$$y_n \longrightarrow \boxed{H(z)} \longrightarrow \hat{x}_{n/n} \qquad H(z) = \frac{G}{1 - 0.6(1 - G)z^{-1}}$$

Insert this expression into the mean-squared estimation error $\mathcal{E} = E[e_n^2]$, where $e_n = x_n - \hat{x}_{n/n}$, and minimize it with respect to the parameter $G$. This should give $G = 0.5$.

Alternatively, $G$ should be such that to *render* the innovations residual (4.6.4) a white noise signal. In requiring this, it is useful to use the spectral factorization model for $y_n$; that is, the fact that $y_n$ is the output of $B(z)$ when driven by the white noise signal $\epsilon_n$. Let us work with $z$-transforms

$$\alpha(z) = Y(z) - 0.6z^{-1}\hat{X}(z) = Y(z) - 0.6z^{-1}H(z)Y(z)$$

$$= \left[1 - 0.6z^{-1}\frac{G}{1 - 0.6(1 - G)z^{-1}}\right] Y(z)$$

$$= \left[\frac{1 - 0.6z^{-1}}{1 - 0.6(1 - G)z^{-1}}\right] Y(z)$$

$$= \left[\frac{1 - 0.6z^{-1}}{1 - 0.6(1 - G)z^{-1}}\right] \left[\frac{1 - 0.3z^{-1}}{1 - 0.6z^{-1}}\right] \epsilon(z)$$

$$= \frac{1 - 0.3z^{-1}}{1 - 0.6(1 - G)z^{-1}} \epsilon(z)$$

Since $\epsilon_n$ is white, it follows that the transfer function relationship between $\alpha_n$ and $\epsilon_n$ must be trivial; otherwise, there will be sequential correlations present in $\alpha_n$. Thus, we must have $0.6(1 - G) = 0.3$, or $G = 0.5$; and in this case, $\alpha_n = \epsilon_n$. It is also possible to set $0.6(1 - G) = 1/0.3$, but this would correspond to an unstable filter.

We have obtained a most interesting result; namely, that when the Wiener filtering problem is recast into its Kalman filter form given by Eq. (4.6.1), then the innovations residual $\alpha_n$, which is computable on line with the estimate $\hat{x}_{n/n}$, is *identical* to the whitening sequence $\epsilon_n$ of the signal model of $y_n$. In other words, the Kalman filter can be thought of as *the whitening filter* for the *observation signal* $y_n$.

To appreciate further the connection between Wiener and Kalman filters and between Kalman filters and the whitening filters of signal models, we consider a generalized version of the above example and cast it in standard Kalman filter notation.

It is desired to estimate $x_n$ from $y_n$. The signal model for $x_n$ is taken to be first-order Markov model

$$x_{n+1} = ax_n + w_n \quad \text{(state model)} \tag{4.6.6}$$

with $|a| < 1$. The observation signal $y_n$ is related to $x_n$ by

$$y_n = cx_n + v_n \quad \text{(measurement model)} \tag{4.6.7}$$

It is further assumed that the state and measurement noises, $w_n$ and $v_n$, are zero-mean, mutually uncorrelated, white noises of variances $Q$ and $R$, respectively; that is,

$$E[w_n w_i] = Q\delta_{ni}, \qquad E[v_n v_i] = R\delta_{ni}, \qquad E[w_n v_i] = 0 \tag{4.6.8}$$

We also assume that $v_n$ is uncorrelated with $x_n$. The parameters $a,c,Q,R$ are assumed to be known. Let $x_1(n)$ be the signal defined by

$$x_1(n) = x_{n+1}$$

and consider the two related Wiener filtering problems of estimating $x_n$ and $x_1(n)$ on the basis of $Y_n = \{y_i;\ -\infty < i \leq n\}$, depicted below

$y_n \longrightarrow \boxed{H(z)} \longrightarrow \hat{x}_{n/n}$      $y_n \longrightarrow \boxed{H_1(z)} \longrightarrow \widehat{x_1(n)} = \hat{x}_{n+1/n}$

The problem of estimating $x_1(n) = x_{n+1}$ is equivalent to the problem of *one-step prediction* into the future on the basis of the past and present. Therefore, we will denote this estimate by $\widehat{x_1(n)} = \hat{x}_{n+1/n}$. The state equation (4.6.6) determines the spectral density of $x_n$:

$$S_{xx}(z) = S_{ww}(z)\frac{1}{(z-a)(z^{-1}-a)} = \frac{Q}{(1-az^{-1})(1-az)}$$

The observation equation (4.6.7) determines the cross-densities

$$S_{xy}(z) = cS_{xx}(z) + S_{xv}(z) = cS_{xx}(z)$$

$$S_{x_1 y}(z) = zS_{xy}(z) = zcS_{xx}(z)$$

where we used the filtering equation $X_1(z) = zX(z)$. The spectral density of $y_n$ can be factored as follows:

$$S_{yy}(z) = c^2 S_{xx}(z) + S_{vv}(z) = \frac{c^2 Q}{(1 - az^{-1})(1 - az)} + R$$

$$= \frac{c^2 Q + R(1 - az^{-1})(1 - az)}{(1 - az^{-1})(1 - az)} = \sigma_\epsilon^2 \left(\frac{1 - fz^{-1}}{1 - az^{-1}}\right)\left(\frac{1 - fz}{1 - az}\right)$$

where $f$ and $\sigma_\epsilon^2$ satisfy the equations

$$\sigma_\epsilon^2 f = Ra \tag{4.6.9}$$

$$\sigma_\epsilon^2(1 + f^2) = c^2 Q + R(1 + a^2) \tag{4.6.10}$$

and $f$ has magnitude less than one. Thus, the corresponding signal model for $y_n$ is the filter

$$B(z) = \frac{1 - fz^{-1}}{1 - az^{-1}} \tag{4.6.11}$$

Next, we compute the causal parts as required by Eq. (4.4.6):

$$[S_{xy}(z)/B(z^{-1})]_+ = \left[\frac{cQ}{(1 - az^{-1})(1 - fz)}\right]_+ = \frac{cQ}{1 - fa}\frac{1}{1 - az^{-1}}$$

$$[S_{x_1y}(z)/B(z^{-1})]_+ = \left[\frac{cQz}{(1 - az^{-1})(1 - fz)}\right]_+ = \frac{cQa}{1 - fa}\frac{1}{1 - az^{-1}}$$

Using Eq. (4.4.6), we determine the optimal Wiener filters $H(z)$ and $H_1(z)$ as follows:

$$H(z) = \frac{1}{\sigma_\epsilon^2 B(z)}\left[\frac{S_{xy}(z)}{B(z^{-1})}\right]_+ = \frac{\dfrac{cQ/(1 - fa)}{(1 - az^{-1})}}{\sigma_\epsilon^2\left(\dfrac{1 - fz^{-1}}{1 - az^{-1}}\right)} = \frac{\left(\dfrac{cQ}{\sigma_\epsilon^2(1 - fa)}\right)}{1 - fz^{-1}}$$

Or, defining the gain $G$ by

$$G = \frac{cQ}{\sigma_\epsilon^2(1 - fa)} \tag{4.6.12}$$

we finally find

$$H(z) = \frac{G}{1 - fz^{-1}} \tag{4.6.13}$$

$$H_1(z) = aH(z) = \frac{K}{1 - fz^{-1}} \tag{4.6.14}$$

where in Eq. (4.6.14) we defined a related gain, also called the Kalman gain, as follows:

$$K = aG = \frac{acQ}{\sigma_\epsilon^2(1 - fa)} \tag{4.6.15}$$

Eq. (4.6.14) immediately implies that

$$\hat{x}_{n+1/n} = a\hat{x}_{n/n} \tag{4.6.16}$$

which is the precise justification of Eq. (4.6.2). The difference equations of the two filters are

$$\hat{x}_{n+1/n} = f\hat{x}_{n/n-1} + Ky_n \tag{4.6.17}$$
$$\hat{x}_{n/n} = f\hat{x}_{n-1/n-1} + Gy_n$$

Using the results of Problem 3.7, we may express all the quantities $f$, $\sigma_\epsilon^2$, $K$, and $G$ in terms of a single positive quantity $P$ which satisfies the *algebraic Riccati* equation:

$$Q = P - \frac{PRa^2}{R + c^2P} \tag{4.6.18}$$

Then, we find the interrelationships

$$K = aG = \frac{acP}{R + c^2P}, \qquad \sigma_\epsilon^2 = R + c^2P,$$

$$f = a - cK = \frac{Ra}{R + c^2P} \tag{4.6.19}$$

It is left as an exercise to show that the minimized mean-squared estimation errors are given in terms of $P$ by

$$E[e_{n/n-1}^2] = P, \qquad E[e_{n/n}^2] = \frac{PR}{R + c^2P}$$

where

$$e_{n/n-1} = x_n - \hat{x}_{n/n-1} \qquad \text{and} \qquad e_{n/n} = x_n - \hat{x}_{n/n}$$

are the corresponding estimation errors for the optimally predicted and filtered estimates, respectively.

Using Eq. (4.6.19), we may rewrite the filtering equation (4.6.17) as

$$\hat{x}_{n+1/n} = (a - cK)\,\hat{x}_{n/n-1} + Ky_n$$

or

$$\hat{x}_{n+1/n} = a\hat{x}_{n/n-1} + K(y_n - c\hat{x}_{n/n-1}) \tag{4.6.20}$$

or

$$\hat{x}_{n+1/n} = a\hat{x}_{n/n-1} + K(y_n - \hat{y}_{n/n-1})$$

where we set

$$\hat{y}_{n/n-1} = c\hat{x}_{n/n-1} \qquad (4.6.21)$$

A realization of the estimation filter based on (4.6.20) is shown below:

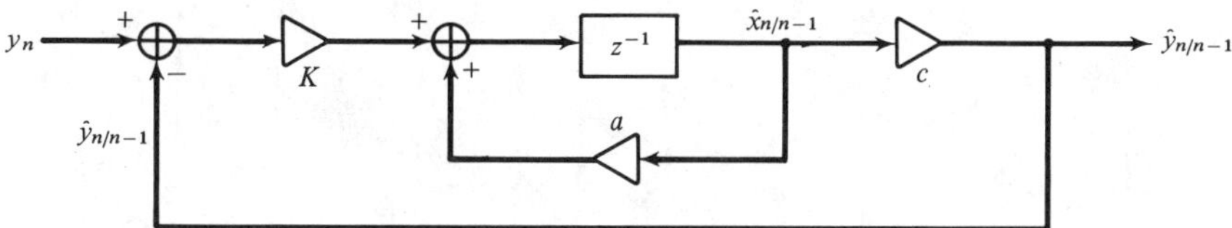


Inserting $K = aG$ and using Eq. (4.6.16) we also find

$$\hat{x}_{n/n} = \hat{x}_{n/n-1} + G(y_n - \hat{y}_{n/n-1}) \qquad (4.6.22)$$

The quantity $\hat{y}_{n/n-1}$ defined in Eq. (4.6.21) is the best estimate of $y_n$ based on its past $Y_{n-1}$. This can be seen in two ways: First, using the results of Problem 1.4 on the linearity of the estimates, we find

$$\hat{y}_{n/n-1} = \widehat{cx_n + v_n} = c\hat{x}_{n/n-1} + \hat{v}_{n/n-1} = c\hat{x}_{n/n-1}$$

where the term $\hat{v}_{n/n-1}$ was dropped. This term represents the estimate of $v_n$ on the basis of the past $ys$; that is, $Y_{n-1}$. Since $v_n$ is white and also uncorrelated with $x_n$, it follows that it will be uncorrelated with all past $ys$; therefore, $\hat{v}_{n/n-1} = 0$. The second way to show that $\hat{y}_{n/n-1}$ is the best prediction of $y_n$ is to show that the innovations residual

$$\alpha_n = y_n - \hat{y}_{n/n-1} = y_n - c\hat{x}_{n/n-1} \qquad (4.6.23)$$

is a white-noise sequence and coincides with the whitening sequence $\epsilon_n$ of $y_n$. Indeed, working in the $z$-domain and using Eq. (4.6.17) and the signal model of $y_n$ we find

$$\alpha(z) = Y(z) - cz^{-1}\hat{X}_1(z) = Y(z) - cz^{-1} H_1(z)Y(z)$$

$$= \left[1 - cz^{-1} \frac{K}{1 - fz^{-1}}\right] Y(z) = \left[\frac{1 - (f + cK)z^{-1}}{1 - fz^{-1}}\right] Y(z)$$

$$= \left[\frac{1 - az^{-1}}{1 - fz^{-1}}\right] Y(z) = \frac{1}{B(z)} Y(z) = \epsilon(z)$$

which implies that

$$\alpha_n = \epsilon_n$$

Finally, we note that the recursive updating of the estimate of $x_n$ given by Eq. (4.6.22) is identical to the result of Problem 1.8.

Our purpose in presenting this example was to tie together a number of ideas from Chapter 1 (correlation canceling, estimation, Gram-Schmidt orthogonalization, linear prediction, and signal modeling) to ideas from this chapter on

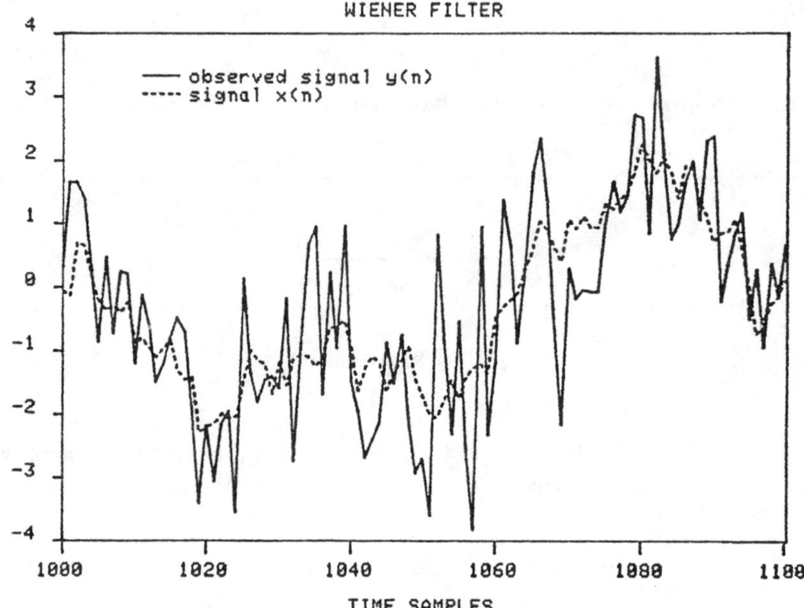


**Figure 4.2**  Desired Signal and Its Noisy Observation

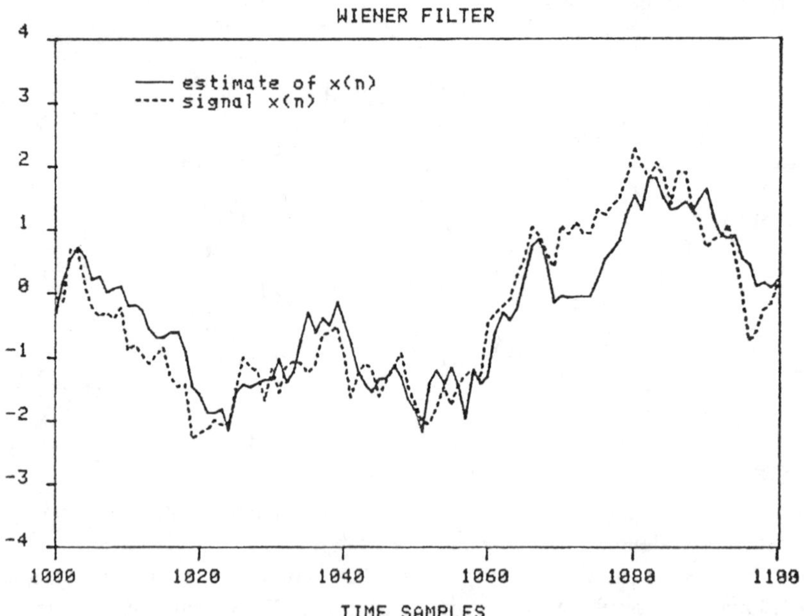


**Figure 4.3**  Best Estimate of Desired Signal

Wiener filtering and its recursive reformulation as a Kalman filter [8–10, 12–17].

We conclude this section by presenting a simulation of this example defined by the following choice of parameters:

$$a = 0.95, \quad c = 1, \quad Q = 1 - a^2, \quad R = 1$$

the above choice for $Q$ normalizes the variance of $x_n$ to unity. Solving the Riccati equation (4.6.18) and using Eq. (4.6.19), we find

$$P = 0.312, \quad K = 0.226, \quad G = 0.238, \quad f = a - cK = 0.724$$

Figure 4.2 shows 100 samples of the observed signal $y(n)$ together with the desired signal $x(n)$. The signal $y(n)$ processed through the Wiener filter $H(z)$ defined by the above parameters is shown in Fig. 4.3 together with $x(n)$. The tracking properties of the filter are evident from the graph. It should be emphasized that this is the best one can do by means of ordinary causal linear filtering!

## 4.7  *Construction of the Wiener Filter by the Gapped Function*

Next, we would like to give an alternative construction of the optimal Wiener filter based on the concept of the gapped function. The gapped function is especially useful in linear prediction. It is defined as the cross-correlation between the estimation error $e_n$ and the observation sequence $y_n$ as follows:

$$g(k) = R_{ey}(k) = E[e_n y_{n-k}] \qquad \text{for all } k \tag{4.7.1}$$

This definition is motivated by the orthogonality equations which state that the prediction error $e_n$ must be orthogonal to all of the available observations; namely, $Y_n = \{y_i; \ -\infty < i \leq n\}$. That is, for the optimal set of filter weights we have

$$g(k) = R_{ey}(k) = E[e_n y_{n-k}] = 0, \qquad \text{for } k \geq 0 \tag{4.7.2}$$

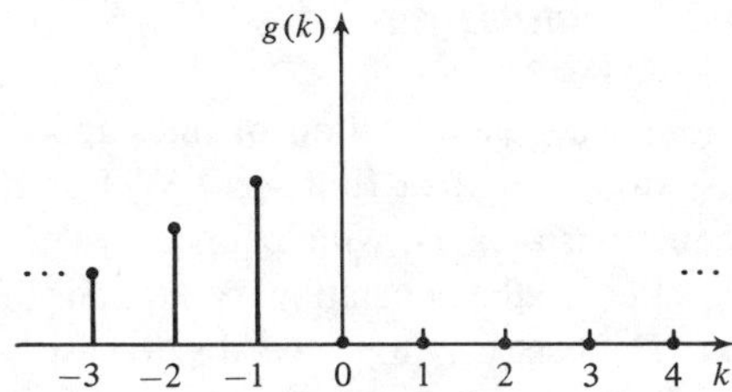

and $g(k)$ develops a right-hand gap. On the other hand, $g(k)$ may be written in the alternative form

$$g(k) = E[e_n y_{n-k}] = E\left[\left(x_n - \sum_{i=0}^{\infty} h_i y_{n-i}\right) y_{n-k}\right]$$

$$= R_{xy}(k) - \sum_{i=0}^{\infty} h_i R_{yy}(k - i)$$

or

$$g(k) = R_{ey}(k) = R_{xy}(k) - \sum_{i=0}^{\infty} h_i R_{yy}(k - i) \qquad (4.7.3)$$

Taking $z$-transforms we find

$$G(z) = S_{xy}(z) - H(z)S_{yy}(z)$$

Due to the gap conditions, the left-hand side contains only positive powers of $z$, whereas the right-hand side contains both positive and negative powers of $z$. Thus, the nonpositive powers of $z$ must drop out of the right side. This condition precisely determines $H(z)$. Introducing the spectral factorization of $S_{yy}(z)$ and dividing both sides by $B(z^{-1})$ we find

$$G(z) = S_{xy}(z) - H(z)S_{yy}(z) = S_{xy}(z) - H(z)B(z)B(z^{-1})\sigma_\epsilon^2$$

$$\frac{G(z)}{B(z^{-1})} = \frac{S_{xy}(z)}{B(z^{-1})} - H(z)B(z)\sigma_\epsilon^2$$

The $z$-transform $B(z^{-1})$ is anticausal and, because of the gap conditions, so is the ratio $G(z)/B(z^{-1})$. Therefore, taking causal parts of both sides we find

$$0 = \left[\frac{S_{xy}(z)}{B(z^{-1})}\right]_+ - H(z)B(z)\sigma_\epsilon^2$$

which can be solved for $H(z)$ to give Eq. (4.4.6).

## 4.8 Construction of the Wiener Filter by Covariance Factorization

In this section, we present a generalization of the gapped-function method to the more general nonstationary and/or finite-past Wiener filter. This is defined by the Wiener-Hopf equations (4.2.7), which are equivalent to the orthogonality equations (4.2.5). The latter are the nonstationary versions of the gapped function of the previous section. The best way to proceed is to cast Eqs. (4.2.5) in matrix form as follows: Without loss of generality we may take the starting point $n_a = 0$. The final point $n_b$ is left arbitrary. Introduce the vectors

$$\mathbf{x} = \begin{bmatrix} x_0 \\ x_1 \\ \vdots \\ x_{n_b} \end{bmatrix} \qquad \mathbf{y} = \begin{bmatrix} y_0 \\ y_1 \\ \vdots \\ y_{n_b} \end{bmatrix}$$

and the corresponding correlation matrices

$$R_{xy} = E[\mathbf{x}\mathbf{y}^T] \qquad \text{and} \qquad R_{yy} = E[\mathbf{y}\mathbf{y}^T]$$

The filtering equation (4.2.4) may be written in vector form as

$$\hat{\mathbf{x}} = H\mathbf{y} \tag{4.8.1}$$

where $H$ is the matrix of optimal weights $\{h(n,i)\}$. The *causality* of the filtering operation (4.8.1), requires $H$ to be *lower-triangular*. The minimization problem becomes equivalent to the problem of minimizing the mean-squared estimation error subject to the constraint that $H$ be lower-triangular. The minimization conditions are the normal equations (4.2.5) which, in this matrix notation, state that the matrix $R_{ey}$ has no lower-triangular (causal) part; or, equivalently, that $R_{ey}$ is *strictly* upper-triangular (i.e., even the main diagonal of $R_{ey}$ is zero), therefore

$$R_{ey} = \text{strictly upper triangular} \qquad\qquad\qquad \tag{4.8.2}$$

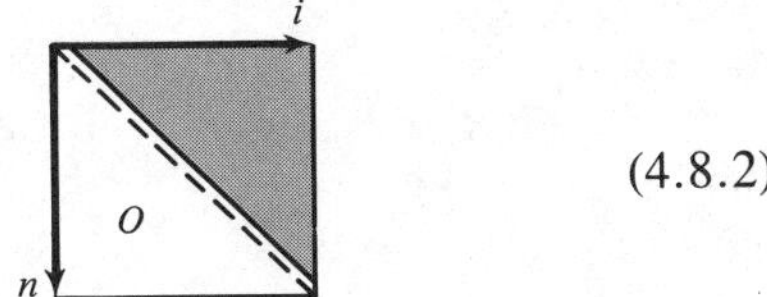

Inserting Eq. (4.8.1) into $R_{ey}$ we find

$$R_{ey} = E[\mathbf{e}\mathbf{y}^T] = E[(\mathbf{x} - H\mathbf{y})\mathbf{y}^T]$$

or

$$R_{ey} = R_{xy} - HR_{yy} \tag{4.8.3}$$

The minimization conditions (4.8.2) require $H$ to be that lower-triangular matrix which renders the combination (4.8.3) upper-triangular. In other words, $H$ should be such that the lower triangular part of the right hand side must vanish. To solve Eqs. (4.8.2) and (4.8.3), we introduce the *LU Cholesky factorization* of the covariance matrix $R_{yy}$ given by

$$R_{yy} = BR_{\epsilon\epsilon}B^T \tag{4.8.4}$$

where $B$ is unit *lower*-triangular, and $R_{\epsilon\epsilon}$ is diagonal. This was discussed in Section 1.5. Inserting in Eq. (4.8.3) we find

$$R_{ey} = R_{xy} - HBR_{\epsilon\epsilon}B^T \tag{4.8.5}$$

Multiplying by the inverse transpose of $B$ we obtain

$$R_{ey}B^{-T} = R_{xy}B^{-T} - HBR_{\epsilon\epsilon} \qquad (4.8.6)$$

Now, the matrix $B^{-T}$ is unit upper-triangular, but $R_{ey}$ is *strictly* upper, therefore, the product $R_{ey}B^{-T}$ will be strictly upper. This can be verified easily for any two such matrices. Extracting the lower-triangular parts of both sides of Eq. (4.8.6) we find

$$0 = [R_{xy}B^{-T}]_{+} - HBR_{\epsilon\epsilon}$$

where we used the fact that the left-hand side was strictly upper and that the term $HBR_{\epsilon\epsilon}$ was already lower-triangular. The notation $[\ ]_{+}$ denotes the lower-triangular part of a matrix including the diagonal. We find finally

$$H = [R_{xy}B^{-T}]_{+} R_{\epsilon\epsilon}^{-1}B^{-1} \qquad (4.8.7)$$

This is the most general solution of the Wiener filtering problem [17,18]. It includes the results of the stationary case, as a special case. Indeed, if all the signals are stationary, then the matrices $R_{xy}$ and $B$ and $B^T$ become Toeplitz and have a $z$-transform associated with them as discussed in Problem 3.8. Using the results of that problem, it is easily seen that Eq. (4.8.7) is the *time domain analog* of Eq. (4.4.6).

The prewhitening approach of Section 4.4 can also be understood in the present matrix framework. Making the change of variables

$$\mathbf{y} = B\boldsymbol{\epsilon}$$

we find that $R_{xy} = E[\mathbf{x}\mathbf{y}^T] = E[\mathbf{x}\boldsymbol{\epsilon}^T]B^T = R_{x\epsilon}B^T$, and therefore $R_{xy}B^{-T} = R_{x\epsilon}$, and the filter $H$ becomes $H = [R_{x\epsilon}]_{+}R_{\epsilon\epsilon}^{-1}B^{-1}$. The corresponding estimate is then

$$\hat{\mathbf{x}} = H\mathbf{y} = HB\boldsymbol{\epsilon} = F\boldsymbol{\epsilon}, \qquad \text{where} \qquad F = HB = [R_{x\epsilon}]_{+}R_{\epsilon\epsilon}^{-1} \qquad (4.8.8)$$

This is the matrix analog of Eq. (4.4.5). The matrix $F$ is lower-triangular by construction. Therefore, to extract the $n$th component $\hat{x}_n$ of Eq. (4.8.8), it is enough to consider the $n \times n$ submatrices as shown below:

$$\hat{x}_n = n \begin{bmatrix} \ \end{bmatrix} \quad \epsilon_n \qquad f(n)^T$$

The $n$th row of $F$ is $\mathbf{f}(n)^T = E[x_n\boldsymbol{\epsilon}_n^T]E[\boldsymbol{\epsilon}_n\boldsymbol{\epsilon}_n^T]^{-1}$. Therefore, the $n$th estimate becomes

$$\hat{x}_n = E[x_n\boldsymbol{\epsilon}_n^T]E[\boldsymbol{\epsilon}_n\boldsymbol{\epsilon}_n^T]^{-1}\boldsymbol{\epsilon}_n$$

which may also be written in the recursive form

$$\hat{x}_{n/n} = \sum_{i=0}^{n} E[x_n\epsilon_i]E[\epsilon_i\epsilon_i]^{-1}\epsilon_i = \sum_{i=0}^{n-1} E[x_n\epsilon_i]E[\epsilon_i\epsilon_i]^{-1}\epsilon_i + G_n\epsilon_n$$

or

$$\hat{x}_{n/n} = \hat{x}_{n/n-1} + G_n\epsilon_n \tag{4.8.9}$$

where we made an obvious change in notation, and $G_n = E[x_n\epsilon_n]E[\epsilon_n\epsilon_n]^{-1}$. This is identical to Eq. (4.6.22); in the stationary case, $G_n$ is a constant independent of $n$. We can also recast the $n$th estimate in "batch" form, expressed directly in terms of the observation vector $\mathbf{y}_n = [y_0, y_1, \ldots, y_n]^T$. By considering the $nxn$ subblock part of the Gram-Schmidt construction, we may write $\mathbf{y}_n = B_n\boldsymbol{\epsilon}_n$, where $B_n$ is unit lower-triangular. Then, $\hat{x}_n$ can be expressed as

$$\hat{x}_n = E[x_n\boldsymbol{\epsilon}_n^T]E[\boldsymbol{\epsilon}_n\boldsymbol{\epsilon}_n^T]^{-1}\boldsymbol{\epsilon}_n = E[x_n\mathbf{y}_n^T]E[\mathbf{y}_n\mathbf{y}_n^T]^{-1}\mathbf{y}_n$$

which is identical to Eq. (4.2.8).

## Problems

### Problem 4.1:

Let $\mathbf{x} = [x_{n_a}, \ldots, x_{n_b}]^T$ and $\mathbf{y} = [y_{n_a}, \ldots, y_{n_b}]^T$ be the desired and available signal vectors. The relationship between $\mathbf{x}$ and $\mathbf{y}$ is assumed to be linear of the form

$$\mathbf{y} = C\mathbf{x} + \mathbf{v}$$

where $C$ represents a linear degradation and $\mathbf{v}$ is a vector of zero-mean independent gaussian samples with a common variance $\sigma_v^2$. Show that the maximum likelihood (ML) estimation criterion is in this case equivalent to the following least-squares criterion, based on the quadratic vector norm:

$$\mathcal{E} = \|\mathbf{y} - C\mathbf{x}\|^2 = \text{minimum with respect to } \mathbf{x}$$

Show that the resulting estimate is given by

$$\hat{\mathbf{x}} = (C^TC)^{-1}C^T\mathbf{y}$$

### Problem 4.2:

Let $\hat{\mathbf{x}} = H\mathbf{y}$ be the optimal linear smoothing estimate of $\mathbf{x}$ given by Eq. (4.1.5). It is obtained by minimizing the mean-squared estimation error $\mathcal{E}_n = E[e_n^2]$ for each $n$ in the interval $[n_a, n_b]$.

(a) Show that this solution for $H$ also minimizes the error covariance matrix

$$R_{ee} = E[\mathbf{ee}^T]$$

where $\mathbf{e}$ is the vector of estimation errors $\mathbf{e} = [e_{n_a}, \ldots, e_{n_b}]^T$.

(b) Show that $H$ also minimizes every quadratic index of the form

$$E[\mathbf{e}^T Q \mathbf{e}] = \text{minimum}$$

for any positive semi-definite matrix $Q$.

(c) Explain how the minimization of each $E[e_n^2]$ can be understood in terms of part (b).

**Problem 4.3:**
Consider the smoothing problem of estimating the signal vector $\mathbf{x}$ from the signal vector $\mathbf{y}$. Assume that $\mathbf{x}$ and $\mathbf{y}$ are linearly related by

$$\mathbf{y} = C\mathbf{x} + \mathbf{v}$$

and that $\mathbf{v}$ and $\mathbf{x}$ are uncorrelated from each other, and that the covariance matrices of $\mathbf{x}$ and $\mathbf{v}$, $R_{xx}$ and $R_{vv}$, are known. Show that the smoothing estimate of $\mathbf{x}$ is in this case

$$\hat{\mathbf{x}} = R_{xx}C^T[CR_{xx}C^T + R_{vv}]^{-1}\mathbf{y}$$

**Problem 4.4:** A stationary random signal has autocorrelation function $R_{xx}(k) = \sigma_x^2 a^{|k|}$, for all $k$. The observation signal is $y_n = x_n + v_n$, where $v_n$ is a zero-mean, white noise sequence of variance $\sigma_v^2$, uncorrelated from $x_n$.

(a) Determine the optimal FIR Wiener filter of order $M = 1$ for estimating $x_n$ in terms of $y_n$.

(b) Repeat for the optimal linear predictor of order $M = 2$ for predicting $x_n$ on the basis of the past two samples $y_{n-1}$ and $y_{n-2}$.

**Problem 4.5:**
A stationary random signal $x_n$ has autocorrelation function $R_{xx}(k) = \sigma_x^2 a^{|k|}$, for all $k$. Consider a time interval $[n_a, n_b]$. The random signal $x_n$ is known only at the end-points of that interval; that is, the only available observations are

$$y(n_a) = x(n_a), \qquad y(n_b) = x(n_b)$$

Determine the optimal estimate of $x(n)$ based on just these two samples in the form

$$\hat{x}(n) = h(n,n_a)y(n_a) + h(n,n_b)y(n_b)$$

for the following values of $n$: (a) $n_a \leq n \leq n_b$ (b) $n \leq n_a$ (c) $n \geq n_b$

***Problem 4.6:***

A stationary random signal $x_n$ is to be estimated on the basis of the noisy observations

$$y_n = x_n + v_n$$

It is given that

$$S_{xx}(z) = \frac{1}{(1 - 0.5z^{-1})(1 - 0.5z)}, \qquad S_{vv}(z) = 5, \qquad S_{xv}(z) = 0$$

(a) Determine the optimal realizable Wiener filter for estimating the signal $x_n$ on the basis of the observations $Y_n = \{y_i; i \leqslant n\}$. Write the difference equation of this filter. Compute the mean-squared estimation error.

(b) Determine the optimal realizable Wiener filter for predicting one step into the future; that is, estimate $x_{n+1}$ on the basis of $Y_n$.

(c) Cast the results of (a) and (b) in a predictor/corrector Kalman filter form, and show explicitly that the innovations residual of the observation signal $y_n$ is identical to the corresponding whitening sequence $\epsilon_n$ driving the signal model of $y_n$.

***Problem 4.7:***

Repeat the previous problem for the following choice of state and measurement models

$$x_{n+1} = x_n + w_n, \qquad y_n = x_n + v_n$$

where $w_n$ and $v_n$ have variances $Q = 0.5$ and $R = 1$, respectively.

***Problem 4.8:***

Consider the state and measurement equations

$$x_{n+1} = ax_n + w_n \qquad y_n = cx_n + v_n$$

as discussed in Section 4.6. For *any* value of the Kalman gain $K$, consider the Kalman predictor/corrector algorithm defined by the equation

$$\hat{x}_{n+1/n} = a\hat{x}_{n/n-1} + K(y_n - c\hat{x}_{n/n-1}) = f\hat{x}_{n/n-1} + Ky_n \tag{P.1}$$

where $f = a - cK$. The stability requirement of this estimation filter requires further that $K$ be such that $|f| < 1$.

(a) Let $e_{n/n-1} = x_n - \hat{x}_{n/n-1}$ be the corresponding estimation error. Assuming all signals are stationary, and working with $z$-transforms, show that the power spectral density of $e_{n/n-1}$ is given by

$$S_{ee}(z) = \frac{Q + K^2R}{(1 - fz^{-1})(1 - fz)}$$

(b) Integrating (a) around the unit circle, show that the mean-squared value of the estimation error is given by

$$\mathcal{E} = E[e_{n/n-1}^2] = \frac{Q + K^2 R}{1 - f^2} = \frac{Q + K^2 R}{1 - (a - cK)^2} \tag{P.2}$$

(c) To select the *optimal value* of the Kalman gain $K$, differentiate $\mathcal{E}$ with respect to $K$ and set the derivative to zero. Show that the resulting equation for $K$ can be expressed in the form

$$K = \frac{caP}{R + c^2 P}$$

where $P$ stands for the minimized value of $\mathcal{E}$; that is,

$$P = \mathcal{E}_{min}$$

(d) Inserting this expression for $K$ back into the expression (P.2) for $\mathcal{E}$, show that the quantity $P$ must satisfy the algebraic Riccati equation

$$Q = P - \frac{PRa^2}{R + c^2 P}$$

Thus, the resulting estimator filter is identical to the optimal one-step prediction filter discussed in Section 4.6.

### Problem 4.9:

(a) Show that Eq. (P.2) of Problem 4.8 can be derived without using $z$-transforms, by using only stationarity, as suggested below: Using the state and measurement model equations and Eq. (P.1), show that the estimation error $e_{n/n-1}$ satisfies the difference equation

$$e_{n+1/n} = f e_{n/n-1} + w_n - K v_n$$

Then, invoking stationarity, derive Eq. (P.2).

(b) Using similar methods, show that the mean-squared estimation error is given by

$$E[e_{n/n}^2] = \frac{PR}{R + c^2 P}$$

where $e_{n/n} = x_n - \hat{x}_{n/n}$ is the estimation error of the optimal filter (4.6.13).

### Problem 4.10:

Consider the general example of Section 4.6. It was shown there that the innovations residual was the same as the whitening sequence $\epsilon_n$ driving the signal model of $y_n$

$$\epsilon_n = y_n - \hat{y}_{n/n-1} = y_n - c\hat{x}_{n/n-1}$$

Show that it can be written as

$$\epsilon_n = ce_{n/n-1} + v_n$$

where $e_{n/n-1} = x_n - \hat{x}_{n/n-1}$ is the estimation error. Then, show that

$$\sigma_\epsilon^2 = E[\epsilon_n^2] = c^2 P + R$$

***Problem 4.11:*** Computer experiment
Consider the signal and measurement model defined by Eqs. (4.6.6) through (4.6.8), with the choices $a = 0.9$, $c = 1$, $Q = 1 - a^2$, and $R = 1$. Generate 1500 samples of the random noises $w_n$ and $v_n$. Generate the corresponding signals $x_n$ and $y_n$ according to the state and measurement equations. Determine the optimal Wiener filter of the form (4.6.13) for estimating $x_n$ on the basis of $y_n$. Filter the sequence $y_n$ through the Wiener filter to generate the sequence $\hat{x}_{n/n}$.

(a) On the same graph, plot the desired signal $x_n$ and the available noisy version $y_n$ for $n$ ranging over the last 100 values (i.e., $n = 1400$–$1500$).

(b) On the same graph, plot the recovered signal $\hat{x}_{n/n}$ together with the original signal $x_n$ for $n$ ranging over the last 100 values.

(c) Repeat (a) and (b) using a different realization of $w_n$ and $v_n$.

(d) Repeat (a), (b), and (c) for the choice $a = -0.9$.

***Problem 4.12:***
Consider the optimal Wiener filtering problem in its matrix formulation of Section 4.8. Let $\mathbf{e} = \mathbf{x} - \hat{\mathbf{x}} = \mathbf{x} - H\mathbf{y}$ be the estimation error corresponding to a particular choice of the lower-triangular matrix $H$. Minimize the error covariance matrix $R_{ee} = E[\mathbf{e}\mathbf{e}^T]$ with respect to $H$ subject to the *constraint* that $H$ be lower-triangular. These constraints are $H_{ni} = 0$ for $n < i$. To do this, introduce a set of Lagrange multipliers $\Lambda_{ni}$; $n < i$, one for each constraint equation, and incorporate them into an effective performance index

$$J = E[\mathbf{e}\mathbf{e}^T] + \Lambda H^T + H\Lambda^T$$

where the matrix $\Lambda$ is strictly upper-triangular. Show that this formulation of the minimization problem yields exactly the same solution as Eq. (4.8.7).

## References


1. N. Wiener, *Extrapolation, Interpolation and Smoothing of Stationary Time Series with Engineering Applications*, New York, Wiley, 1949.

2. A. N. Kolmogorov, Sur l'Interpolation et Extrapolation des Suites Stationnaires, *C. R. Acad. Sci*, **208**, 2043–2045 (1939). See also Interpolation and Extrapolation of Stationary Random Sequences, and Stationary Sequences in Hilbert Space, reprinted in T. Kailath, Ed., *Linear Least-Squares Estimation*, Stroudsburg, PA, Dowden, Hutchinson, and Ross, 1977.

3. H. W. Bode and C. E. Shannon, A Simplified Derivation of Linear Least-Squares Smoothing and Prediction Theory, *Proc. IRE*, **38,** 417–425 (1950).

4. P. Whittle, *Prediction and Regulation*, New York: Van Nostrand Reinhold, 1963.

5. A. M. Yaglom, *Theory of Stationary Random Functions*, Englewood Cliffs, Prentice-Hall, 1962.

6. E. A. Robinson, *Multichannel Time-Series Analysis with Digital Computer Programs*, San Francisco, Holden-Day, 1967.

7. E. A. Robinson, *Statistical Communication and Detection*, New York, Hafner, 1967.

8. A. P. Sage and J. L. Melsa, *Estimation Theory with Applications to Communication and Control*, New York, McGraw-Hill, 1971.

9. S. Tretter, *Introduction to Discrete-Time Signal Processing*, New York, Wiley, 1976.

10. M. Srinath and P. Rajasekaran, *Introduction to Statistical Signal Processing*, New York, Wiley, 1979.

11. E. Robinson and S. Treitel, *Geophysical Signal Analysis*, Englewood Cliffs, Prentice-Hall, 1980.

12. T. Kailath, "A View of Three Decades of Linear Filtering Theory," *IEEE Trans. Info. Theory*, **IT-20,** 146 (1974).

13. R. E. Kalman, A New Approach to Linear Filtering and Prediction Problems, *Trans. ASME*, Ser. D, *J. Basic Eng.*, **82,** 34–45 (1960).

14. R. E. Kalman and R. S. Bucy, New Results in Linear Filtering and Prediction Theory, *Trans. ASME*, Ser. D, *J. Basic Eng.*, **83,** 95–107 (1961).

15. B. Anderson and J. Moore, *Optimal Filtering*, Englewood Cliffs, Prentice-Hall, 1979.

16. A. Gelb, *Applied Optimal Estimation*, Cambridge, MA, MIT Press, 1974.

17. T. Kailath, An Innovations Approach to Least-Squares Estimation. Part I: Linear Filtering in Additive White Noise, *IEEE Trans. Autom. Control*, **AC-13,** 646–655 (1968).

18. T. Kailath, Some Topics in Linear Estimation, in M. Hazewinkel and J. C. Willems, Eds., *Stochastic Systems: The Mathematics of Filtering and Identification*, Boston, D. Reidel Publications, 1981, pp. 307–350.

# 5

# Linear Prediction and Related Topics

## 5.1  *Pure Prediction and Signal Modeling*

In Sections 1.10 and 1.15, we discussed the connection between linear prediction and signal modeling. Here, we rederive the same results by considering the linear prediction problem as a special case of the Wiener filtering problem, given by Eq. (4.4.6). Our aim is to cast the results in a form that will suggest a practical way to solve the prediction problem and hence also the modeling problem. Consider a stationary signal $y_n$ having a signal model

$$S_{yy}(z) = \sigma_\epsilon^2 B(z) B(z^{-1}) \qquad \epsilon_n \longrightarrow \boxed{B(z)} \longrightarrow y_n \qquad (5.1.1)$$

as guaranteed by the spectral factorization theorem. Let $R_{yy}(k)$ denote the autocorrelation of $y_n$:

$$R_{yy}(k) = E[y_{n+k}y_n]$$

The linear prediction problem is to predict the current value $y_n$ on the basis of all the past values $Y_{n-1} = \{y_i;\ -\infty < i \leq n - 1\}$. If we define the delayed signal $y_1(n) = y_{n-1}$, then the linear prediction problem is equivalent to the optimal Wiener filtering problem of estimating $y_n$ from the related signal $y_1(n)$. The optimal estimation filter $H(z)$ is given by Eq. (4.4.6), where we must identify

**145**

$x_n$ and $y_n$ with $y_n$ and $y_1(n)$ of the present notation. Using the filtering equation $Y_1(z) = z^{-1}Y(z)$, we find that $y_n$ and $y_1(n)$ have the same spectral factor $B(z)$

$$S_{y_1 y_1}(z) = (z^{-1})(z^{-1})^{-1}S_{yy}(z) = S_{yy}(z) = \sigma_\epsilon^2 B(z)B(z^{-1})$$

and also that

$$S_{yy_1}(z) = S_{yy}(z)z = z\sigma_\epsilon^2 B(z)B(z^{-1})$$

Inserting these into Eq. (4.4.6) we find for the optimal filter $H(z)$

$$H(z) = \frac{1}{\sigma_\epsilon^2 B(z)}\left[\frac{S_{yy_1}(z)}{B(z^{-1})}\right]_+ = \frac{1}{\sigma_\epsilon^2 B(z)}\left[\frac{z\sigma_\epsilon^2 B(z)B(z^{-1})}{B(z^{-1})}\right]_+$$

or

$$H(z) = \frac{1}{B(z)}[zB(z)]_+ \tag{5.1.2}$$

The causal instruction can be removed as follows: Noting that $B(z)$ is a causal and stable filter, we may expand it in the power series

$$B(z) = 1 + b_1 z^{-1} + b_2 z^{-2} + b_3 z^{-3} + \cdots .$$

The causal part of $zB(z)$ is then

$$\begin{aligned}
[zB(z)]_+ &= [z + b_1 + b_2 z^{-1} + b_3 z^{-2} + \cdots]_+ \\
&= b_1 + b_2 z^{-1} + b_3 z^{-2} + \cdots \\
&= z(b_1 z^{-1} + b_2 z^{-2} + b_3 z^{-3} + \cdots) \\
&= z[B(z) - 1]
\end{aligned}$$

The prediction filter $H(z)$ then becomes

$$H(z) = \frac{1}{B(z)}z[B(z) - 1] = z\left[1 - \frac{1}{B(z)}\right] \tag{5.1.3}$$

The input to this filter is $y_1(n)$ and the output is the prediction $\hat{y}_{n/n-1}$.

*Example 5.1.1:*  Suppose $y_n$ is generated by driving the all-pole filter

$$y_n = 0.9y_{n-1} - 0.2y_{n-2} + \epsilon_n$$

by zero-mean white noise $\epsilon_n$. Find the best predictor $\hat{y}_{n/n-1}$. The signal model in this case is $B(z) = 1/(1 - 0.9z^{-1} + 0.2z^{-2})$ and Eq. (5.1.3) gives

$$z^{-1}H(z) = 1 - \frac{1}{B(z)} = 1 - (1 - 0.9z^{-1} + 0.2z^{-2})$$

$$= 0.9z^{-1} - 0.2z^{-2}$$

The I/O equation for the prediction filter is obtained by

$$\hat{Y}(z) = H(z)Y_1(z) = z^{-1}H(z)Y(z) = [0.9z^{-1} - 0.2z^{-2}]Y(z)$$

and in the time domain

$$\hat{y}_{n/n-1} = 0.9y_{n-1} - 0.2y_{n-2}$$

*Example 5.1.2:*   Suppose

$$S_{yy}(z) = \frac{(1 - 0.25z^{-2})(1 - 0.25z^2)}{(1 - 0.8z^{-1})(1 - 0.8z)}$$

Determine the best predictor $\hat{y}_{n/n-1}$. Here, the minimal-phase spectral factor
is

$$B(z) = \frac{1 - 0.25z^{-2}}{1 - 0.8z^{-1}}$$

and therefore the prediction filter is

$$z^{-1}H(z) = 1 - \frac{1 - 0.8z^{-1}}{1 - 0.25z^{-2}}$$
$$= \frac{1 - 0.25z^{-2} - 1 + 0.8z^{-1}}{1 - 0.25z^{-2}} = \frac{0.8z^{-1} - 0.25z^{-2}}{1 - 0.25z^{-2}}$$

The I/O equation of this filter is conveniently given recursively by the
difference equation

$$\hat{y}_{n/n-1} = 0.25\hat{y}_{n-2/n-3} + 0.8y_{n-1} - 0.25y_{n-2}$$

The prediction error

$$e_{n/n-1} = y_n - \hat{y}_{n/n-1}$$

is *identical* to the whitening sequence $\epsilon_n$ driving the signal model (5.1.1) of $y_n$;
indeed,

$$E(z) = Y(z) - \hat{Y}(z) = Y(z) - H(z)Y_1(z) = Y(z) - H(z)z^{-1}Y(z)$$

$$= [1 - H(z)z^{-1}]Y(z) = \frac{1}{B(z)} Y(z) = \epsilon(z)$$

Thus,

$$e_{n/n-1} = y_n - \hat{y}_{n/n-1} = \epsilon_n \tag{5.1.4}$$

in accordance with the results of Sections 1.10 and 1.15. An overall realization
of the linear predictor is shown in Fig. 5.1.

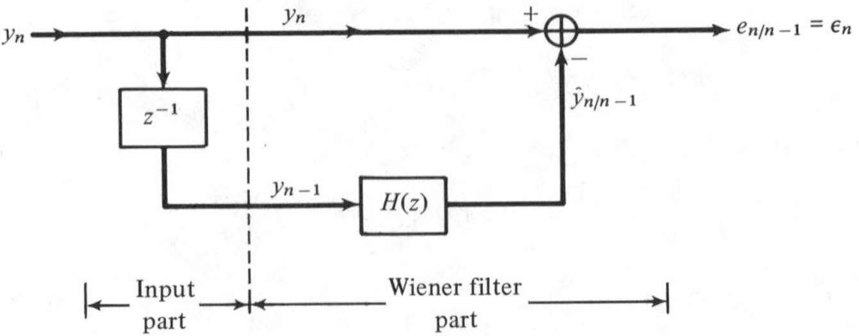



**Figure 5.1**   Linear Predictor

The indicated dividing line separates the linear predictor into the Wiener filtering part and the input part which provides the proper input signals to the Wiener part. The transfer function from $y_n$ to $e_{n/n-1}$ is the *whitening inverse filter*

$$A(z) = \frac{1}{B(z)} = 1 - z^{-1}H(z)$$

which is *stable and causal* by the minimal-phase property of the spectral factorization (5.1.1). In the $z$-domain we have

$$E(z) = \epsilon(z) = A(z)Y(z)$$

and in the time domain

$$e_{n/n-1} = \epsilon_n = \sum_{m=0}^{\infty} a_m y_{n-m} = y_n + a_1 y_{n-1} + a_2 y_{n-2} + \cdots$$

The predicted estimate $\hat{y}_{n/n-1} = y_n - e_{m/n-1}$ is

$$\hat{y}_{n/n-1} = -[a_1 y_{n-1} + a_2 y_{n-2} + \cdots]$$

These results are identical to Eqs. (1.15.2) and (1.15.3). The relationship noted above between linear prediction and signal modeling can also be understood in terms of the gapped-function approach of Section 4.7. Rewriting Eq. (5.1.1) in terms of the prediction-error filter $A(z)$ we have

$$S_{yy}(z) = \frac{\sigma_\epsilon^2}{A(z)A(z^{-1})} \tag{5.1.5}$$

from which we obtain

$$A(z)S_{yy}(z) = \frac{\sigma_\epsilon^2}{A(z^{-1})} = \sigma_\epsilon^2 B(z^{-1}) \tag{5.1.6}$$

Since we have the filtering equation $\epsilon(z) = A(z)Y(z)$, it follows that

$$S_{\epsilon y}(z) = A(z)S_{yy}(z)$$

and in the time domain

$$R_{\epsilon y}(k) = E[\epsilon_n y_{n-k}] = \sum_{i=0}^{\infty} a_i R_{yy}(k - i) \qquad (5.1.7)$$

which is recognized as the *gapped* function (4.7.1). By construction, $\epsilon_n$ is the orthogonal complement of $y_n$ with respect to the *entire* past subspace $Y_{n-1} = \{y_{n-k};\ k = 1,2, \ldots\}$, therefore, $\epsilon_n$ will be orthogonal to each $y_{n-k}$ for $k \geq 1$. These are precisely the gap conditions. Because the prediction is based on the entire past, the gapped function develops an infinite right-hand side gap. Thus, Eq. (5.1.7.) implies

$$R_{\epsilon y}(k) = E[\epsilon_n y_{n-k}] = \sum_{i=0}^{\infty} a_i R_{yy}(k - i) = 0 \quad \text{for all } k = 1,2, \ldots \qquad (5.1.8)$$

The same result, of course, also follows from the $z$-domain equation (5.1.6). Both sides of the equation are stable, but since $A(z)$ is minimal-phase, $A(z^{-1})$ will be maximum phase, and therefore it will have a stable but anticausal inverse $1/A(z^{-1})$. Thus, the right-hand side of Eq. (5.1.6) has no strictly causal part. Equating to zero all the coefficients of positive powers of $z^{-1}$, results in Eq. (5.1.8). The value of the gapped function at $k = 0$ is equal to $\sigma_\epsilon^2$. Indeed, using Eq. (5.1.8) we find

$$\begin{aligned} \sigma_\epsilon^2 = E[\epsilon_n^2] &= E[\epsilon_n(y_n + a_1 y_{n-1} + a_2 y_{n-2} + \cdots)] \\ &= R_{\epsilon y}(0) + a_1 R_{\epsilon y}(1) + a_2 R_{\epsilon y}(2) + \cdots \\ &= R_{\epsilon y}(0) = E[\epsilon_n y_n] \end{aligned}$$

Using Eq. (5.1.7) with $k = 0$ and the symmetry property $R_{yy}(i) = R_{yy}(-i)$, we find

$$\begin{aligned} \sigma_\epsilon^2 = E[\epsilon_n^2] &= E[\epsilon_n y_n] \\ &= R_{yy}(0) + a_1 R_{yy}(1) + a_2 R_{yy}(2) + \cdots \end{aligned} \qquad (5.1.9)$$

Equations (5.1.8) and (5.1.9) may be combined into one,

$$\sum_{i=0}^{\infty} a_i R_{yy}(k - i) = \sigma_\epsilon^2 \delta(k) \qquad \text{for all } k \geq 0 \qquad (5.1.10)$$

which can be cast in the matrix form

$$\begin{bmatrix} R_{yy}(0) & R_{yy}(1) & R_{yy}(2) & R_{yy}(3) & \cdots \\ R_{yy}(1) & R_{yy}(0) & R_{yy}(1) & R_{yy}(2) & \cdots \\ R_{yy}(2) & R_{yy}(1) & R_{yy}(0) & R_{yy}(1) & \cdots \\ R_{yy}(3) & R_{yy}(2) & R_{yy}(1) & R_{yy}(0) & \cdots \\ \vdots & \vdots & \vdots & \vdots & \end{bmatrix} \begin{bmatrix} 1 \\ a_1 \\ a_2 \\ a_3 \\ \vdots \end{bmatrix} = \begin{bmatrix} \sigma_\epsilon^2 \\ 0 \\ 0 \\ 0 \\ \vdots \end{bmatrix} \qquad (5.1.11)$$

These equations are known as the *normal equations* of linear prediction [1–11]. They provide the solution to both *signal modeling* and linear *prediction* problems. They determine the *model parameters* $\{a_1, a_2, \ldots ; \sigma_\epsilon^2\}$ of the signal $y_n$ directly in terms of the *experimentally accessible* quantities $R_{yy}(k)$. To render them computationally manageable, the infinite matrix equation (5.1.11) must be reduced to a *finite* one, and furthermore, the quantities $R_{yy}(k)$ must be *estimated* from actual data samples of $y_n$. We discuss these matters next.

## 5.2  *Autoregressive Models*

In general, the number of prediction coefficients $\{a_1, a_2, \ldots\}$ is infinite since the predictor is based on the infinite past. However, there is an important exception to this; namely, when the process $y_n$ is *autoregressive*. In this case, the signal model $B(z)$ is an *all-pole* filter of the type

$$B(z) = \frac{1}{A(z)} = \frac{1}{1 + a_1 z^{-1} + a_2 z^{-2} + \cdots + a_p z^{-p}} \tag{5.2.1}$$

which implies that the prediction-error filter is a polynomial

$$A(z) = 1 + a_1 z^{-1} + a_2 z^{-2} + \cdots + a_p z^{-p} \tag{5.2.2}$$

The signal generator for $y_n$ is the following difference equation, driven by the uncorrelated sequence $\epsilon_n$:

$$y_n + a_1 y_{n-1} + a_2 y_{n-2} + \cdots + a_p y_{n-p} = \epsilon_n \tag{5.2.3}$$

and the optimal prediction of $y_n$ is simply given by

$$\hat{y}_{n/n-1} = -[a_1 y_{n-1} + a_2 y_{n-2} + \cdots + a_p y_{n-p}] \tag{5.2.4}$$

In this case, the best prediction of $y_n$ depends *only* on the past $p$ samples $\{y_{n-1}, y_{n-2}, \ldots, y_{n-p}\}$. The infinite set of equations (5.1.10) or (5.1.11) are still satisfied even though only the first $p + 1$ coefficients $1, a_1, a_2, \ldots, a_p$ are nonzero. The $(p + 1) \times (p + 1)$ portion of Eq. (5.1.11) is *sufficient* to determine the $(p + 1)$ model parameters $\{a_1, a_2, \ldots, a_p; \sigma_\epsilon^2\}$:

$$\begin{bmatrix} R_{yy}(0) & R_{yy}(1) & R_{yy}(2) & \cdots & R_{yy}(p) \\ R_{yy}(1) & R_{yy}(0) & R_{yy}(1) & \cdots & R_{yy}(p-1) \\ R_{yy}(2) & R_{yy}(1) & R_{yy}(0) & \cdots & R_{yy}(p-2) \\ \vdots & \vdots & \vdots & & \vdots \\ R_{yy}(p) & R_{yy}(p-1) & R_{yy}(p-2) & \cdots & R_{yy}(0) \end{bmatrix} \begin{bmatrix} 1 \\ a_1 \\ a_2 \\ \vdots \\ a_p \end{bmatrix} = \begin{bmatrix} \sigma_\epsilon^2 \\ 0 \\ 0 \\ \vdots \\ 0 \end{bmatrix} \tag{5.2.5}$$

Such equations may be solved efficiently by Levinson's algorithm, which requires $O(p^2)$ operations and $O(p)$ storage locations to obtain the $a_i$s, instead of

$O(p^3)$ and $O(p^2)$, respectively, that would be required if the inverse of the autocorrelation matrix $R_{yy}$ were to be computed. The finite set of model parameters $\{a_1, a_2, \ldots, a_p; \sigma_\epsilon^2\}$ determines the signal model of $y_n$ completely. Setting $z = \exp(j\omega)$ into Eq. (5.1.5) we find a simple *parametric representation* of the power spectrum of the AR signal $y_n$:

$$S_{yy}(\omega) = \frac{\sigma_\epsilon^2}{|A(\omega)|^2}$$

$$= \frac{\sigma_\epsilon^2}{|1 + a_1 e^{-j\omega} + a_2 e^{-2j\omega} + \cdots + a_p e^{-j\omega p}|^2} \quad \text{(AR spectrum)}$$

$$(5.2.6)$$

In practice, the normal equations (5.2.5) provide a means of determining approximate estimates for the model parameters $\{a_1, a_2, \ldots, a_p; \sigma_\epsilon^2\}$. Typically, a block of length $N$ of recorded data is available

$$\boxed{y_0, y_1, y_2 \cdots y_{N-1}}$$

There are many different methods of extracting reasonable estimates of the model parameters using this block of data. We mention: (1) the *autocorrelation* or *Yule-Walker* method, (2) the *covariance* method, and (3) *Burg's* method. There are also some variations of these methods. The first method, the Yule-Walker method, is perhaps the most obvious and straightforward one. In the normal equations (5.2.5), one simply replaces the ensemble autocorrelations $R_{yy}(k)$ by the corresponding *sample* autocorrelations computed from the given block of data; that is,

$$\hat{R}_{yy}(k) = \frac{1}{N} \sum_{n=0}^{N-1-k} y_{n+k} y_n \qquad \text{for } 0 \leq k \leq p \qquad (5.2.7)$$

where only the first $p + 1$ lags are needed in Eq. (5.2.5). We must have, of course, $p \leq N - 1$. As discussed in Section 1.10, the resulting *estimates* of the model parameters $\{\hat{a}_1, \hat{a}_2, \ldots, \hat{a}_p; \hat{\sigma}_\epsilon^2\}$ may be used now in a number of ways; examples include obtaining a *spectral estimate* of the power spectrum of the sequence $y_n$

$$\hat{S}_{yy}(\omega) = \frac{\hat{\sigma}_\epsilon^2}{|1 + \hat{a}_1 e^{-j\omega} + \hat{a}_2 e^{-2j\omega} + \cdots + \hat{a}_p e^{-j\omega p}|^2} \quad \text{(YW spectral estimate)}$$

and *representing* the block of $N$ samples $y_n$ in terms of a few (i.e., $p + 1$) filter parameters. To synthesize the original samples one would generate white noise $\epsilon_n$ of variance $\hat{\sigma}_\epsilon^2$, and send it through the generator filter whose coefficients are the estimated values; that is, the filter

$$\hat{B}(z) = \frac{1}{\hat{A}(z)} = \frac{1}{1 + \hat{a}_1 z^{-1} + \hat{a}_2 z^{-2} + \cdots + \hat{a}_p z^{-p}}$$

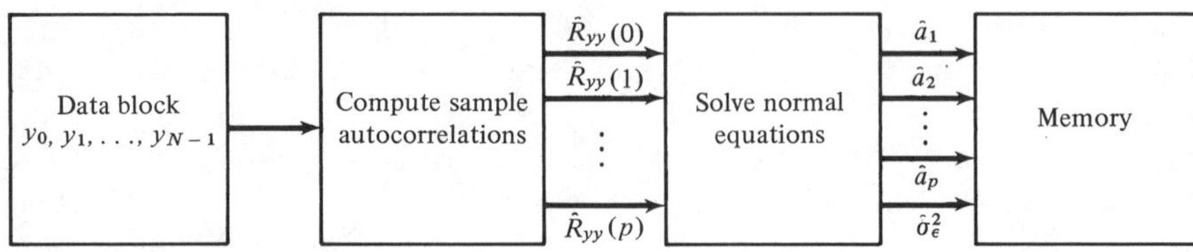

**Figure 5.2**  Yule-Walker Analysis Algorithm

The Yule-Walker analysis procedure, also referred to as the autocorrelation method of linear prediction [3], is summarized in Fig. 5.2.

## 5.3  *Linear Prediction Based on the Finite Past— Levinson Recursion*

In the last section, we saw that if the signal being predicted is autoregressive of order $p$, then the optimal linear predictor collapses to a $p$th order predictor. The infinite dimensional Wiener filtering problem collapses to a *finite* dimensional one. A geometrical way to understand this property is to say that the projection of $y_n$ on the subspace spanned by the entire past $\{y_{n-i}; i = 1,2, \ldots\}$ is the same as the projection of $y_n$ onto the subspace spanned only by the past $p$ samples; namely, $\{y_{n-i}; i = 1,2, \ldots ,p\}$. This is a consequence of the difference equation (5.2.3) generating $y_n$. If the process $y_n$ is not autoregressive, these two projections will be different. For any given $p$, the projection of $y_n$ onto the past $p$ samples will still provide the *best linear prediction* of $y_n$ that can be made on the basis of these *past p samples*. As $p$ increases, more and more past information is taken into account, and we expect the prediction of $y_n$ to become better and better in the sense of yielding a smaller mean-squared prediction error. In this section, we consider the finite-past prediction problem and discuss its efficient solution via the Levinson recursion [1–11]. For *sufficiently large* values of $p$, it may be considered an adequate *approximation* to the *full* prediction problem and hence also to the *modeling* problem.

Consider a stationary time series $y_n$ with autocorrelation function $R(k) = E[y_{n+k}y_n]$. For any given $p$, we seek the best linear predictor of order $p$ of the form

$$\hat{y}_n = -[a_1 y_{n-1} + a_2 y_{n-2} + \cdots + a_p y_{n-p}] \tag{5.3.1}$$

The $p$ prediction coefficients $a_1, a_2, \ldots ,a_p$ are chosen to minimize the mean-squared prediction error

$$\mathscr{E} = E[e_n^2] = \min \tag{5.3.2}$$

where $e_n$ is the prediction error

$$e_n = y_n - \hat{y}_n = y_n + a_1 y_{n-1} + a_2 y_{n-2} + \cdots + a_p y_{n-p} \tag{5.3.3}$$

Differentiating Eq. (5.3.2) with respect to each coefficient $a_i$; $i = 1,2, \ldots ,p$ yields the orthogonality equations

$$E[e_n y_{n-i}] = 0, \qquad \text{for } i = 1,2, \ldots ,p \qquad (5.3.4)$$

which express the fact that the optimal predictor $\hat{y}_n$ is the projection onto the span of the past $p$ samples; that is, $\{y_{n-i}; i = 1,2, \ldots ,p\}$. Inserting the expression (5.3.3) for $e_n$ into Eq. (5.3.4), we obtain $p$ linear equations for the coefficients

$$\sum_{j=0}^{p} a_j E[y_{n-j} y_{n-i}] = \sum_{j=0}^{p} R(i - j)a_j = 0, \qquad \text{for } i = 1,2, \ldots ,p \qquad (5.3.5)$$

Using the conditions (5.3.4) we also find for the *minimized* value of

$$\sigma_e^2 = \mathcal{E} = E[e_n^2] = E[e_n y_n] = \sum_{i=0}^{p} R(i)a_i \qquad (5.3.6)$$

Equations (5.3.5) and (5.3.6) can be combined into the $(p + 1) \times (p + 1)$ matrix equation

$$\begin{bmatrix} R(0) & R(1) & R(2) & \cdots & R(p) \\ R(1) & R(0) & R(1) & \cdots & R(p-1) \\ R(2) & R(1) & R(0) & \cdots & R(p-2) \\ \vdots & \vdots & \vdots & & \vdots \\ R(p) & R(p-1) & R(p-2) & \cdots & R(0) \end{bmatrix} \begin{bmatrix} 1 \\ a_1 \\ a_2 \\ \vdots \\ a_p \end{bmatrix} = \begin{bmatrix} \sigma_e^2 \\ 0 \\ 0 \\ \vdots \\ 0 \end{bmatrix} \qquad (5.3.7)$$

which is identical to Eq. (5.2.5) for the autoregressive case. It is also the *truncated* version of the infinite matrix equation (5.1.11) for the full prediction problem.

Instead of solving the normal equations (5.3.7) directly, we would like to imbed this problem into a whole class of similar problems; namely, of determining the best linear predictors of orders $p = 1$, $p = 2$, $p = 3$, $\ldots$ and so on. This approach will lead to Levinson's algorithm and to the so-called lattice realizations of linear prediction filters. Pictorially this class of problems is illustrated below

$$e_1(n) = y_n + a_{11} y_{n-1}$$
$$e_2(n) = y_n + a_{21} y_{n-1} + a_{22} y_{n-2}$$
$$e_3(n) = y_n + a_{31} y_{n-1} + a_{32} y_{n-2} + a_{33} y_{n-3}$$
$$e_p(n) = y_n + a_{p1} y_{n-1} + a_{p2} + y_{n-2} + \cdots + a_{pp} y_{n-p}$$

where $(1,a_{11})$, $(1,a_{21},a_{22})$, $(1,a_{31},a_{32},a_{33})$, $\ldots$ represent the best predictors of orders $p = 1,2,3, \ldots$ , respectively. It was necessary to tack on an extra index indicating the order of the predictor. Levinson's algorithm is an iterative procedure that constructs the *next* predictor from the *previous* one. In the process, *all* optimal predictors of lower order are also computed.

Consider the predictors of orders $p$ and $p + 1$, below.

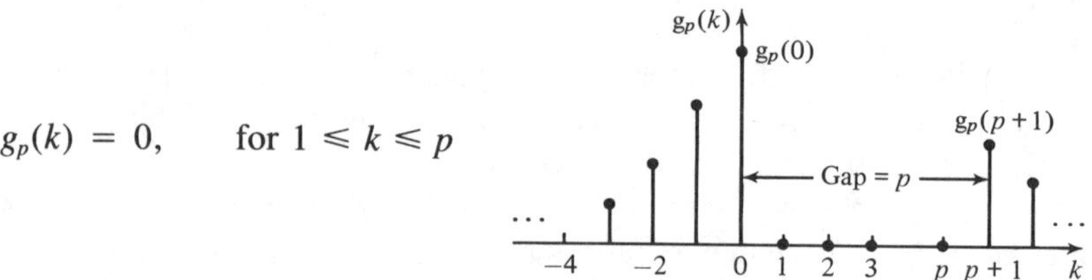

$$g_p(k) = E[e_p(n)y_{n-k}] = E\left[\left(\sum_{i=0}^{p} a_{pi}y_{n-i}\right)y_{n-k}\right]$$

$$= \sum_{i=0}^{p} a_{pi}R(k - i)$$

(5.3.8)

It has a gap of length $p$ as shown; that is,

$$g_p(k) = 0, \qquad \text{for } 1 \leq k \leq p$$

These gap conditions are the same as the orthogonality equations (5.3.4). Using $g_p(k)$ we now construct a new gapped function $g_{p+1}(k)$ of gap $p + 1$. To do this, first we reflect $g_p(k)$ about the origin; that is, $g_p(k) \rightarrow g_p(-k)$. The reflected function has a gap of length $p$ but at negatives times. A delay of $(p + 1)$ units of time will *realign* this gap with the original gap. This follows because if $1 \leq k \leq p$, then $1 \leq p + 1 - k \leq p$. The reflected-delayed function will be $g_p(p + 1 - k)$. These operations are shown in the accompanying figure.

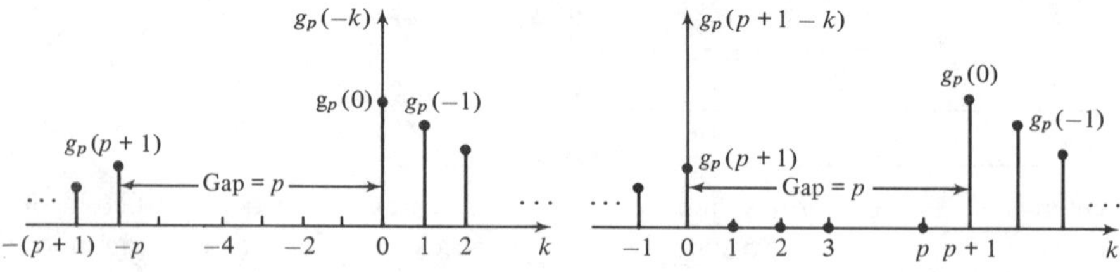

Since both $g_p(k)$ and $g_p(p + 1 - k)$ have exactly the same gap, it follows that so will any linear combination of these two functions. Therefore, if we write

$$g_{p+1}(k) = g_p(k) - \gamma_{p+1} g_p(p + 1 - k) \qquad (5.3.9)$$

there will be a gap of length at least $p$. We now select the parameter $\gamma_{p+1}$ so that $g_{p+1}(k)$ acquires an *extra* gap point; its gap is now of length $p + 1$. The extra gap condition is

$$g_{p+1}(p + 1) = g_p(p + 1) - \gamma_{p+1} g_p(0) = 0$$

which may be solved for

$$\gamma_{p+1} = \frac{g_p(p + 1)}{g_p(0)}$$

Evaluating Eq. (5.3.8) at $k = p + 1$, and using the fact that the value of the gapped function at $k = 0$ is the *minimized* value of the mean-squared error, that is,

$$E_p = E[e_p(n)^2] = E[e_p(n)y_n] = g_p(0) \qquad (5.3.10)$$

we finally find

$$
\gamma_{p+1} = \frac{\displaystyle\sum_{i=0}^{p} a_{pi} R(p + 1 - i)}{\displaystyle\sum_{i=0}^{p} a_{pi} R(i)}
$$

$$
= \frac{R(p + 1) + a_{p1}R(p) + a_{p2}R(p - 1) + \cdots + a_{pp}R(1)}{R(0) + a_{p1}R(1) + a_{p2}R(2) + \cdots + a_{pp}R(p)} \qquad (5.3.11)
$$

The coefficient $\gamma_{p+1}$ is called the *reflection* or *PARCOR* coefficient. This terminology will become clear later. Evaluating Eq. (5.3.9) at $k = 0$ and using $\gamma_{p+1} = g_p(p + 1)/g_p(0)$, we also find a recursion for the quantity $E_{p+1} = g_{p+1}(0)$

$$E_{p+1} = g_{p+1}(0) = g_p(0) - \gamma_{p+1} g_p(p + 1) = (1 - \gamma_{p+1}^2) g_p(0)$$

or

$$E_{p+1} = (1 - \gamma_{p+1}^2) E_p \qquad (5.3.12)$$

This represents the minimum value of the mean-squared prediction error $E[e_{p+1}(n)^2]$ for the predictor of order $p + 1$. Since both $E_p$ and $E_{p+1}$ are nonnegative, it follows that the factor $(1 - \gamma_{p+1}^2)$ will be positive and less than one. It represents the *improvement* in the prediction afforded by using a predictor of order $p + 1$ instead of a predictor of order $p$. It also follows that $\gamma_{p+1}$ has *magnitude less* than one. To find the new prediction coefficients, we use the fact that the gapped functions are equal to the convolution of the corresponding prediction-error filters with the autocorrelation function of $y_n$:

$$g_p(k) = \sum_{m=0}^{p} a_{pm} R(k - m) \rightarrow G_p(z) = A_p(z) S_{yy}(z)$$

$$g_{p+1}(k) = \sum_{m=0}^{p+1} a_{p+1,m} R(k - m) \rightarrow G_{p+1}(z) = A_{p+1}(z) S_{yy}(z)$$

where $S_{yy}(z)$ is the power spectral density of $y_n$. Taking the $z$-transforms of both sides of Eq. (5.3.9), we find

$$G_{p+1}(z) = G_p(z) - \gamma_{p+1} z^{-(p+1)} G_p(z^{-1})$$

or

$$A_{p+1}(z) S_{yy}(z) = A_p(z) S_{yy}(z) - \gamma_{p+1} z^{-(p+1)} A_p(z^{-1}) S_{yy}(z^{-1})$$

where we used the fact that the reflected gapped function has $z$-transform $G_p(z^{-1})$, and therefore the delayed (by $p + 1$) as well as reflected gapped function $g_p(p + 1 - k)$ has $z$-transform $z^{-(p+1)} G_p(z^{-1})$. Since $S_{yy}(z) = S_{yy}(z^{-1})$ because of the symmetry relations $R(k) = R(-k)$, it follows that $S_{yy}(z)$ is a common factor in all terms. Therefore, we obtain a relationship between the new best prediction-error filter $A_{p+1}(z)$ and the old one $A_p(z)$

$$A_{p+1}(z) = A_p(z) - \gamma_{p+1} z^{-(p+1)} A_p(z^{-1}) \qquad \text{(Levinson recursion)} \qquad (5.3.13)$$

Taking inverse $z$-transforms, we find

$$\begin{bmatrix} 1 \\ a_{p+1,1} \\ a_{p+1,2} \\ \vdots \\ a_{p+1,p} \\ a_{p+1,p+1} \end{bmatrix} = \begin{bmatrix} 1 \\ a_{p1} \\ a_{p2} \\ \vdots \\ a_{pp} \\ 0 \end{bmatrix} - \gamma_{p+1} \begin{bmatrix} 0 \\ a_{pp} \\ a_{p,p-1} \\ \vdots \\ a_{p1} \\ 1 \end{bmatrix} \qquad (5.3.14)$$

which can also be written as

$$a_{p+1,m} = a_{pm} - \gamma_{p+1} a_{p,p+1-m} \qquad \text{for } 1 \leq m \leq p$$

$$a_{p+1,p+1} = -\gamma_{p+1}$$

Introducing the *reverse* polynomial $A_p^R(z) = z^{-p} A_p(z^{-1})$, we may write Eq. (5.3.13) as

$$A_{p+1}(z) = A_p(z) - \gamma_{p+1} z^{-1} A_p^R(z) \qquad (5.3.15)$$

Taking the reverse of both sides we find

$$A_{p+1}(z^{-1}) = A_p(z^{-1}) - \gamma_{p+1} z^{p+1} A_p(z)$$

$$A_{p+1}^R(z) = z^{-(p+1)} A_{p+1}(z^{-1}) = z^{-(p+1)} A_p(z^{-1}) - \gamma_{p+1} A_p(z)$$

or

$$A^R_{p+1}(z) = z^{-1}A^R_p(z) - \gamma_{p+1}A_p(z) \qquad (5.3.16)$$

Equation (5.3.16) is, in a sense, redundant, but it will prove convenient to think of the Levinson recursion as a recursion on *both* the forward, $A_p(z)$, and the reverse, $A^R_p(z)$, polynomials. Both Eq. (5.3.15) and Eq. (5.3.16) may be combined into a $2 \times 2$ matrix recursion equation

$$\begin{bmatrix} A_{p+1}(z) \\ A^R_{p+1}(z) \end{bmatrix} = \begin{bmatrix} 1 & -\gamma_{p+1}z^{-1} \\ -\gamma_{p+1} & z^{-1} \end{bmatrix} \begin{bmatrix} A_p(z) \\ A^R_p(z) \end{bmatrix} \qquad \text{(forward recursion)}$$

$$(5.3.17)$$

The recursion is initialized at $p = 0$, by setting

$$A_0(z) = 1, \qquad A^R_0(z) = 1, \qquad \text{and} \qquad E_0 = R(0) = E[y_n^2] \qquad (5.3.18)$$

which correspond to no prediction at all.

We summarize the computational sequence of the Levinson algorithm:

0. Initialize at $p = 0$ using Eq. (5.3.18).
1. At stage $p$, the filter $A_p(z)$ and error $E_p$ are available.
2. Using Eq. (5.3.11), compute $\gamma_{p+1}$.
3. Using Eq. (5.3.13) or Eq. (5.3.17), determine the new polynomial $A_{p+1}(z)$.
4. Using Eq. (5.3.12), update the mean-squared prediction error to $E_{p+1}$.
5. Go to stage $p + 1$.

The iteration may be continued until the final desired order is reached. The dependence on the autocorrelation $R(k)$ of the signal $y_n$ is entered through Eq. (5.3.11) and $E_0 = R(0)$. To reach stage $p$, only the first $p + 1$ autocorrelation lags $R(0), R(1), \ldots, R(p)$ are required. At the $p$th stage, the iteration already has provided *all* the prediction filters of lower order, and all the previous reflection coefficients. Thus, an alternative parametrization of the $p$th order predictor is in terms of the sequence of reflection coefficients $\{\gamma_1, \gamma_2, \ldots, \gamma_p\}$ and the prediction error $E_p$

$$\{E_p, a_{p1}, a_{p2}, \ldots, a_{pp}\} \leftrightarrow \{E_p, \gamma_1, \gamma_2, \ldots, \gamma_p\}$$

One may pass from one parameter set to another. And both sets are equivalent to the autocorrelation set $\{R(0), R(1), \ldots, R(p)\}$. The alternative parametrization of the autocorrelation function $R(k)$ of a stationary random sequence in terms of the equivalent set of reflection coefficients is a general result [12], and has also been extended to the multichannel case [13]. If the process $y_n$ is autoregressive of order $p$, then as soon as the Levinson recursion reaches this order, it will provide the autoregressive coefficients $a_1, a_2, \ldots, a_p$ which are also the best prediction coefficients for the full prediction problem. Further continuation of the Levinson recursion will produce nothing new; all prediction coefficients of order higher than $p$ will be zero.

The four FORTRAN 77 subroutines LEV, FRWLEV, BKWLEV, and RLEV (see appendix) allow the passage from one parameter set to another. The subroutine LEV is an implementation of the computational sequence outlined above. The input to the subroutine is the final desired order of the predictor, say $M$, and the vector of autocorrelation lags $\{R(0), R(1), \ldots, R(M)\}$. Its output is the lower-triangular matrix $L$ whose rows are the *reverse* of all the lower order prediction-error filters. For example, for $M = 4$ the matrix $L$ would be

$$L = \begin{bmatrix} 1 & 0 & 0 & 0 & 0 \\ a_{11} & 1 & 0 & 0 & 0 \\ a_{22} & a_{21} & 1 & 0 & 0 \\ a_{33} & a_{32} & a_{31} & 1 & 0 \\ a_{44} & a_{43} & a_{42} & a_{41} & 1 \end{bmatrix}$$

The first column of $L$ contains the negatives of all the reflection coefficients. This follows from the Levinson recursion (5.3.14) which implies that the *negative* of the highest coefficient of the $p$th prediction-error filter is the $p$th reflection coefficient; namely,

$$\gamma_p = -a_{pp}; \qquad p = 1, 2, \ldots, M \tag{5.3.19}$$

This choice for $L$ is justified in Section 5.7. The subroutine LEV also produces the vector of mean-squared prediction errors $\{E_0, E_1, \ldots, E_M\}$ according to the recursion (5.3.12). The subroutine FRWLEV is an implementation of the forward Levinson recursion (5.3.17) or (5.3.14). Its input is the set of reflection coefficients $\{\gamma_1, \gamma_2, \ldots, \gamma_M\}$ and its output is the set of all prediction-error filters up to the order $M$; that is, $A_p(z); p = 1, 2, \ldots, M$. Again, this output is arranged into the matrix $L$. The subroutine BKWLEV is the inverse operation to FRWLEV. Its input is the prediction-error filter coefficients $[1, a_{M1}, a_{M2}, \ldots, a_{MM}]$ of the final order $M$, and its output is the matrix $L$ containing all the lower order prediction-error filters. The set of reflection coefficients are extracted from the first column of $L$. This subroutine is based on the inverse of the matrix equation (5.3.17). Shifting $p$ down by one unit, we write Eq. (5.3.17) as

$$\begin{bmatrix} A_p(z) \\ A_p^R(z) \end{bmatrix} = \begin{bmatrix} 1 & -\gamma_p z^{-1} \\ -\gamma_p & z^{-1} \end{bmatrix} \begin{bmatrix} A_{p-1}(z) \\ A_{p-1}^R(z) \end{bmatrix} \tag{5.3.20}$$

Its inverse is

$$\begin{bmatrix} A_{p-1}(z) \\ A_{p-1}^R(z) \end{bmatrix} = \frac{1}{1 - \gamma_p^2} \begin{bmatrix} 1 & \gamma_p \\ \gamma_p z & z \end{bmatrix} \begin{bmatrix} A_p(z) \\ A_p^R(z) \end{bmatrix} \qquad \text{(backward recursion)} \tag{5.3.21}$$

At each stage $p$, start with $A_p(z)$ and extract $\gamma_p = -a_{pp}$ from the highest coefficient of $A_p(z)$. Then, use Eq. (5.3.21) to obtain the polynomial $A_{p-1}(z)$. The iteration begins at the given order $M$ and proceeds downwards to $p =$

$M - 1, M - 2, \ldots, 1, 0$. Finally, the subroutine RLEV generates the set of autocorrelation lags $\{R(0), R(1), \ldots, R(M)\}$ from the knowledge of the *final* prediction-error filter $A_M(z)$ and *final* prediction error $E_M$. It calls BKWLEV to generate all the lower order prediction-error filters, and then it reconstructs the autocorrelation lags using the gapped function condition $g_p(p) = \sum_{i=0}^{p} a_{pi} R(p - i) = 0$, which may be solved for $R(p)$ in terms of $R(p - i); i = 1, 2, \ldots, p$, as follows:

$$R(p) = - \sum_{i=1}^{p} a_{pi} R(p - i) \qquad p = 1, 2, \ldots, M \qquad (5.3.22)$$

For example, the first few iterations of Eq. (5.3.22) will be

$$R(1) = -a_{11} R(0)$$
$$R(2) = -[a_{21} R(1) + a_{22} R(0)]$$
$$R(3) = -[a_{31} R(2) + a_{32} R(1) + a_{33} R(0)]$$

To get this recursion started, the value of $R(0)$ may be obtained from Eq. (5.3.12). Using Eq. (5.3.12) repeatedly, and $E_0 = R(0)$ we find

$$E_M = (1 - \gamma_1^2)(1 - \gamma_2^2) \cdots (1 - \gamma_M^2) R(0) \qquad (5.3.23)$$

Since the reflection coefficients are already known and $E_M$ is given, this equation provides the right value for $R(0)$.

*Example 5.3.1:*  Given the five autocorrelation lags

$$\{R(0), R(1), R(2), R(3), R(4)\} = \{128, -64, 80, -88, 89\}$$

Find all the prediction-error filters $A_p(z)$ up to order four, the four reflection coefficients, and the corresponding mean-squared prediction errors.

It is recommended that the reader go through the iteration of the Levinson algorithm by hand. Below, we simply state the results obtained using the subroutine LEV:

$$A_1(z) = 1 + 0.5z^{-1}$$
$$A_2(z) = 1 + 0.25z^{-1} - 0.5z^{-2}$$
$$A_3(z) = 1 - 0.375z^{-2} + 0.5z^{-3}$$
$$A_4(z) = 1 - 0.25z^{-1} - 0.1875z^{-2} + 0.5z^{-3} - 0.5z^{-4}$$

The reflection coefficients are the negatives of the highest coefficients; thus,

$$\{\gamma_1, \gamma_2, \gamma_3, \gamma_4\} = \{-0.5, 0.5, -0.5, 0.5\}$$

The vector of mean-squared prediction errors is given by

$$\{E_0, E_1, E_2, E_3, E_4\} = \{128, 96, 72, 54, 40.5\}$$

Sending the above vector of reflection coefficients through the subroutine FRWLEV would generate the above set of polynomials. Sending the coefficients of $A_4(z)$ through BKWLEV would generate the same set of polynomials. Sending the coefficients of $A_4(z)$ and $E_4 = 40.5$ through RLEV would recover the original autocorrelation lags $R(k)$; $k = 0,1, \ldots ,4$.

The *Yule-Walker method* (see Section 5.2) can be used to extract the linear prediction parameters from a given set of signal samples. From a given block of data

$$\boxed{y_0, y_1, \ldots, y_N}$$

compute the sample autocorrelations $\{\hat{R}(0),\hat{R}(1), \ldots ,\hat{R}(M)\}$ using, for example, Eq. (5.2.7), and send them through the Levinson recursion. The subroutine YW (see appendix) implements the Yule-Walker method. The input to the subroutine is the data vector of samples $\{y_0, y_1, \ldots , y_N\}$ (note that the actual length of this sequence is $N + 1$) and the desired final order $M$ of the predictor. Its output is the set of all prediction-error filters up to order $M$, arranged in the matrix $L$, and the vector of mean-squared prediction errors up to order $M$, $\{E_0, E_1, \ldots ,E_M\}$.

*Example 5.3.2:*   Given the signal samples

$$\{y_0, y_1, y_2, y_3, y_4\} = \{1,1,1,1,1\}$$

determine all the prediction-error filters up to order four. Using the fourth order predictor, predict the sixth value in the above sequence.

The sample autocorrelation of the above signal is easily computed using the methods of Chapter 1. We find (ignoring the $1/N$ normalization factor):

$$\{\hat{R}(0),\hat{R}(1),\hat{R}(2),\hat{R}(3),\hat{R}(4)\} = \{5,4,3,2,1\}$$

Sending these lags through the subroutine LEV we find the prediction-error filters

$$A_1(z) = 1 - 0.8z^{-1}$$

$$A_2(z) = 1 - 0.889z^{-1} + 0.111z^{-2}$$

$$A_3(z) = 1 - 0.875z^{-1} + 0.125z^{-3}$$

$$A_4(z) = 1 - 0.857z^{-1} + 0.143z^{-4}$$

Therefore, the fourth order prediction of $y_n$ given by Eq. (5.3.1) is

$$\hat{y}_n = 0.857y_{n-1} - 0.143y_{n-4}$$

which gives $\hat{y}_5 = 0.857 - 0.143 = 0.714$

## 5.4  *Levinson's Algorithm in Matrix Form*

The objective is to solve matrix equations of the type

$$\begin{bmatrix} R_0 & R_1 & R_2 & R_3 \\ R_1 & R_0 & R_1 & R_2 \\ R_2 & R_1 & R_0 & R_1 \\ R_3 & R_2 & R_1 & R_0 \end{bmatrix} \begin{bmatrix} 1 \\ a_{31} \\ a_{32} \\ a_{33} \end{bmatrix} = \begin{bmatrix} E_3 \\ 0 \\ 0 \\ 0 \end{bmatrix}$$

for the unknowns $(1, a_{31}, a_{32}, a_{33})$ and also indirectly for $E_3$. The corresponding prediction-error filter will be

$$A_3(z) = 1 + a_{31}z^{-1} + a_{32}z^{-2} + a_{33}z^{-3}$$

and the minimum value of the prediction error will be $E_3$.

The solution is obtained in an iterative manner, by solving a family of similar matrix equations of lower dimensionality. Starting at the upper left corner, the

$$\begin{bmatrix} R_0 & R_1 & R_2 & R_3 \\ R_1 & R_0 & R_1 & R_2 \\ R_2 & R_1 & R_0 & R_1 \\ R_3 & R_2 & R_1 & R_0 \end{bmatrix}$$

$R$ matrices are successively enlarged until the desired dimension is reached ($4 \times 4$ in this example). Therefore, one successively solves

$$[R_0]\,[1] = [E_0]; \qquad \begin{bmatrix} R_0 & R_1 \\ R_1 & R_0 \end{bmatrix} \begin{bmatrix} 1 \\ a_{11} \end{bmatrix} = \begin{bmatrix} E_1 \\ 0 \end{bmatrix};$$

$$\begin{bmatrix} R_0 & R_1 & R_2 \\ R_1 & R_0 & R_1 \\ R_2 & R_1 & R_0 \end{bmatrix} \begin{bmatrix} 1 \\ a_{21} \\ a_{22} \end{bmatrix} = \begin{bmatrix} E_2 \\ 0 \\ 0 \end{bmatrix}; \text{ etc.}$$

The solution of each problem is obtained *in terms of* the solution of the previous one. In this manner, the final solution is gradually built up. In the process, one also finds *all* the lower order prediction-error filters.

The iteration is based on two key properties of the autocorrelation matrix: first, the autocorrelation matrix of a given size contains as *subblocks* all the lower order autocorrelation matrices; and second, the autocorrelation matrix is *reflection invariant*. That is, it remains invariant under interchange of its columns and then its rows. This interchanging operation is equivalent to the similarity transformation by the "reversing" matrix $J$ defined by

$$J = \begin{bmatrix} 0 & 0 & 0 & 1 \\ 0 & 0 & 1 & 0 \\ 0 & 1 & 0 & 0 \\ 1 & 0 & 0 & 0 \end{bmatrix} \qquad (5.4.1)$$

The invariance property means that the autocorrelation matrix commutes with the matrix $J$

$$JRJ^{-1} = R \qquad (5.4.2)$$

This property immediately implies that if

$$\begin{bmatrix} R_0 & R_1 & R_2 & R_3 \\ R_1 & R_0 & R_1 & R_2 \\ R_2 & R_1 & R_0 & R_1 \\ R_3 & R_2 & R_1 & R_0 \end{bmatrix} \begin{bmatrix} a_0 \\ a_1 \\ a_2 \\ a_3 \end{bmatrix} = \begin{bmatrix} b_0 \\ b_1 \\ b_2 \\ b_3 \end{bmatrix}$$

then also
$$\begin{bmatrix} R_0 & R_1 & R_2 & R_3 \\ R_1 & R_0 & R_1 & R_2 \\ R_2 & R_1 & R_0 & R_1 \\ R_3 & R_2 & R_1 & R_0 \end{bmatrix} \begin{bmatrix} a_3 \\ a_2 \\ a_1 \\ a_0 \end{bmatrix} = \begin{bmatrix} b_3 \\ b_2 \\ b_1 \\ b_0 \end{bmatrix}$$

The steps of the algorithm are explicitly as follows:

### Step 0

Solve $R_0 \cdot 1 = E_0$. This defines $E_0$. Then, enlarge to the next size by padding a zero; that is,

$$\begin{bmatrix} R_0 & R_1 \\ R_1 & R_0 \end{bmatrix} \begin{bmatrix} 1 \\ 0 \end{bmatrix} = \begin{bmatrix} E_0 \\ D_0 \end{bmatrix} ; \qquad \text{this defines } D_0. \text{ Then also,}$$

$$\begin{bmatrix} R_0 & R_1 \\ R_1 & R_0 \end{bmatrix} \begin{bmatrix} 0 \\ 1 \end{bmatrix} = \begin{bmatrix} D_0 \\ E_0 \end{bmatrix} ; \qquad \text{by reversal invariance.}$$

These are the preliminaries to Step 1.

### Step 1

We wish to solve

$$\begin{bmatrix} R_0 & R_1 \\ R_1 & R_0 \end{bmatrix} \begin{bmatrix} 1 \\ a_{11} \end{bmatrix} = \begin{bmatrix} E_1 \\ 0 \end{bmatrix}$$

Try an expression of the form

$$\begin{bmatrix} 1 \\ a_{11} \end{bmatrix} = \begin{bmatrix} 1 \\ 0 \end{bmatrix} - \gamma_1 \begin{bmatrix} 0 \\ 1 \end{bmatrix} \qquad \text{with } \gamma_1 \text{ to be determined.}$$

Acting on both sides by

$$\begin{bmatrix} R_0 & R_1 \\ R_1 & R_0 \end{bmatrix} \qquad \text{and using the results of Step 0, we obtain}$$

$$\begin{bmatrix} E_1 \\ 0 \end{bmatrix} = \begin{bmatrix} E_0 \\ D_0 \end{bmatrix} - \gamma_1 \begin{bmatrix} D_0 \\ E_0 \end{bmatrix}$$

or

$$E_0 - \gamma_1 D_0 = E_1$$

$$D_0 - \gamma_1 E_0 = 0$$

or

$$\gamma_1 = \frac{D_0}{E_0}, \qquad E_1 = E_0 - \gamma_1 D_0 = (1 - \gamma_1^2) E_0$$

These define $\gamma_1$ and $E_1$. As a preliminary step to Step 2, enlarge to the next size by padding a zero

$$\begin{bmatrix} R_0 & R_1 & R_2 \\ R_1 & R_0 & R_1 \\ R_2 & R_1 & R_0 \end{bmatrix} \begin{bmatrix} 1 \\ a_{11} \\ 0 \end{bmatrix} = \begin{bmatrix} E_1 \\ 0 \\ D_1 \end{bmatrix}; \qquad \text{this defines } D_1. \text{ Then also,}$$

$$\begin{bmatrix} R_0 & R_1 & R_2 \\ R_1 & R_0 & R_1 \\ R_2 & R_1 & R_0 \end{bmatrix} \begin{bmatrix} 0 \\ a_{11} \\ 1 \end{bmatrix} = \begin{bmatrix} D_1 \\ 0 \\ E_1 \end{bmatrix}; \qquad \text{by reversal invariance.}$$

### Step 2

We wish to solve

$$\begin{bmatrix} R_0 & R_1 & R_2 \\ R_1 & R_0 & R_1 \\ R_2 & R_1 & R_0 \end{bmatrix} \begin{bmatrix} 1 \\ a_{21} \\ a_{22} \end{bmatrix} = \begin{bmatrix} E_2 \\ 0 \\ 0 \end{bmatrix}$$

Try expression

$$\begin{bmatrix} 1 \\ a_{21} \\ a_{22} \end{bmatrix} = \begin{bmatrix} 1 \\ a_{11} \\ 0 \end{bmatrix} - \gamma_2 \begin{bmatrix} 0 \\ 1 \\ a_{11} \end{bmatrix}; \qquad \text{with } \gamma_2 \text{ to be determined.}$$

From Step 1, we find

$$
\begin{bmatrix} E_2 \\ 0 \\ 0 \end{bmatrix} = \begin{bmatrix} E_1 \\ 0 \\ D_1 \end{bmatrix} - \gamma_2 \begin{bmatrix} D_1 \\ 0 \\ E_1 \end{bmatrix}
$$

or

$$
E_1 - \gamma_2 D_1 = E_2
$$

$$
D_1 - \gamma_2 E_1 = 0
$$

or

$$
\gamma_2 = \frac{D_1}{E_1} = \frac{R_2 + a_{11}R_1}{R_0 + a_{11}R_1}, \quad \text{and} \quad E_2 = (1 - \gamma_2^2)E_1
$$

These define $\gamma_2$ and $E_2$. Enlarge to next size by padding a zero

$$
\begin{bmatrix} R_0 & R_1 & R_2 & R_3 \\ R_1 & R_0 & R_1 & R_2 \\ R_2 & R_1 & R_0 & R_1 \\ R_3 & R_2 & R_1 & R_0 \end{bmatrix} \begin{bmatrix} 1 \\ a_{21} \\ a_{22} \\ 0 \end{bmatrix} = \begin{bmatrix} E_2 \\ 0 \\ 0 \\ D_2 \end{bmatrix} ; \quad \text{this defines } D_2. \text{ Then also,}
$$

$$
\begin{bmatrix} R_0 & R_1 & R_2 & R_3 \\ R_1 & R_0 & R_1 & R_2 \\ R_2 & R_1 & R_0 & R_1 \\ R_3 & R_2 & R_1 & R_0 \end{bmatrix} \begin{bmatrix} 0 \\ a_{22} \\ a_{21} \\ 1 \end{bmatrix} = \begin{bmatrix} D_2 \\ 0 \\ 0 \\ E_2 \end{bmatrix} ; \quad \text{by reversal invariance.}
$$

### Step 3

We wish to solve

$$
\begin{bmatrix} R_0 & R_1 & R_2 & R_3 \\ R_1 & R_0 & R_1 & R_2 \\ R_2 & R_1 & R_0 & R_1 \\ R_3 & R_2 & R_1 & R_0 \end{bmatrix} \begin{bmatrix} 1 \\ a_{31} \\ a_{32} \\ a_{33} \end{bmatrix} = \begin{bmatrix} E_3 \\ 0 \\ 0 \\ 0 \end{bmatrix}
$$

Try expression

$$
\begin{bmatrix} 1 \\ a_{31} \\ a_{32} \\ a_{33} \end{bmatrix} = \begin{bmatrix} 1 \\ a_{21} \\ a_{22} \\ 0 \end{bmatrix} - \gamma_3 \begin{bmatrix} 0 \\ a_{22} \\ a_{21} \\ 1 \end{bmatrix} \quad \text{with } \gamma_3 \text{ to be determined}
$$

From Step 2 we find

$$
\begin{bmatrix} E_3 \\ 0 \\ 0 \\ 0 \end{bmatrix} = \begin{bmatrix} E_2 \\ 0 \\ 0 \\ D_2 \end{bmatrix} - \gamma_3 \begin{bmatrix} D_2 \\ 0 \\ 0 \\ E_2 \end{bmatrix}
$$

or

$$
E_2 - \gamma_3 D_2 = E_3
$$

$$
D_2 - \gamma_3 E_2 = 0
$$

or

$$
\gamma_3 = \frac{D_2}{E_2} = \frac{R_3 + a_{21}R_2 + a_{22}R_1}{R_0 + a_{21}R_1 + a_{22}R_2}, \qquad \text{and} \qquad E_3 = (1 - \gamma_3^2)E_2
$$

Clearly, the procedure can be continued to higher and higher dimensions, as required in each problem.


## 5.5   *Analysis and Synthesis Lattice Filters*

The Levinson recursion, expressed in the $2 \times 2$ matrix form of Eq. (5.3.17), forms the basis of the so-called *lattice,* or ladder, realizations of the prediction-error filters and their inverses [3,6]. Remembering that the prediction-error sequence $e_p(n)$ is the convolution of the prediction-error filter $(1, a_{p1}, a_{p2}, \ldots, a_{pp})$ with the original data sequence $y_n$; that is,

$$
e_p^+(n) = y_n + a_{p1}y_{n-1} + a_{p2}y_{n-2} + \cdots + a_{pp}y_{n-p} \tag{5.5.1}
$$

we find in the $z$-domain

$$
E_p^+(z) = A_p(z)Y(z) \tag{5.5.2}
$$

where we changed the notation slightly and denoted $e_p(n)$ by $e_p^+(n)$.

At this point, it proves convenient to introduce the *backward* prediction-error sequence, defined in terms of the *reverse* of the prediction-error filter, as follows:

$$
E_p^-(z) = A_p^R(z)Y(z) \tag{5.5.3}
$$

$$
e_p^-(n) = y_{n-p} + a_{p1}y_{n-p+1} + a_{p2}y_{n-p+2} + \cdots + a_{pp}y_n \tag{5.5.4}
$$

where $A_p^R(z)$ is the reverse of $A_p(z)$; namely,

$$
A_p^R(z) = z^{-p}A_p(z^{-1})
$$

$$
= a_{pp} + a_{p,p-1}z^{-1} + a_{p,p-2}z^{-2} + \cdots + a_{p1}z^{-(p-1)} + z^{-p}
$$

The signal sequence $e_p^-(n)$ may be interpreted as the *postdiction* error in post-dicting the value of $y_{n-p}$ on the basis of the *future $p$ samples* $\{y_{n-p+1}, y_{n-p+2}, \ldots, y_n\}$, as shown below

| 1 | $a_{p1}$ | $a_{p2}$ | $\cdots$ | $a_{p,p-1}$ | $a_{pp}$ |

$y_{n-p-1}$   $y_{n-p}$   $y_{n-p+1}$  $y_{n-p+2}$   $\cdots$   $y_{n-1}$   $y_n$

Actually, the above choice of postdiction coefficients is the *optimal* one that minimizes the *mean-squared postdiction error*

$$E[e_p^-(n)^2] = \min \tag{5.5.5}$$

This is easily shown by inserting Eq. (5.5.4) into Eq. (5.5.5) and using stationarity

$$E[e_p^-(n)^2] = E\left[\left(\sum_{m=0}^{p} a_{pm} y_{n-p+m}\right)^2\right] = \sum_{m,k=0}^{p} a_{pm} E[y_{n-p+m} y_{n-p+k}] a_{pk}$$

$$= \sum_{m,k=0}^{p} a_{pm} R(m-k) a_{pk} = E[e_p^+(n)^2]$$

which shows that the forward and the backward prediction error criteria are the same, thus having the *same* solution for the optimal coefficients. Multiplying both sides of the Levinson recursion (5.3.17) by $Y(z)$, we cast it in the equivalent form in terms of the forward and backward prediction-error sequences:

$$\begin{bmatrix} E_{p+1}^+(z) \\ E_{p+1}^-(z) \end{bmatrix} = \begin{bmatrix} 1 & -\gamma_{p+1} z^{-1} \\ -\gamma_{p+1} & z^{-1} \end{bmatrix} \begin{bmatrix} E_p^+(z) \\ E_p^-(z) \end{bmatrix} \tag{5.5.6}$$

and in the time domain

$$e_{p+1}^+(n) = e_p^+(n) - \gamma_{p+1} e_p^-(n-1) \tag{5.5.7}$$

$$e_{p+1}^-(n) = e_p^-(n-1) - \gamma_{p+1} e_p^+(n)$$

These are to be initialized at $p = 0$ by

$$E_0^{\pm}(z) = A_0(z) Y(z) = Y(z) \quad \text{and} \quad e_0^{\pm}(n) = y_n \tag{5.5.8}$$

The lattice realization of the prediction-error filter is based on the recursion (5.5.7). Starting at $p = 0$ the output of the $p$th stage of (5.5.7) becomes the input of the $(p + 1)$th stage, up to the final desired order $p = M$. This is depicted in Fig. 5.3.

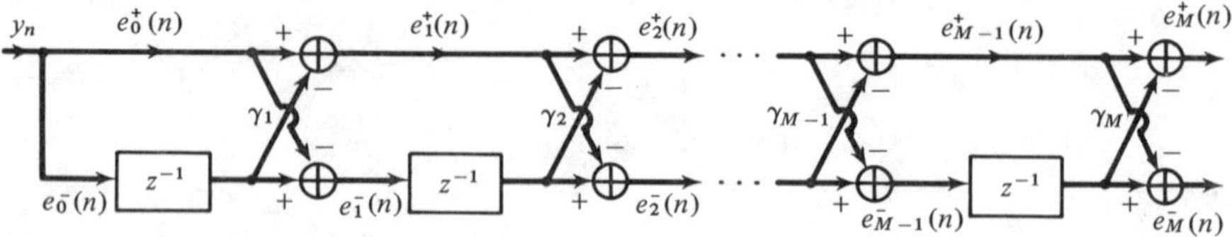


**Figure 5.3**   Analysis Lattice Filter

Equations (5.5.2) and (5.5.3) imply that the transfer function from the input $y_n$ to the output $e_M^+(n)$ is the desired prediction-error filter $A_M(z)$, whereas the transfer function from $y_n$ to $e_M^-(n)$ is the reversed filter $A_M^R(z)$. The lattice realization is therefore equivalent to the direct form realization

$$e_M^+(n) = y_n + a_{M1}y_{n-1} + \cdots + a_{MM}y_{n-M}$$

realized directly in terms of the prediction coefficients.

The synthesis filter $1/A_M(z)$ can also be realized in a lattice form. The input to the synthesis filter is the prediction error sequence $e_M^+(n)$ and its output is the original sequence $y_n$:

Since $y_n$ corresponds to $e_0^+(n)$, we must write Eq. (5.5.7) in an order-decreasing form, starting at $e_M^+(n)$ and ending with $e_0^+(n) = y_n$. Rearranging the terms of the first of Eqs. (5.5.7) we have

$$e_p^+(n) = e_{p+1}^+(n) + \gamma_{p+1}e_p^-(n-1)$$
$$e_{p+1}^-(n) = e_p^-(n-1) - \gamma_{p+1}e_p^+(n)$$

$$(5.5.9)$$

which can be realized as shown below:

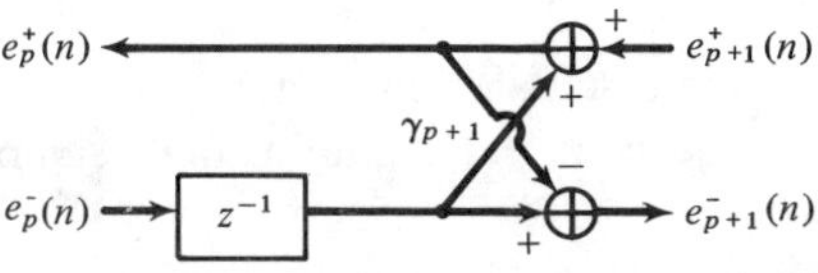

Note the difference in signs in the upper and lower adders. Putting together the stages from $p = M$ to $p = 0$, we obtain the synthesis lattice filter shown in Fig. 5.4.

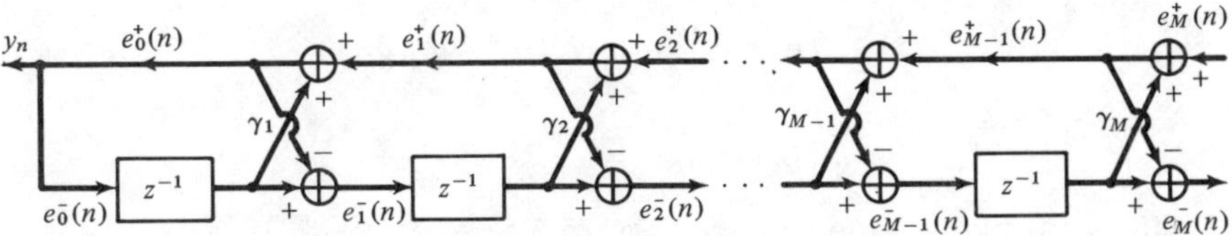

**Figure 5.4**   Synthesis Lattice Filter

## 5.6 *Alternative Proof of the Minimal Phase Property*

The synthesis filter $1/A_M(z)$ must be stable and causal, which requires all the $M$ zeros of the prediction-error filter $A_M(z)$ to lie inside the unit circle in the complex $z$-plane. We have already presented a proof of this fact which was based on the property that the coefficients of $A_M(z)$ minimized the mean-squared prediction error $E[e_M^+(n)^2]$. Here, we present an alternative proof based on the Levinson recursion and the fact that all reflection coefficients $\gamma_p$ have magnitude less than one [6,11].

From the definition (5.5.4), it follows that

$$e_p^-(n-1) = y_{n-p-1} + a_{p1}y_{n-p} + a_{p2}y_{n-p+1} + \cdots + a_{pp}y_{n-1} \qquad (5.6.1)$$

This quantity represents the estimation error of postdicting $y_{n-p-1}$ on the basis of the $p$ future samples $y_{n-p}, y_{n-p+1}, \ldots, y_{n-1}$. Another way to say this is that the linear combination of these $p$ samples is the projection of $y_{n-p-1}$ onto the subspace of random variables spanned by $\{y_{n-p}, y_{n-p+1}, \ldots, y_{n-1}\}$; that is,

$$e_p^-(n-1) = y_{n-p-1} - (\text{projection of } y_{n-p-1} \text{ onto}$$

$$\{y_{n-p}, y_{n-p+1}, \ldots, y_{n-1}\}) \qquad (5.6.2)$$

On the other hand, $e_p^+(n)$ given in Eq. (5.5.1) is the estimation error of $y_n$ based on the *same* set of samples. Therefore,

$$e_p^+(n) = y_n - (\text{projection of } y_n \text{ onto } \{y_{n-p}, y_{n-p+1}, \ldots, y_{n-1}\}) \qquad (5.6.3)$$

The samples $y_{n-p}, y_{n-p+1}, \ldots, y_{n-1}$ are the *intermediate* set of samples between $y_{n-p-1}$ and $y_n$, as shown below:

$$y_{n-p-1} \quad y_{n-p} \quad y_{n-p+1} \quad \cdots \quad y_{n-2} \quad y_{n-1} \quad y_n$$

Therefore, according to the discussion in Section 1.6, the PARCOR coefficient between $y_{n-p-1}$ and $y_n$ with the effect of intermediate samples removed is given by

$$\text{PARCOR} = \frac{E[e_p^+(n)e_p^-(n-1)]}{E[e_p^-(n-1)^2]}$$

This is precisely the reflection coefficient $\gamma_{p+1}$ of Eq. (5.3.11). Indeed, using Eq. (5.6.1) and the gapped conditions we find

$$E[e_p^+(n)e_p^-(n-1)] = E[e_p^+(n)(y_{n-p-1} + a_{p1}y_{n-p} + \cdots + a_{pp}y_{n-1})]$$

$$= g_p(p+1) + a_{p1}g_p(p) + a_{p2}g_p(p-1)$$

$$+ \cdots + a_{pp}g_p(1)$$

$$= g_p(p+1)$$

Similarly, invoking stationarity and Eq. (5.5.5),

$$E[e_p^-(n-1)^2] = E[e_p^-(n)^2] = E[e_p^+(n)^2] = g_p(0)$$

Thus, the reflection coefficient $\gamma_{p+1}$ is really a *PARCOR coefficient*:

$$\gamma_{p+1} = \frac{E[e_p^+(n)e_p^-(n-1)]}{E[e_p^-(n-1)^2]} = \frac{E[e_p^+(n)e_p^-(n-1)]}{E[e_p^+(n)^2]} \tag{5.6.4}$$

Using the Schwarz inequality with respect to the inner product $E[uv]$, that is,

$$(E[uv])^2 \leq E[u^2]E[v^2]$$

Eq. (5.6.4) implies that $\gamma_{p+1}$ has magnitude less than one:

$$|\gamma_{p+1}| \leq 1 \qquad \text{for each } p = 0,1, \ldots \tag{5.6.5}$$

To prove the minimal phase property of $A_M(z)$ we must show that all of its $M$ zeros are inside the unit circle. We will do this by induction. Let $Z_p$ and $N_p$ denote the number of zeros and poles of $A_p(z)$ that lie *inside* the unit circle. Levinson's recursion, Eq. (5.3.13), expresses $A_{p+1}(z)$ as the sum of $A_p(z)$ and a correction term $F(z) = -\gamma_{p+1}z^{-1}A_p^R(z)$

$$A_{p+1}(z) = A_p(z) + F(z)$$

Using the inequality (5.6.5) and the fact that $A_p^R(z)$ has the *same* magnitude spectrum as $A_p(z)$, we find the inequality

$$|F(z)| = |-\gamma_{p+1}z^{-1}A_p^R(z)| = |\gamma_{p+1}A_p(z)| \leq |A_p(z)|$$

for $z = \exp(j\omega)$ on the unit circle. Then, the argument principle and Rouche's theorem imply that the addition of the function $F(z)$ will not affect the difference $N_p - Z_p$ of poles and zeros contained inside the unit circle. Thus,

$$N_{p+1} - Z_{p+1} = N_p - Z_p$$

Since the only pole of $A_p(z)$ is the multiple pole of order $p$ at the origin arising from the term $z^{-p}$, it follows that $N_p = p$. Therefore,

$$(p + 1) - Z_{p+1} = p - Z_p$$

or

$$Z_{p+1} = Z_p + 1$$

Starting at $p = 0$ with $A_0(z) = 1$, we have $Z_0 = 0$. It follows that

$$Z_p = p$$

which states that for each $p$, all the $p$ zeros of the polynomial $A_p(z)$ lie inside the unit circle.

Another way to state this result is: "A necessary and sufficient condition for a polynomial $A_M(z)$ to have all of its $M$ zeros strictly inside the unit circle is that all reflection coefficients $\{\gamma_1, \gamma_2, \ldots, \gamma_M\}$ resulting from $A_M(z)$ via the backward recursion (5.3.21) have magnitude strictly less than one." This is essentially equivalent to the classic *Schur-Cohn* test of stability [14–16]. The subroutine BKWLEV can be used in this regard to obtain the sequence of reflection coefficients.

*Example 5.6.1:*   Test the minimal phase property of the polynomials

(a) $A(z) = 1 - 2.60z^{-1} + 2.55z^{-2} - 2.80z^{-3} + 0.50z^{-4}$
(b) $A(z) = 1 - 1.40z^{-1} + 1.47z^{-2} - 1.30z^{-3} + 0.50z^{-4}$

Sending the coefficients of each through the subroutine BKWLEV we find the set of reflection coefficients

(a) $\{0.4, -0.5, 2.0, -0.5\}$
(b) $\{0.4, -0.5, 0.8, -0.5\}$

Since among (a) there is one reflection coefficient of magnitude greater than one, case (a) will not be minimal phase, whereas case (b) is.

## 5.7   *Orthogonality of Backward Prediction Errors— Cholesky Factorization*

Another interesting structural property of the lattice realizations is that, in a certain sense, the *backward* prediction errors $e_p^-(n)$ are orthogonal to each other [3,6]. To see this, consider the case $M = 3$, and form the matrix product

$$\underbrace{\begin{bmatrix} R_0 & R_1 & R_2 & R_3 \\ R_1 & R_0 & R_1 & R_2 \\ R_2 & R_1 & R_0 & R_1 \\ R_3 & R_2 & R_1 & R_0 \end{bmatrix}}_{R} \underbrace{\begin{bmatrix} 1 & a_{11} & a_{22} & a_{33} \\ 0 & 1 & a_{21} & a_{32} \\ 0 & 0 & 1 & a_{31} \\ 0 & 0 & 0 & 1 \end{bmatrix}}_{L^T} = \underbrace{\begin{bmatrix} E_0 & 0 & 0 & 0 \\ * & E_1 & 0 & 0 \\ * & * & E_2 & 0 \\ * & * & * & E_3 \end{bmatrix}}_{L_1}$$

Because the normal equations (written upside down) are satisfied by each prediction-error filter, the right-hand side will be a lower-triangular matrix. The "don't care" entries have been denoted by *s. Multiply from the left by $L$ to get

$$LRL^T = LL_1 = \begin{bmatrix} E_0 & 0 & 0 & 0 \\ * & E_1 & 0 & 0 \\ * & * & E_2 & 0 \\ * & * & * & E_3 \end{bmatrix}$$

Since $L$ is by definition lower-triangular, the right-hand side will still be lower-triangular. But the left-hand side is symmetric. Thus, so is the right-hand side and as a result it must be diagonal. We have shown

$$LRL^T = D = \text{diag}\{E_0, E_1, E_2, E_3\} \tag{5.7.1}$$

or, written explicitly,

$$
\begin{bmatrix} 1 & 0 & 0 & 0 \\ a_{11} & 1 & 0 & 0 \\ a_{22} & a_{21} & 1 & 0 \\ a_{33} & a_{32} & a_{31} & 1 \end{bmatrix}
\begin{bmatrix} R_0 & R_1 & R_2 & R_3 \\ R_1 & R_0 & R_1 & R_2 \\ R_2 & R_1 & R_0 & R_1 \\ R_3 & R_2 & R_1 & R_0 \end{bmatrix}
\begin{bmatrix} 1 & a_{11} & a_{22} & a_{33} \\ 0 & 1 & a_{21} & a_{32} \\ 0 & 0 & 1 & a_{31} \\ 0 & 0 & 0 & 1 \end{bmatrix}
$$

$$
= \begin{bmatrix} E_0 & 0 & 0 & 0 \\ 0 & E_1 & 0 & 0 \\ 0 & 0 & E_2 & 0 \\ 0 & 0 & 0 & E_3 \end{bmatrix}
$$

The $pq$th element of this matrix equation is then

$$
\mathbf{a}_p^T R \mathbf{a}_q = \delta_{pq} E_p \qquad 0 \leq p,q \leq 3 \tag{5.7.2}
$$

where $\mathbf{a}_p$ and $\mathbf{a}_q$ denote the $p$th and $q$th columns of $L^T$. These are recognized as the *backward prediction-error filters* of orders $p$ and $q$. Eq. (5.7.2) implies then the orthogonality of the backward prediction-error filters with respect to an inner product $\mathbf{x}^T R \mathbf{y}$.

The backward prediction errors $e_p^-(n)$ can be expressed in terms of the $\mathbf{a}_p$s and the vector of samples $\mathbf{y}(n) = [y_n, y_{n-1}, y_{n-2}, y_{n-3}]^T$, as follows:

$$
e_0^-(n) = [1,\ 0,\ 0,\ 0]\mathbf{y}(n) = \mathbf{a}_0^T \mathbf{y}(n) = y_n
$$

$$
e_1^-(n) = [a_{11}, 1,\ 0,\ 0]\mathbf{y}(n) = \mathbf{a}_1^T \mathbf{y}(n) = a_{11} y_n + y_{n-1}
$$

$$
e_2^-(n) = [a_{22}, a_{21}, 1,\ 0]\mathbf{y}(n) = \mathbf{a}_2^T \mathbf{y}(n) = a_{22} y_n + a_{21} y_{n-1} + y_{n-2}
$$

$$
e_3^-(n) = [a_{33}, a_{32}, a_{31}, 1]\mathbf{y}(n) = \mathbf{a}_3^T \mathbf{y}(n) = a_{33} y_n + a_{32} y_{n-1} + a_{31} y_{n-2} + y_{n-3}
$$

$$
\tag{5.7.3}
$$

which can be arranged into the vector form

$$
\mathbf{e}^-(n) = \begin{bmatrix} e_0^-(n) \\ e_1^-(n) \\ e_2^-(n) \\ e_3^-(n) \end{bmatrix}
= \begin{bmatrix} 1 & 0 & 0 & 0 \\ a_{11} & 1 & 0 & 0 \\ a_{22} & a_{21} & 1 & 0 \\ a_{33} & a_{32} & a_{31} & 1 \end{bmatrix}
\begin{bmatrix} y_n \\ y_{n-1} \\ y_{n-2} \\ y_{n-3} \end{bmatrix} = L\mathbf{y}(n) \tag{5.7.4}
$$

Using Eq. (5.7.1), it follows now that the covariance matrix of $\mathbf{e}^-(n)$ is *diagonal*; indeed, since $R = E[\mathbf{y}(n)\mathbf{y}(n)^T]$,

$$
R_{e^- e^-} = E[\mathbf{e}^-(n)\mathbf{e}^-(n)^T] = LRL^T = D \tag{5.7.5}
$$

which can also be expressed component-wise as the zero lag cross-correlation

$$
R_{e_p^- e_q^-}(0) = E[e_p^-(n) e_q^-(n)] = \delta_{pq} E_p \tag{5.7.6}
$$

Thus, at each time instant $n$, the backward prediction errors $e_p^-(n)$ are mutually uncorrelated (orthogonal) with each other. The orthogonality conditions (5.7.6) and the lower-triangular nature of $L$ render the transformation (5.7.4) equivalent to the *Gram-Schmidt* orthogonalization of the data vector $\mathbf{y}(n) = [y_n, y_{n-1}, y_{n-2}, y_{n-3}]^T$. Equation (5.7.1), written as

$$R = L^{-1}DL^{-T}$$

corresponds to an *LU Cholesky factorization* of the covariance matrix $R$.

Since the backward errors $e_p^-(n)$; $p = 0,1,2, \ldots ,M$, for an $M$th order predictor are generated at the output of each successive lattice segment of Fig. 5.3, we may view the *analysis lattice* filter as an *implementation* of the *Gram-Schmidt orthogonalization* of the vector $\mathbf{y}(n) = [y_n, y_{n-1}, y_{n-2}, \ldots ,y_{n-M}]^T$. It is interesting to note, in this respect, that this implementation requires only knowledge of the reflection coefficients $\{\gamma_1, \gamma_2, \ldots ,\gamma_M\}$.

The above orthogonalization may also be understood in the $z$-domain: Since the backward prediction error $e_p^-(n)$ is the output of the reversed prediction-error filter $A_p^R(z)$ driven by the data sequence $y_n$, we have for the cross-density

$$S_{e_p^- e_q^-}(z) = A_p^R(z) S_{yy}(z) A_q^R(z^{-1})$$

Integrating this expression over the unit circle and using Eq. (5.7.6), we find

$$\oint_{\text{u.c.}} A_p^R(z) S_{yy}(z) A_q^R(z^{-1}) \frac{dz}{2\pi jz} = \oint S_{e_p^- e_q^-}(z) \frac{dz}{2\pi jz} = R_{e_p^- e_q^-}(0)$$
$$= E[e_p^-(n) e_q^-(n)] = \delta_{pq} E_p \tag{5.7.7}$$

That is, the reverse polynomials $A_p^R(z)$ are *mutually orthogonal* with respect to the above inner product defined by the (positive-definite) weighting function $S_{yy}(z)$. Equation (5.7.7) is the $z$-domain expression of Eq. (5.7.2). This result establishes an intimate connection between the linear prediction problem and the theory of *orthogonal polynomials* on the unit circle developed by Szegö [17,18].

The LU factorization of $R$ implies a UL factorization of the inverse of $R$; that is, solving Eq. (5.7.1) for $R^{-1}$ we have:

$$R^{-1} = L^T D^{-1} L \tag{5.7.8}$$

Since the Levinson recursion generates all the lower order prediction-error filters, it essentially generates the inverse of $R$.

The computation of this inverse may also be done *recursively* in the *order*, as follows. To keep track of the order let us tack an extra index

$$R_3^{-1} = L_3^T D_3^{-1} L_3 \tag{5.7.9}$$

The matrix $L_3$ contains as a submatrix the matrix $L_2$; in fact

$$L_3 = \begin{bmatrix} 1 & 0 & 0 & 0 \\ a_{11} & 1 & 0 & 0 \\ a_{22} & a_{21} & 1 & 0 \\ a_{33} & a_{32} & a_{31} & 1 \end{bmatrix} = \begin{bmatrix} L_2 & \mathbf{0} \\ \boldsymbol{\alpha}_3^{RT} & 1 \end{bmatrix} \qquad (5.7.10)$$

where $\boldsymbol{\alpha}_3^{RT}$ denotes the transpose of the *reverse* of the vector of prediction coefficients; namely, $\boldsymbol{\alpha}_3^{RT} = [a_{33}, a_{32}, a_{31}]$. The diagonal matrix $D_3^{-1}$ may also be block divided in the same manner:

$$D_3^{-1} = \begin{bmatrix} D_2^{-1} & \mathbf{0} \\ \mathbf{0}^T & E_3^{-1} \end{bmatrix}$$

Inserting these block decompositions into Eq. (5.7.9) and using the lower order result $R_2^{-1} = L_2^T D_2^{-1} L_2$, we find

$$R_3^{-1} = \begin{bmatrix} R_2^{-1} + \dfrac{1}{E_3}\, \boldsymbol{\alpha}_3^R \boldsymbol{\alpha}_3^{RT} & \dfrac{1}{E_3}\, \boldsymbol{\alpha}_3^R \\ \dfrac{1}{E_3}\, \boldsymbol{\alpha}_3^{RT} & \dfrac{1}{E_3} \end{bmatrix} \qquad (5.7.11)$$

Thus, through Levinson's algorithm, as the prediction coefficients $\boldsymbol{\alpha}_3$ and error $E_3$ are obtained, the inverse of $R$ may be *updated* to the next higher order. The result (5.7.11) also suggests an efficient way of solving more general normal equations of the type

$$R_3 \mathbf{h}_3 = \begin{bmatrix} R_0 & R_1 & R_2 & R_3 \\ R_1 & R_0 & R_1 & R_2 \\ R_2 & R_1 & R_0 & R_1 \\ R_3 & R_2 & R_1 & R_0 \end{bmatrix} \begin{bmatrix} h_{30} \\ h_{31} \\ h_{32} \\ h_{33} \end{bmatrix} = \begin{bmatrix} r_0 \\ r_1 \\ r_2 \\ r_3 \end{bmatrix} = \mathbf{r}_3 \qquad (5.7.12)$$

for a given right-hand vector $\mathbf{r}_3$. Such normal equations arise in the design of FIR Wiener filters; for example, Eq. (4.3.9). The solution for $\mathbf{h}_3$ is obtained recursively from the solution of similar linear equations of lower order. For example, let $\mathbf{h}_2$ be the solution of the previous order

$$R_2 \mathbf{h}_2 = \begin{bmatrix} R_0 & R_1 & R_2 \\ R_1 & R_0 & R_1 \\ R_2 & R_1 & R_0 \end{bmatrix} \begin{bmatrix} h_{20} \\ h_{21} \\ h_{22} \end{bmatrix} = \begin{bmatrix} r_0 \\ r_1 \\ r_2 \end{bmatrix} = \mathbf{r}_2$$

where the right-hand side vector $\mathbf{r}_2$ is *part* of $\mathbf{r}_3$. Then, Eq. (5.7.11) implies a recursive relationship between $\mathbf{h}_3$ and $\mathbf{h}_2$:

$$\mathbf{h}_3 = R_3^{-1}\mathbf{r}_3 = \left[\begin{array}{c|c} R_2^{-1} + \dfrac{1}{E_3}\boldsymbol{\alpha}_3^R\boldsymbol{\alpha}_3^{RT} & \dfrac{1}{E_3}\boldsymbol{\alpha}_3^R \\ \hline \dfrac{1}{E_3}\boldsymbol{\alpha}_3^{RT} & \dfrac{1}{E_3} \end{array}\right]\left[\begin{array}{c} \mathbf{r}_2 \\ \hline r_3 \end{array}\right]$$

$$= \left[\begin{array}{c} \mathbf{h}_2 + \dfrac{1}{E_3}\boldsymbol{\alpha}_3^R(r_3 + \boldsymbol{\alpha}_3^{RT}\mathbf{r}_2) \\ \hline \dfrac{1}{E_3}(r_3 + \boldsymbol{\alpha}_3^{RT}\mathbf{r}_2) \end{array}\right]$$

In terms of the reverse prediction-error filter $\mathbf{a}_3^{RT} = [a_{33}, a_{32}, a_{31}, 1] = [\boldsymbol{\alpha}_3^{RT}, 1]$, we may write

$$\mathbf{h}_3 = \left[\begin{array}{c} \mathbf{h}_2 \\ \hline 0 \end{array}\right] + c\mathbf{a}_3^R \qquad \text{where } c = (r_3 + \boldsymbol{\alpha}_3^{RT}\mathbf{r}_2)/E_3 = \mathbf{a}_3^{RT}\mathbf{r}_3/E_3 \qquad (5.7.13)$$

Thus, the recursive updating of the solution $\mathbf{h}$ must be done by carrying out the auxiliary updating of the prediction-error filters. The method requires $O(M^2)$ operations, compared to $O(M^3)$ if the inverse of $R$ were to be computed directly. This recursive method of solving general normal equations, developed by Robinson and Treitel, has been reviewed elsewhere [7,8,19–21].

Some additional insight into the properties of these recursions can be gained by using the Toeplitz property of $R$. This property together with the symmetric nature of $R$ imply that $R$ commutes with the reversing matrix:

$$J_3 = \begin{bmatrix} 0 & 0 & 0 & 1 \\ 0 & 0 & 1 & 0 \\ 0 & 1 & 0 & 0 \\ 1 & 0 & 0 & 0 \end{bmatrix} = J_3^{-1} \qquad J_3 R_3 J_3 = R_3 \qquad (5.7.14)$$

Therefore, even though the inverse $R_3^{-1}$ is not Toeplitz, it still commutes with this reversing matrix; that is,

$$J_3 R_3^{-1} J_3 = R_3^{-1} \qquad (5.7.15)$$

The effect of this symmetry property on the block decomposition (5.7.11) may be seen by decomposing $J_3$ also as

$$J_3 = \left[\begin{array}{c|c} \mathbf{0} & J_2 \\ \hline 1 & \mathbf{0}^T \end{array}\right] = \left[\begin{array}{c|c} \mathbf{0}^T & 1 \\ \hline J_2 & \mathbf{0} \end{array}\right]$$

where $J_2$ is the lower order reversing matrix. Combining Eq. (5.7.15) with Eq. (5.7.11) we find

$$R_3^{-1} = J_3 R_3^{-1} J_3 = \begin{bmatrix} \mathbf{0}^T & 1 \\ \hline J_2 & 0 \end{bmatrix} \begin{bmatrix} R_2^{-1} + \dfrac{1}{E_3}\boldsymbol{\alpha}_3^R \boldsymbol{\alpha}_3^{RT} & \dfrac{1}{E_3}\boldsymbol{\alpha}_3^R \\ \hline \dfrac{1}{E_3}\boldsymbol{\alpha}_3^{RT} & \dfrac{1}{E_3} \end{bmatrix} \begin{bmatrix} 0 & J_2 \\ \hline 1 & \mathbf{0}^T \end{bmatrix}$$

or, since $R_2$ commutes with $J_2$ and $J_2 \boldsymbol{\alpha}_3^R = \boldsymbol{\alpha}_3$,

$$R_3^{-1} = \begin{bmatrix} \dfrac{1}{E_3} & \dfrac{1}{E_3}\boldsymbol{\alpha}_3^T \\ \hline \dfrac{1}{E_3}\boldsymbol{\alpha}_3 & R_2^{-1} + \dfrac{1}{E_3}\boldsymbol{\alpha}_3\boldsymbol{\alpha}_3^T \end{bmatrix} \tag{5.7.16}$$

Both ways of expressing $R_3^{-1}$, given by Eqs.(5.7.16) and (5.7.11), are useful. They may be combined as follows: Eq.(5.7.16) gives for the $ij$th entry

$$(R_3^{-1})_{ij} = (R_2^{-1} + \boldsymbol{\alpha}_3 \boldsymbol{\alpha}_3^T E_3^{-1})_{i-1,j-1} = (R_2^{-1})_{i-1,j-1} + \alpha_{3i}\alpha_{3j}E_3^{-1}$$

which is valid for $1 \le i,j \le 3$. On the other hand, from Eq. (5.7.11) we have

$$(R_3^{-1})_{i-1,j-1} = (R_2^{-1})_{i-1,j-1} + \alpha_{3i}^R \alpha_{3j}^R E_3^{-1}$$

which is valid also for $1 \le i,j \le 3$. Subtracting the two to cancel the common term $(R_2^{-1})_{i-1,j-1}$ we obtain the Goberg-Semencul-Trench-Zohar recursion [22–26]

$$(R_3^{-1})_{ij} = (R_3^{-1})_{i-1,j-1} + (\boldsymbol{\alpha}_3 \boldsymbol{\alpha}_3^T - \boldsymbol{\alpha}_3^R \boldsymbol{\alpha}_3^{RT})_{ij} E_3^{-1}, \; 1 \le i,j \le 3 \tag{5.7.17}$$

which allows the building-up of $R_3^{-1}$ along *each diagonal,* provided one knows the "boundary" values to get these recursions started. But these are:

$$(R_3^{-1})_{00} = E_3^{-1}, \qquad (R_3^{-1})_{i0} = (R_3^{-1})_{0i} = a_{3i}E_3^{-1}, \qquad 1 \le i \le 3 \tag{5.7.18}$$

Thus, from the prediction-error filter $\mathbf{a}_3$ and its reverse, the entire inverse of the autocorrelation matrix may be built up. Computationally, of course, the best procedure is to use Eq. (5.7.8), where $L$ and $D$ are obtained as byproducts of the Levinson recursion. The subroutine LEV of the appendix starts with the $M + 1$ autocorrelation lags $\{R(0), R(1), \ldots, R(M)\}$ and generates the required matrices $L$ and $D$. The new VLSI hardware has motivated the development of efficient computational structures for the implementation of Levinson's algorithm, the inversion and diagonalization of Toeplitz matrices, and other related signal processing computations [27–31].


## 5.8  Lattice Realizations of FIR Wiener Filters

In this section, we combine the results of Sections 4.3 and 5.7 to derive alternative realizations of Wiener filters that are based on the Gram-Schmidt lattice structures. Consider the FIR Wiener filtering problem of estimating a desired signal $x_n$ on the basis of the related signal $y_n$, using an $M$th order filter. The I/O equation of the optimal filter is given by Eq. (4.3.8). The vector of optimal weights is determined by solving the set of normal equations, given by Eq. (4.3.9). The discussion of the previous section suggests that Eq. (4.3.9) can be solved efficiently using the Levinson recursion. Defining the data vector

$$\mathbf{y}(n) = \begin{bmatrix} y_n \\ y_{n-1} \\ \vdots \\ y_{n-M} \end{bmatrix} \tag{5.8.1}$$

we rewrite Eq. (4.3.9) in the compact matrix form

$$R_{yy}\mathbf{h} = \mathbf{r}_{xy} \tag{5.8.2}$$

where $R_{yy}$ is the $(M + 1) \times (M + 1)$ autocorrelation matrix of $\mathbf{y}(n)$, and $\mathbf{r}_{xy}$ the $(M + 1)$-vector of cross-correlations between $x_n$ and $\mathbf{y}(n)$; namely,

$$R_{yy} = E[\mathbf{y}(n)\mathbf{y}(n)^T] \qquad \mathbf{r}_{xy} = E[x_n\mathbf{y}(n)] = \begin{bmatrix} R_{xy}(0) \\ R_{xy}(1) \\ \vdots \\ R_{xy}(M) \end{bmatrix} \tag{5.8.3}$$

and $\mathbf{h}$ is the $(M + 1)$-vector of optimal weights

$$\mathbf{h} = \begin{bmatrix} h(0) \\ h(1) \\ \vdots \\ h(M) \end{bmatrix} \tag{5.8.4}$$

The I/O equation of the filter, Eq. (4.3.8), is

$$\hat{x}_n = \mathbf{h}^T\mathbf{y}(n) = h(0)y_n + h(1)y_{n-1} + \cdots + h(M)y_{n-M} \tag{5.8.5}$$

Next, consider the Gram-Schmidt transformation of Eq. (5.7.4) from the data vector $\mathbf{y}(n)$ to the decorrelated vector $\mathbf{e}^-(n)$:

$$\mathbf{e}^-(n) = L\mathbf{y}(n) \qquad \text{or} \qquad \begin{bmatrix} e_0^-(n) \\ e_1^-(n) \\ \vdots \\ e_M^-(n) \end{bmatrix} = L \begin{bmatrix} y_n \\ y_{n-1} \\ \vdots \\ y_{n-M} \end{bmatrix} \tag{5.8.6}$$

Inserting Eq. (5.8.6) into Eq. (5.8.5), we find

$$\hat{x}_n = \mathbf{h}^T L^{-1} \mathbf{e}^-(n)$$

Defining the $(M + 1)$-vector

$$\mathbf{g} = L^{-T}\mathbf{h} \tag{5.8.7}$$

we obtain the alternative I/O equation for the Wiener filter:

$$\hat{x}_n = \mathbf{g}^T\mathbf{e}^-(n) = \sum_{p=0}^{M} g_p e_p^-(n)$$
$$= g_0 e_0^-(n) + g_1 e_1^-(n) + \cdots + g_M e_M^-(n) \tag{5.8.8}$$

This is easily recognized as the projection of $x_n$ onto the subspace spanned by $\{e_0^-(n), e_1^-(n), \ldots, e_M^-(n)\}$ which is the same as that spanned by the data vector $\{y_n, y_{n-1}, \ldots, y_{n-M}\}$. Indeed, it follows from Eqs. (5.8.7) and (5.8.2) that

$$\begin{aligned} \mathbf{g}^T = \mathbf{h}^T L^{-1} &= E[x_n \mathbf{y}(n)^T] E[\mathbf{y}(n)\mathbf{y}(n)^T]^{-1} L^{-1} \\ &= E[x_n \mathbf{e}^-(n)^T] L^{-T} (L^{-1} E[\mathbf{e}^-(n)\mathbf{e}^-(n)^T] L^{-T})^{-1} L^{-1} \\ &= E[x_n \mathbf{e}^-(n)^T] E[\mathbf{e}^-(n)\mathbf{e}^-(n)^T]^{-1} \\ &= [E[x_n e_0^-(n)]/E_0, E[x_n e_1^-(n)]/E_1, \ldots, E[x_n e_M^-(n)]/E_M] \end{aligned}$$

so that the estimate of $x_n$ can be expressed as

$$\hat{x}_n = E[x_n \mathbf{e}^-(n)^T] E[\mathbf{e}^-(n)\mathbf{e}^-(n)^T]^{-1} \mathbf{e}^-(n) = E[x_n \mathbf{y}(n)^T] E[\mathbf{y}(n)\mathbf{y}(n)^T]^{-1} \mathbf{y}(n)$$

The key to the lattice realization of the optimal filtering equation (5.8.8) is the observation that the *analysis lattice filter* of Fig. (5.3) for the process $y_n$ provides, in its successive lattice stages, the signals $e_p^-(n)$ which are required in the sum (5.8.8). Thus, if the weight vector $\mathbf{g}$ is known, an alternative realization of the optimal filter will be as shown in Fig. 5.5. By comparison, the *direct form* realization using Eq. (5.8.5) operates directly on the vector $\mathbf{y}(n)$, which, at each time instant $n$, is available at the tap registers of the filter. This is depicted in Fig. 5.6.

Both types of realizations can be formulated *adaptively,* without requiring prior knowledge of the filter coefficients or the correlation matrices $R_{yy}$ and $R_{xy}$. We will discuss adaptive implentations in Chapter 6. If $R_{yy}$ and $R_{xy}$ are known,

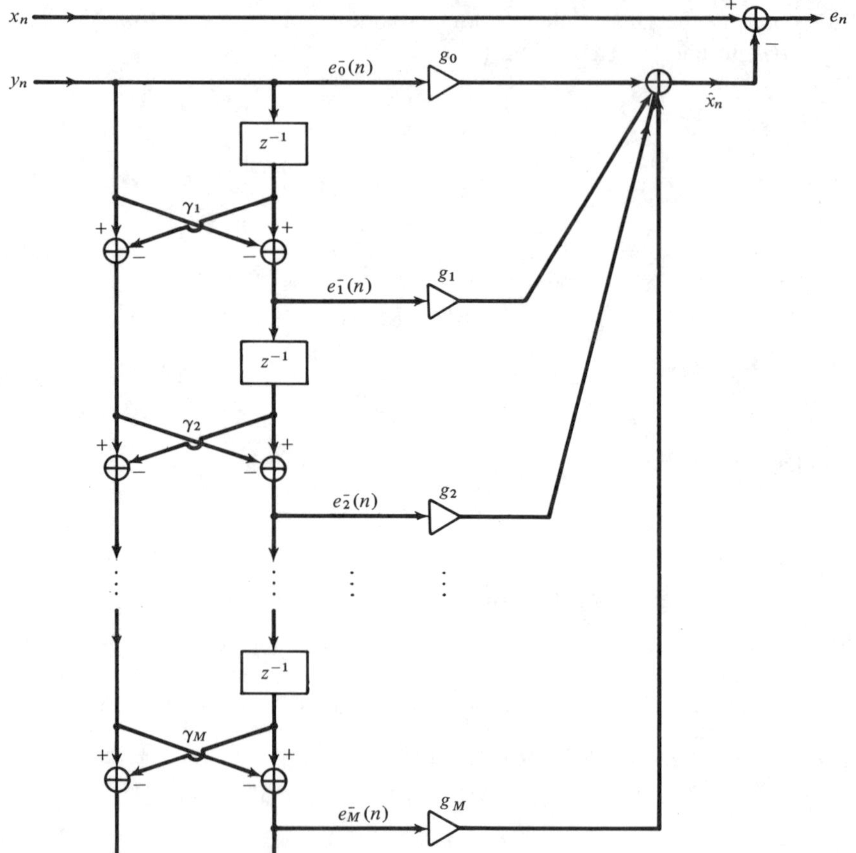



**Figure 5.5** Lattice Realization of the FIR Wiener Filter

or can be estimated, then the design procedure for both the lattice and the direct form realizations is:

1. Using Levinson's algorithm, implemented by the subroutine LEV, perform the LU Cholesky factorization of $R_{yy}$ to determine the matrices $L$ and $D$.

2. The vector of weights $\mathbf{g}$ can be computed in terms of the known quantities $L$, $D$, $\mathbf{r}_{xy}$ as follows:

$$\mathbf{g} = L^{-T}\mathbf{h} = L^{-T}R_{yy}^{-1}\mathbf{r}_{xy} = L^{-T}(L^T D^{-1}L)\mathbf{r}_{xy} = D^{-1}L\mathbf{r}_{xy}$$

3. The vector $\mathbf{h}$ can be recovered from $\mathbf{g}$ by $\mathbf{h} = L^T\mathbf{g}$.

The subroutine FIRW (see appendix) is an implementation of this design procedure. The inputs to the subroutine are the order $M$, and the correlation lags $\{R_{yy}(0), R_{yy}(1), \ldots, R_{yy}(M)\}$ and $\{R_{xy}(0), R_{xy}(1), \ldots, R_{xy}(M)\}$. The outputs are the quantities $L$, $D$, $\mathbf{g}$, and $\mathbf{h}$.

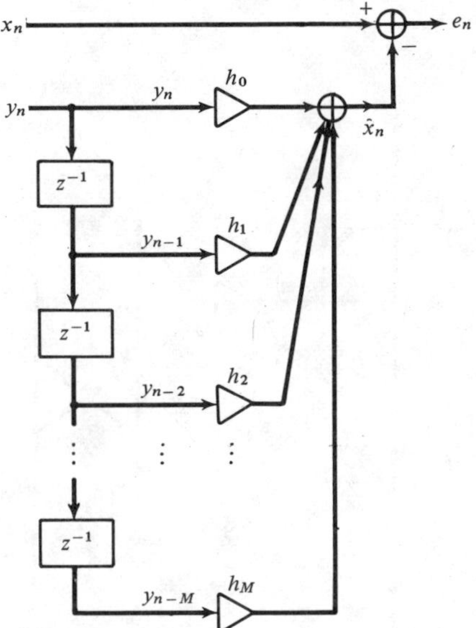

**Figure 5.6**  Direct Form Realization of the FIR Wiener Filter

The estimate (5.8.8) may also be written recursively *in the order* of the filter: If we denote

$$\hat{x}_p(n) = \sum_{i=0}^{p} g_i e_i^-(n) \tag{5.8.9}$$

we obtain the recursion

$$\hat{x}_p(n) = \hat{x}_{p-1}(n) + g_p e_p^-(n), \qquad p = 0,1, \ldots, M \tag{5.8.10}$$

initialized as $\hat{x}_{-1}(n) = 0$. The quantity $\hat{x}_p(n)$ is the *projection* of $x_n$ on the subspace spanned by $\{e_0^-(n), e_1^-(n), \ldots, e_p^-(n)\}$ which by virtue of the lower-triangular nature of the matrix $L$ is the *same* space as that spanned by $\{y_n, y_{n-1}, \ldots, y_{n-p}\}$. Thus, $\hat{x}_p(n)$ represents the optimal estimate of $x_n$ based on a *p*th order filter. Similarly, $\hat{x}_{p-1}(n)$ represents the optimal estimate of $x_n$ based on the $(p-1)$th order filter; that is, based on the past $p-1$ samples $\{y_n, y_{n-1}, \ldots, y_{n-p+1}\}$. These two subspaces differ by $y_{n-p}$. The term $e_p^-(n)$ is by construction the best postdiction error of estimating $y_{n-p}$ from the samples $\{y_n, y_{n-1}, \ldots, y_{n-p+1}\}$; that is, $e_p^-(n)$ is the *orthogonal complement* of $y_{n-p}$ projected on that subspace. Therefore, the term $g_p e_p^-(n)$ in Eq. (5.8.10) represents the improvement in the estimate of $x_n$ that results by taking into account the *additional* past value $y_{n-p}$; it represents that part of $x_n$ that cannot be estimated in terms of the subspace $\{y_n, y_{n-1}, \ldots, y_{n-p+1}\}$. The estimate $\hat{x}_p(n)$ of $x_n$ is *better* than $\hat{x}_{p-1}(n)$ in the sense that it produces a smaller mean-squared estimation error. To see this, define the estimation errors in the two cases

$$e_p(n) = x_n - \hat{x}_p(n) \qquad e_{p-1}(n) = x_n - \hat{x}_{p-1}(n)$$

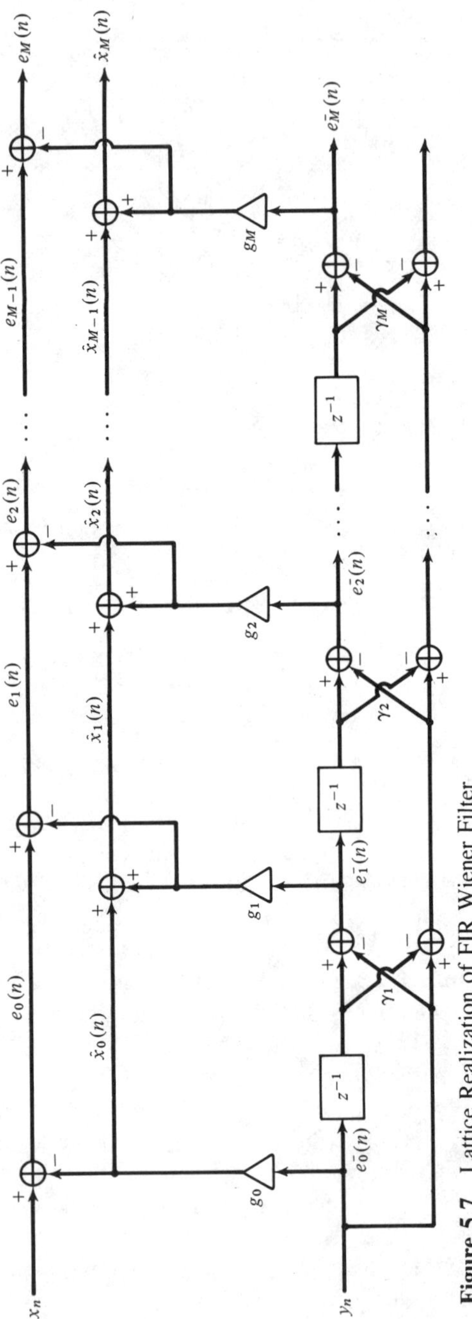


**Figure 5.7**  Lattice Realization of FIR Wiener Filter

Using the recursion (5.8.10) we find

$$e_p(n) = e_{p-1}(n) - g_p e_p^-(n) \qquad (5.8.11)$$

Using $g_p = E[x_n e_p^-(n)]/E_p$, we find for $\mathcal{E}_p = E[e_p(n)^2]$

$$\mathcal{E}_p = E[x_n^2] - \sum_{i=0}^{p} g_i E[e_i^-(n)x_n] = \mathcal{E}_{p-1} - g_p E[e_p^-(n)x_n]$$

$$= \mathcal{E}_{p-1} - (E[x_n e_p^-(n)])^2/E_p = \mathcal{E}_{p-1} - g_p^2 E_p$$

Thus, $\mathcal{E}_p$ is smaller than $\mathcal{E}_{p-1}$. This result shows explicitly how the estimate is constantly improved as the length of the filter is increased. The nice feature of the lattice realization is that the filter length can be increased simply by adding more lattice sections *without* having to recompute the weights $g_p$ of the previous sections. A realization *equivalent* to Fig. 5.5, but which shows explicitly the *recursive* construction (5.8.10) of the estimate of $x_n$ and of the estimation error (5.8.11), is shown in Fig. 5.7.

Next, we present a Wiener filter design example for a *noise canceling* application. The primary and secondary signals $x(n)$ and $y(n)$ are of the form

$$x(n) = s(n) + v_1(n) \qquad y(n) = v_2(n)$$

where $s(n)$ is a desired signal corrupted by noise $v_1(n)$. The signal $v_2(n)$ is correlated with $v_1(n)$ but not with $s(n)$, and provides a *reference* noise signal. The noise canceler is to be implemented as a Wiener filter of order $M$, realized either in the direct or the lattice form. It is shown below:

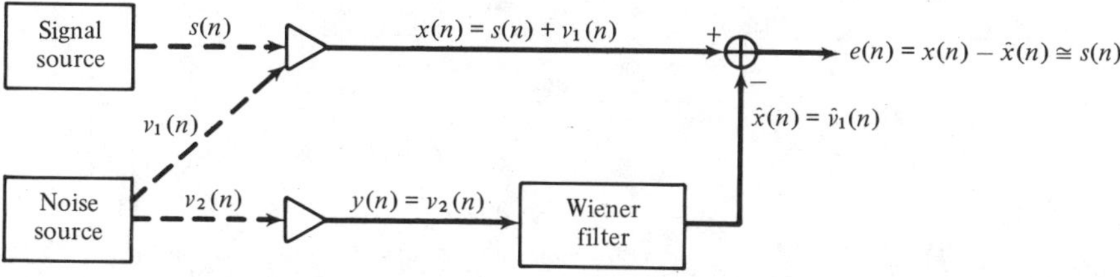


Its basic operation is that of a correlation canceler; that is, the optimally designed filter $H(z)$ will transform the reference noise $v_2(n)$ into the best replica of $v_1(n)$, and then proceed to cancel it from the output, leaving a clean signal $s(n)$. For the purpose of the simulation, we took $s(n)$ to be a simple sinusoid

$$s(n) = \sin(\omega_0 n) \qquad \text{where } \omega_0 = 0.075\pi \ [\text{rads/sample}]$$

and $v_1(n)$ and $v_2(n)$ were generated by the difference equations

$$v_1(n) = -0.5v_1(n-1) + v(n)$$

$$v_2(n) = 0.8v_2(n-1) + v(n)$$

driven by a common, zero-mean, unit-variance, uncorrelated sequence $v(n)$. The difference equations establish a correlation between the two noise components

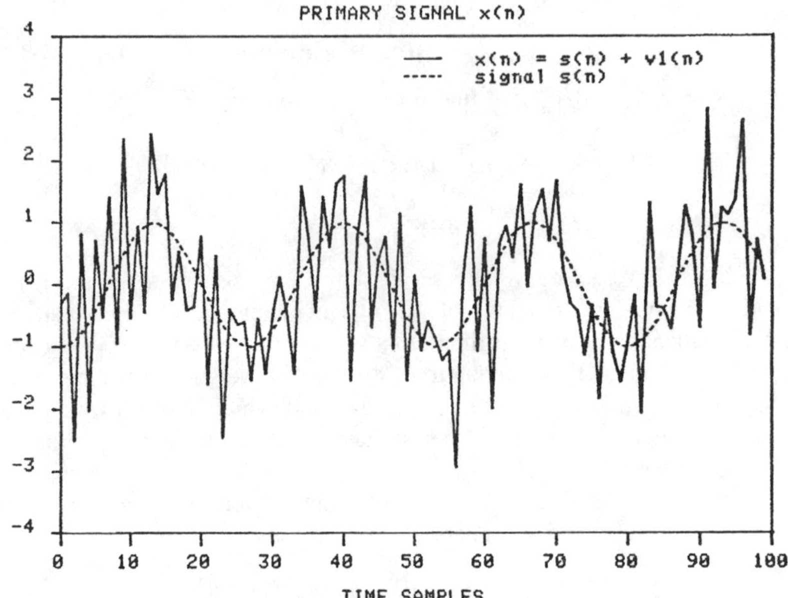


**Figure 5.8**   Noise Corrupted Sinusoid

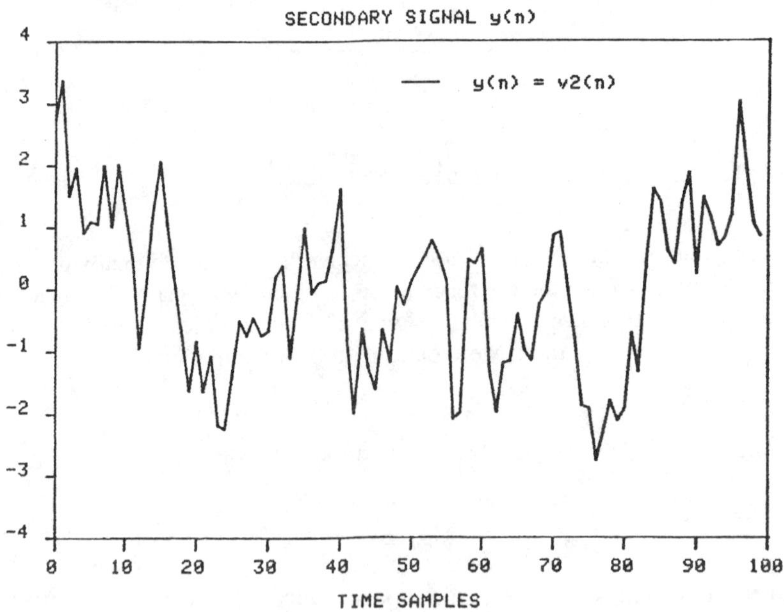


**Figure 5.9**   Reference Noise

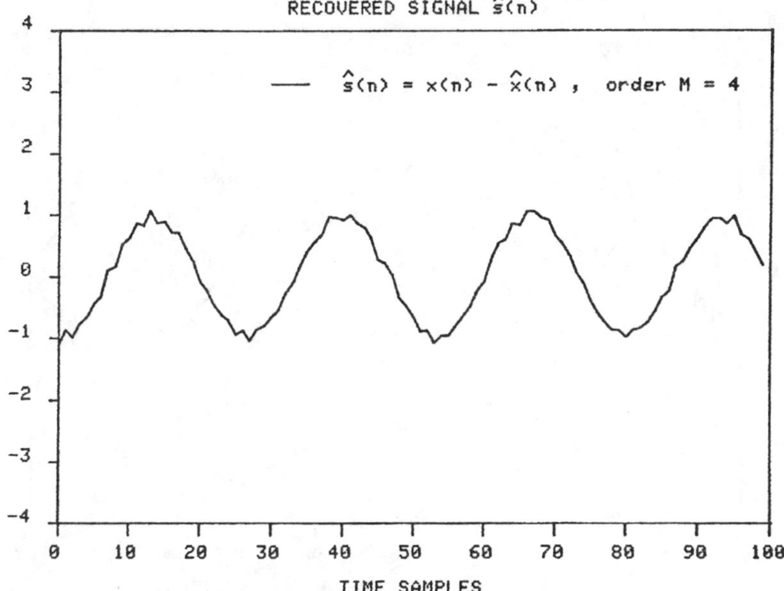


**Figure 5.10**   Output of Noise Canceler ($M = 4$)

$v_1$ and $v_2$, which is exploited by the canceler to effect the noise cancellation. Figures 5.8 and 5.9 show 100 samples of the signals $x(n)$, $s(n)$, and $y(n)$, generated by a particular realization of $v(n)$. For $M = 4$ and $M = 6$, the sample autocorrelation and cross-correlation lags, $R_{yy}(k), R_{xy}(k); k = 0, 1, \ldots, M$, were computed and sent through the routine FIRW to get the filter weights **g** and **h**. The reference signal $y_n$ was filtered through $H$ to get the estimate $\hat{x}_n$—which is really an estimate of $v_1(n)$—and the estimation error $e(n) = x(n) - \hat{x}(n)$, which is really an estimate of $s(n)$. This estimate of $s(n)$ is shown in Figs. 5.10 and 5.11, for the cases $M = 4$ and $M = 6$, respectively. The improvement afforded by using a higher order filter is evident. For the particular realization of $x(n)$ and $y(n)$ that we used, the *sample* correlations $R_{yy}(k), R_{xy}(k), k = 0, 1, 2, \ldots, M$ were:

$$R_{yy} = [2.4540, 1.9265, 1.4636, 1.1031, 0.8425, 0.6557, 0.5120]$$

$$R_{xy} = [0.6857, -0.3097, 0.1496, -0.0714, 0.0626, 0.0105, 0.0315]$$

and the resulting vector of lattice weights $g_p, p = 0, 1, \ldots, M$, reflection coefficients $\gamma_p; p = 1, 2, \ldots, M$, and direct form weights $h_m; m = 0, 1, \ldots, M$ were, for $M = 6$,

$$\mathbf{g} = [0.2794, -0.9005, 0.4693, -0.2186, 0.1220, -0.0521, 0.0290]$$

$$\boldsymbol{\gamma} = [0.7850, -0.0518, -0.0059, 0.0128, 0.0114, 0.0004]$$

$$\mathbf{h} = [1.0084, -1.2964, 0.6545, -0.3212, 0.1663, -0.0760, 0.0290]$$

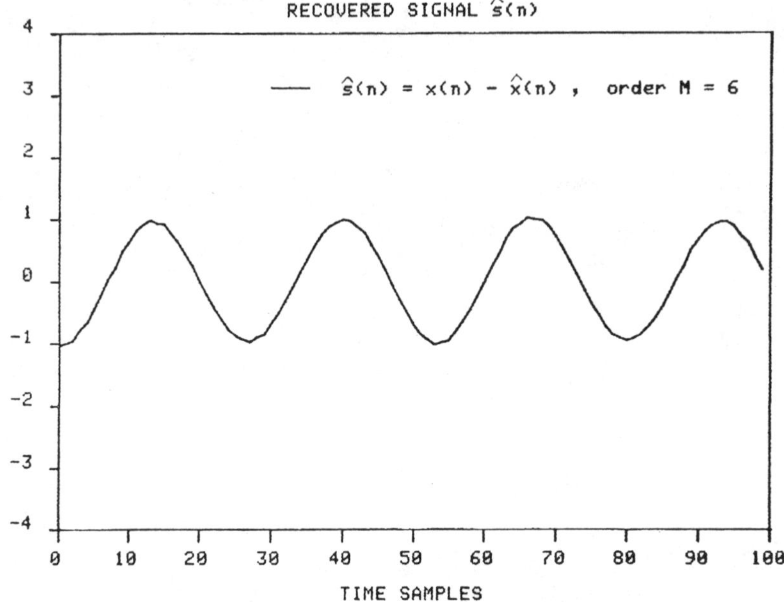


**Figure 5.11**  Output of Noise Canceler ($M = 6$)

To get the **g** and **γ** of the case $M = 4$, simply ignore the last two entries in the above. The corresponding **h** is in this case:

$$\mathbf{h} = [1.0078, \; -1.2963, \; 0.6554, \; -0.3193, \; 0.1220]$$

## 5.9  *Autocorrelation, Covariance, and Burg's Methods*

As mentioned in Section 5.3, the finite order linear prediction problem may be thought of as an *approximation* to the infinite order prediction problem. For large enough order $p$ of the predictor, the prediction-error filter $A_p(z)$ may be considered to be an adequate approximation to the whitening filter $A(z)$ of the process $y_n$. In this case, the prediction-error sequence $e_p^+(n)$ is approximately white, and the inverse synthesis filter $1/A_p(z)$ is an approximation to the signal model $B(z)$ of $y_n$. Thus, we have obtained an approximate solution to the signal modeling problem depicted below:

$$e_p^+(n) \longrightarrow \boxed{1/A_p(z)} \longrightarrow y_n$$

The variance of $e_p^+(n)$ is $E_p$. Depending on the realization one uses, the model parameters are either the set $\{a_{p1}, a_{p2}, \ldots, a_{pp}; E_p\}$ or $\{\gamma_1, \gamma_2, \ldots, \gamma_p; E_p\}$. Because these model parameters can be determined by solving a simple *linear* system of equations—that is, the normal equations (5.3.7)—this approach to the modeling problem has become widespread.

In this section, we present three widely used methods of extracting the model parameters from a given block of measured signal values $y_n$ [3,6,10,11,32–37]. These methods are:

1. The autocorrelation, or Yule-Walker, method
2. The covariance method
3. Burg's method

We have already discussed the Yule-Walker method, which consists simply of replacing the theoretical autocorrelations $R_{yy}(k)$ with the corresponding sample autocorrelations $\hat{R}_{yy}(k)$ computed from the given frame of data. This method, like the other two, can be justified on the basis of an appropriate *least-squares* minimization criterion obtained by replacing the ensemble averages $E[e_p^+(n)^2]$ by appropriate time averages.

The theoretical minimization criteria for the optimal forward and backward predictors are

$$E[e_p^+(n)^2] = \min \qquad E[e_p^-(n)^2] = \min \qquad (5.9.1)$$

where $e_p^+(n)$ and $e_p^-(n)$ are the result of filtering $y_n$ through the prediction-error filter $\mathbf{a} = [1, a_{p1}, \ldots, a_{pp}]^T$ and its reverse $\mathbf{a}^R = [a_{pp}, a_{p,p-1}, \ldots, a_{p1}, 1]^T$, respectively; namely,

$$e_p^+(n) = y_n + a_{p1}y_{n-1} + a_{p2}y_{n-2} + \cdots + a_{pp}y_{n-p} \qquad (5.9.2)$$
$$e_p^-(n) = y_{n-p} + a_{p1}y_{n-p+1} + a_{p2}y_{n-p+2} + \cdots + a_{pp}y_n$$

Note that in both cases the mean-squared value of $e_p^\pm(n)$ can be expressed in terms of the $(p + 1) \times (p + 1)$ autocorrelation matrix

$$R(i,j) = R(i - j) = E[y_{n+i-j}y_n] = E[y_{n-j}y_{n-i}], \qquad 0 \leq i,j \leq p$$

as follows

$$E[e_p^+(n)^2] = E[e_p^-(n)^2] = \mathbf{a}^T R \mathbf{a} \qquad (5.9.3)$$

Consider a frame of length $N$ of measured values of $y_n$

$$\boxed{y_0, y_1, \cdots, y_{N-1}}$$

1. The *Yule-Walker*, or *autocorrelation*, method replaces the ensemble average (5.9.1) by the least-squares time-average criterion

$$\mathcal{E} = \sum_{n=0}^{N+p-1} e_p^+(n)^2 = \min \qquad (5.9.4)$$

where $e_p^+(n)$ is obtained by convolving the length-$(p + 1)$ prediction-error sequence $\mathbf{a} = [1, a_{p1}, a_{p2}, \ldots, a_{pp}]^T$ with the length-$N$ data sequence $y_n$. The length of the sequence $e_p^+(n)$ is, therefore, $N + (p + 1) - 1 = N + p$, which justifies the upper-limit in the summation of Eq. (5.9.4). This convolution operation is equivalent to assuming that the block of data $y_n$ has been extended both to the left and to the right by padding it with zeros and running the filter over this *extended* sequence. The last $p$ output samples $e_p^+(n); N \leq n \leq N + p - 1$ correspond to running the filter off the ends of the data sequence to the right. These terms arise because the prediction-error filter has memory of $p$ samples. This is depicted below:

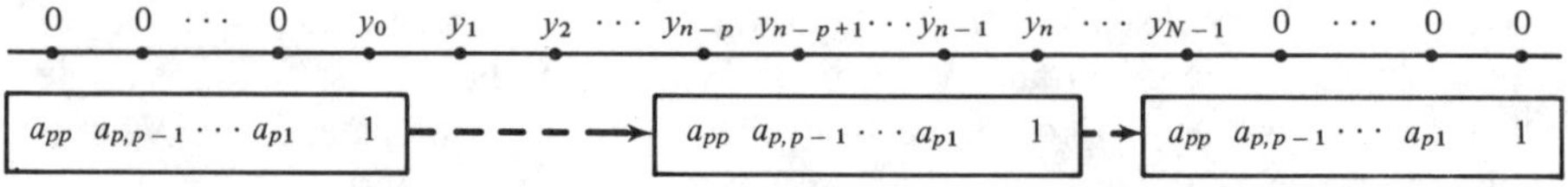

$$
\begin{array}{ccccccccccccccc}
0 & 0 & \cdots & 0 & y_0 & y_1 & y_2 & \cdots & y_{n-p} & y_{n-p+1} & \cdots y_{n-1} & y_n & \cdots & y_{N-1} & 0 & \cdots & 0 & 0
\end{array}
$$

$$
\boxed{a_{pp}\ a_{p,p-1}\cdots a_{p1} \quad 1} \text{ - - - - } \rightarrow \boxed{a_{pp}\ a_{p,p-1}\cdots a_{p1} \quad 1} \mapsto \boxed{a_{pp}\ a_{p,p-1}\cdots a_{p1} \quad 1}
$$

Inserting Eq. (5.9.2) into Eq. (5.9.4), it is easily shown that $\mathcal{E}$ can be expressed in the equivalent form

$$
\mathcal{E} = \sum_{n=0}^{N+p-1} e_p^+(n)^2 = \sum_{i,j=0}^{p} a_i \hat{R}(i-j)\, a_j = \mathbf{a}^T \hat{R} \mathbf{a} \tag{5.9.5}
$$

where $\hat{R}(k)$ denotes the sample autocorrelation of the length-$N$ data sequence $y_n$; namely

$$
\hat{R}(k) = \hat{R}(-k) = \sum_{n=0}^{N-1-k} y_{n+k} y_n \qquad 0 \leq k \leq N - 1
$$

where the usual normalization factor $1/N$ has been ignored. This equation is identical to Eq. (5.9.3) with $R$ replaced by $\hat{R}$. Thus, the minimization of the time-average index (5.9.5) with respect to the prediction coefficients will lead exactly to the *same set* of normal equations (5.3.7) with $R$ replaced by $\hat{R}$. The *positive definiteness* of the sample autocorrelation matrix also guarantees that the resulting prediction-error filter be *minimal phase*, and thus also that all reflection coefficients have magnitude *less* than one.

2. The *covariance method* replaces Eq. (5.9.1) by the time average

$$
\mathcal{E} = \sum_{n=p}^{N-1} e_p^+(n)^2 = \min \tag{5.9.6}
$$

where the summation in $n$ is such that the filter *does not* run off the ends of the data block, as shown below:

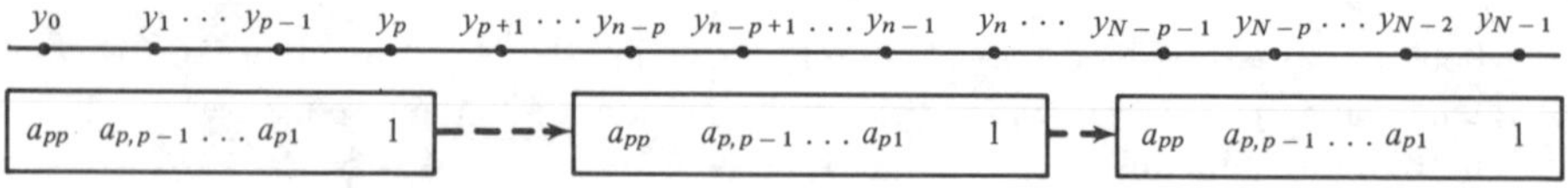

$$
\begin{array}{ccccccccccccccc}
y_0 & y_1 \cdots y_{p-1} & y_p & y_{p+1} \cdots y_{n-p} & y_{n-p+1} \ldots y_{n-1} & y_n & \cdots & y_{N-p-1} & y_{N-p} \cdots y_{N-2} & y_{N-1}
\end{array}
$$

$$
\boxed{a_{pp}\ a_{p,p-1} \ldots a_{p1} \quad 1} \text{ - - } \rightarrow \boxed{a_{pp} \quad a_{p,p-1} \ldots a_{p1} \quad 1} \mapsto \boxed{a_{pp} \quad a_{p,p-1} \ldots a_{p1} \quad 1}
$$

To explain the method and to see its potential problems with stability, consider a simple example of a length-three sequence and a first order predictor:

$$\mathscr{E} = \sum_{n=1}^{2} e_1^+(n)^2 = e_1^+(1)^2 + e_1^+(2)^2 = (y_1 + a_{11}y_0)^2 + (y_2 + a_{11}y_1)^2$$

Differentiating with respect to $a_{11}$ and setting the derivative to zero gives

$$(y_1 + a_{11}y_0)y_0 + (y_2 + a_{11}y_1)y_1 = 0$$

$$a_{11} = -\frac{y_1 y_0 + y_2 y_1}{y_0^2 + y_1^2}$$

Note that the denominator does not depend on the variable $y_2$ and therefore it is possible, if $y_2$ is large enough, for $a_{11}$ to have magnitude greater than one, making the prediction-error filter nonminimal phase. Although this potential stability problem exists, this method has been used with good success in speech processing, with few, if any, such stability problems. The autocorrelation method is sometimes preferred in speech processing because the resulting normal equations have a Toeplitz structure and their solution can be obtained efficiently using Levinson's algorithm. However, similar ways of solving the covariance equations have been developed recently that are just as efficient [37].

  3. Although the autocorrelation method is implemented efficiently, and the resulting prediction-error filter is guaranteed to be minimal phase, it suffers from the effect of prewindowing the data sequence $y_n$ by padding it with zeros to the left and to the right. This reduces the accuracy of the method somewhat, especially when the data record $N$ is short. In this case, the effect of prewindowing is felt more strongly. The proper way to extend the sequence $y_n$, if it must be extended, is a way *compatible* with the signal model generating this sequence. Since we are trying to determine this model, the fairest way of proceeding is to try to use the available data block in a way which is maximally noncommittal to what the sequence is like beyond the ends of the block. *Burg's method,* also known as the *maximum entropy method* (MEM), arose from the desire on the one hand not to run off the ends of the data, and, on the other, to always result in a minimal-phase filter. Burg's minimization criterion is to minimize the sum-squared of both the *forward* and the *backward* prediction errors:

$$\mathscr{E} = \sum_{n=p}^{N-1} [e_p^+(n)^2 + e_p^-(n)^2] = \min \tag{5.9.7}$$

where the summation range is the same as in the covariance method, but with both the forward and the reversed filters running over the data, as shown:

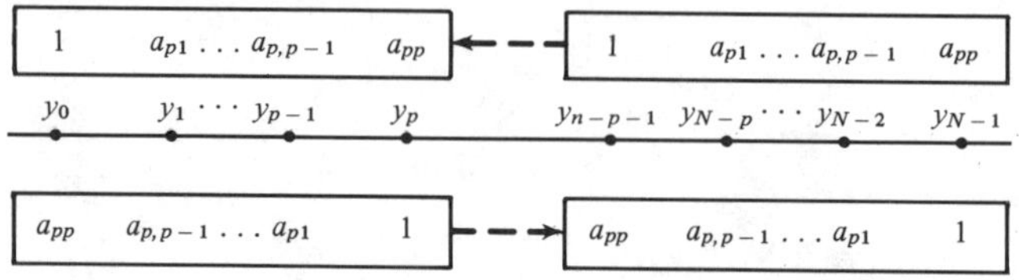

If the minimization is performed with respect to the coefficients $a_{pi}$, it is still possible for the resulting prediction-error filter not to be minimal phase. Instead, Burg suggests an *iterative* procedure: Suppose the prediction-error filter $[1,a_{p-1,1},a_{p-1,2}, \ldots ,a_{p-1,p-1}]$ of order $(p-1)$ has already been determined. Then, to determine the prediction-error filter of order $p$, one needs to know the reflection coefficient $\gamma_p$ and to apply the Levinson recursion:

$$
\begin{bmatrix} 1 \\ a_{p,1} \\ a_{p,2} \\ \vdots \\ a_{p,p-1} \\ a_{p,p} \end{bmatrix} = \begin{bmatrix} 1 \\ a_{p-1,1} \\ a_{p-1,2} \\ \vdots \\ a_{p-1,p-1} \\ 0 \end{bmatrix} - \gamma_p \begin{bmatrix} 0 \\ a_{p-1,p-1} \\ a_{p-1,p-2} \\ \vdots \\ a_{p-1,1} \\ 1 \end{bmatrix} \tag{5.9.8}
$$

To guarantee the minimal-phase property, the reflection coefficient $\gamma_p$ must have magnitude less than one. The best choice for $\gamma_p$ is that which minimizes the performance index (5.9.7). Differentiating with respect to $\gamma_p$ and setting the derivative to zero we find

$$
\frac{\partial \mathcal{E}}{\partial \gamma_p} = 2 \sum_{n=p}^{N-1} \left[ e_p^+(n) \frac{\partial e_p^+(n)}{\partial \gamma_p} + e_p^-(n) \frac{\partial e_p^-(n)}{\partial \gamma_p} \right] = 0
$$

Using the lattice relationships

$$
\begin{aligned}
e_p^+(n) &= e_{p-1}^+(n) - \gamma_p e_{p-1}^-(n-1) \\
e_p^-(n) &= e_{p-1}^-(n-1) - \gamma_p e_{p-1}^+(n)
\end{aligned} \tag{5.9.9}
$$

both valid for $p \leq n \leq N - 1$ if the filter is not to run off the data, we find the conditions

$$
\sum_{n=p}^{N-1} [e_p^+(n)e_{p-1}^-(n-1) + e_p^-(n)e_{p-1}^+(n)] = 0
$$

$$
\sum_{n=p}^{N-1} [(e_{p-1}^+(n) - \gamma_p e_{p-1}^-(n-1))e_{p-1}^-(n-1)
$$

$$
+ (e_{p-1}^-(n-1) - \gamma_p e_{p-1}^+(n))e_{p-1}^+(n)] = 0
$$

which can be solved for $\gamma_p$ to give

$$\gamma_p = \frac{2 \sum\limits_{n=p}^{N-1} e_{p-1}^+(n) e_{p-1}^-(n-1)}{\sum\limits_{n=p}^{N-1} [e_{p-1}^+(n)^2 + e_{p-1}^-(n-1)^2]} \tag{5.9.10}$$

This expression for $\gamma_p$ is of the form

$$\gamma_p = \frac{2\mathbf{a} \cdot \mathbf{b}}{|\mathbf{a}|^2 + |\mathbf{b}|^2}$$

where $\mathbf{a}$ and $\mathbf{b}$ are vectors. Using the Schwarz inequality, it is easily verified that $\gamma_p$ has magnitude less than one.

Equations (5.9.8) through (5.9.10) define *Burg's method*. The computational steps are summarized below:

0. Initialize as follows:

$$e_0^+(n) = e_0^-(n) = y_n; \text{ for } 0 \le n \le N - 1 \text{ and } E_0 = \frac{1}{N} \sum_{n=0}^{N-1} y_n^2$$

1. At stage $(p - 1)$, we have available the quantities:

$$A_{p-1}(z), \; E_{p-1} \text{ and } e_{p-1}^\pm(n); \text{ for } p - 1 \le n \le N - 1.$$

2. Using Eq. (5.9.10), compute the reflection coefficient $\gamma_p$.
3. Using (5.9.8), compute $A_p(z)$.
4. Using (5.9.9), compute $e_p^\pm(n)$; for $p \le n \le N - 1$.
5. Update the mean-squared error by $E_p = (1 - \gamma_p^2)E_{p-1}$.
6. Go to stage $p$.

The subroutine BURG (see appendix) is an implementation of this method. The inputs to the subroutine are the vector of data samples $\{y_0, y_1, \ldots, y_{N-1}\}$ and the desired final order $M$ of the predictor. The outputs are all the prediction-error filters of order up to $M$, arranged as usual into the lower triangular matrix $L$, and the corresponding mean-squared prediction errors $\{E_0, E_1, \ldots, E_M\}$.

*Example 5.9.1:*  The length-six block of data

$$y_n = [4.684, \; 7.247, \; 8.423, \; 8.650, \; 8.640, \; 8.392]$$

for $n = 0,1,2,3,4,5$, is known to have been generated by sending zero-mean, unit-variance, white noise $\epsilon_n$ through the difference equation

$$y_n - 1.70 y_{n-1} + 0.72 y_{n-2} = \epsilon_n$$

Thus, the theoretical prediction-error filter and mean-squared error are $A_2(z) = 1 - 1.70z^{-1} + 0.72z^{-2}$, and $E_2 = 1.00$. Using Burg's method, extract the model parameters for a second order model. The reader is urged to go

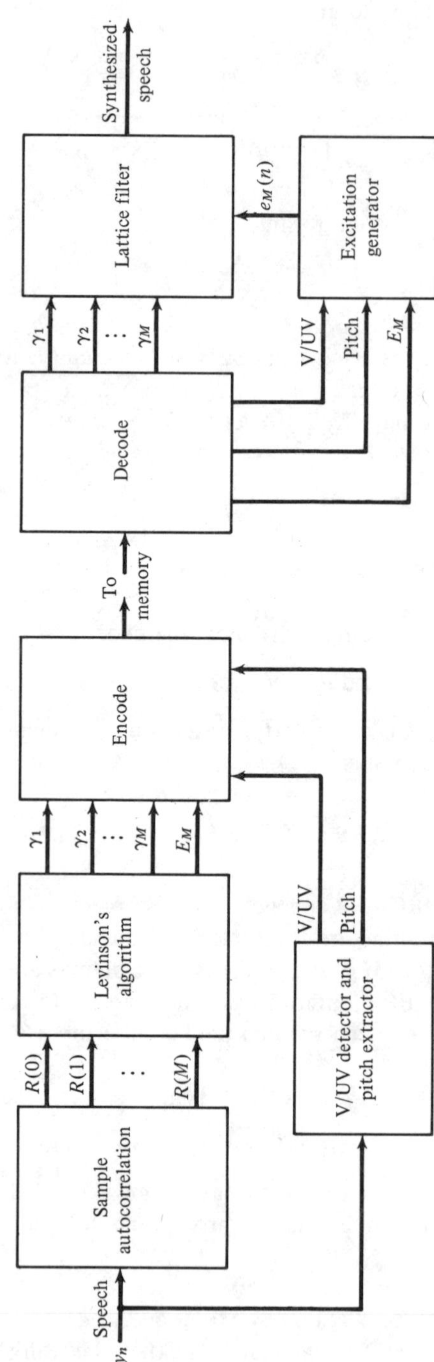



**Figure 5.12** LPC Analysis and Synthesis of Speech

through the algorithm by hand. Sending the above six $y_n$ samples through the routine BURG, we find the first and second order prediction-error filters and the corresponding errors:

$$A_1(z) = 1 - 0.989z^{-1} \quad \text{and} \quad E_1 = 1.529$$
$$A_2(z) = 1 - 1.757z^{-1} + 0.779z^{-2} \quad \text{and} \quad E_2 = 0.60$$

The resulting set of LPC model parameters, from any of the above analysis methods, can be used in a number of ways as suggested in Section 1.10. One of the most successful applications has been to analysis and synthesis of speech [6,38–41]. Each frame of speech, of duration of the order of 20 msec, is subjected to the Yule-Walker analysis method to extract the corresponding set of model parameters. The order of the predictor is typically $M = 10$–$15$. Pitch and voiced/unvoiced information are also extracted. The resulting set of parameters represents that speech segment. To synthesize the segment, the set of model parameters are recalled from memory and used in the synthesizer to drive the synthesis filter. The latter is commonly realized as a *lattice filter*. Lattice realizations are preferred because they are much better well-behaved under quantization of their coefficients (i.e., the reflection coefficients) than the direct form realizations [6,42,43]. A typical speech analysis and synthesis system is shown in Fig. 5.12.

Linear predictive modeling techniques have also been applied to EEG signal processing in order to *model* EEG spectra, to *classify* EEGs automatically, to detect EEG *transients* that might have diagnostic significance, and to *predict* the onset of epileptic seizures [44–51].

LPC methods have been applied successfully to *signal classification* problems such as speech recognition [41,52–57] or the automatic classification of EEGs [48]. *Distance measures* between two sets of model parameters extracted from two signal frames can be used as measures of similarity between the frames. Itakura's *LPC distance measure* can be introduced as follows: Consider two *autoregressive* signal sequences, the test sequence $y_T(n)$ to be compared against the reference sequence $y_R(n)$. Let $A_T(z)$ and $A_R(z)$ be the two whitening filters, both of order $M$. The two signal models are

$$\epsilon_T(n) \longrightarrow \boxed{1/A_T(z)} \longrightarrow y_T(n) \qquad \epsilon_R(n) \longrightarrow \boxed{1/A_R(z)} \longrightarrow y_R(n)$$

Now, suppose the sequence to be tested, $y_T(n)$, is filtered through the whitening filter of the reference signal

$$y_T(n) \longrightarrow \boxed{A_R(z)} \longrightarrow e_T(n)$$

resulting in the output signal $e_T(n)$. The mean output power is easily expressed as

$$E[e_T(n)^2] = \mathbf{a}_R^\dagger R_T \mathbf{a}_R = \int_{-\pi}^{\pi} S_{e_T e_T}(\omega) \frac{d\omega}{2\pi}$$

$$= \int_{-\pi}^{\pi} |A_R(\omega)|^2 S_{y_T y_T}(\omega) \frac{d\omega}{2\pi}$$

$$= \int_{-\pi}^{\pi} |A_R(\omega)|^2 \frac{\sigma_{\epsilon_T}^2}{|A_T(\omega)|^2} \frac{d\omega}{2\pi}$$

where $R_T$ is the correlation matrix of $y_T(n)$. On the other hand, if $y_T(n)$ is filtered through its own whitening filter it will produce $\epsilon_T$. Thus, in this case

$$\sigma_{\epsilon_T}^2 = E[\epsilon_T(n)^2] = \mathbf{a}_T^\dagger R_T \mathbf{a}_T$$

It follows that

$$\frac{E[e_T(n)^2]}{E[\epsilon_T(n)^2]} = \frac{\mathbf{a}_R^\dagger R_T \mathbf{a}_R}{\mathbf{a}_T^\dagger R_T \mathbf{a}_T} = \int_{-\pi}^{\pi} \left| \frac{A_R(\omega)}{A_T(\omega)} \right|^2 \frac{d\omega}{2\pi} \tag{5.9.11}$$

The log of this quantity is Itakura's LPC distance measure

$$d(\mathbf{a}_T, \mathbf{a}_R) = \log\left(\frac{E[e_T^2]}{E[\epsilon_T^2]}\right) = \log\left(\frac{\mathbf{a}_R^\dagger R_T \mathbf{a}_R}{\mathbf{a}_T^\dagger R_T \mathbf{a}_T}\right) = \log\left[\int_{-\pi}^{\pi} \left| \frac{A_R(\omega)}{A_T(\omega)} \right|^2 \frac{d\omega}{2\pi}\right]$$

In practice, the quantities $\mathbf{a}_T$, $R_T$, and $\mathbf{a}_R$ are extracted from a frame of $y_T(n)$ and a frame of $y_R(n)$. If the model parameters are equal, the distance is zero. This distance measure effectively provides a comparison between the two spectra of the processes $y_T$ and $y_R$, but instead of comparing them directly, a *prewhitening* of $y_T(n)$ is attempted by sending it through the whitening filter of the other signal. If the two spectra are close, the filtered signal $e_T(n)$ will be close to white— that is, with a spectrum close to being flat; a measure of this flatness is precisely the above integrated spectrum of Eq. (5.9.11).

## 5.10 *Spectrum Estimation by Autoregressive Modeling*

When a block of signal samples is available, it may be too short to provide enough frequency resolution in the periodogram spectrum. Often, it may not even be correct to extend the length by collecting more samples, since this might come into conflict with the stationarity of the segment. In cases such as these, parametric representation of the spectra by means of autoregressive models can provide much better frequency resolution than the classical periodogram method [10,11,32,33,35,36,58–62]. This approach was discussed briefly in Section 1.10.

The spectrum estimation procedure is as follows: First, the given data segment $\{y_0, y_1, \ldots, y_{N-1}\}$ is subjected to one of the analysis methods discussed in Section 5.9 to extract estimates of the LPC model parameters $\{a_1, a_2, \ldots, a_M; E_M\}$. The choice of order $M$ is an important consideration. There are a number of criteria for model order selection [10], but there is no single one that works well under all circumstances. In fact, selecting the right order $M$ is more of an art than science. As an example, we mention Akaike's *final prediction error* (FPE) criterion which selects the $M$ that minimizes the quantity

$$E_M \cdot \frac{N + M + 1}{N - M - 1} = \min$$

where $E_M$ is the estimate of the mean-squared prediction error for the $M$th order predictor, and $N$ is the length of the sequence $y_n$. As $M$ increases, the factor $E_M$ decreases and the second factor increases, thus, there is a minimum value. The spectrum estimate is then given by

$$S_{AR}(\omega) = \frac{E_M}{|A_M(\omega)|^2} = \frac{E_M}{|1 + a_1 e^{-j\omega} + a_2 e^{-2j\omega} + \ldots + a_M e^{-jM\omega}|^2} \qquad (5.10.1)$$

Note that this would be the exact spectrum if $y_n$ were autoregressive with the above set of model parameters. Generally, spectra that have a few dominant spectral peaks can be modeled quite successfully by such all-pole autoregressive models. One can also fit the given block of data to more general ARMA models. The decision to model a spectrum by ARMA, AR, or MA models should ultimately depend on some prior information regarding the physics of the process $y_n$. The reader is referred to the exhaustive review article of Kay and Marple [10], to the special issue [61], and to [11,33,62], for the discussion of essentially all currently available spectrum estimation techniques, and to Robinson's interesting historical account [63].

Next, we compare by means of a simulation example the classical periodogram method, the Yule-Walker method, and Burg's method of computing spectrum estimates. Generally, the rule of thumb to follow is that Burg's method should work better than the other methods on short records of data, and that all three methods tend to improve as the data record becomes longer. For our simulation example, we chose a fourth order autoregressive model characterized by two very sharp peaks in its spectrum. The signal generator for the sequence $y_n$ was

$$y_n + a_1 y_{n-1} + a_2 y_{n-2} + a_3 y_{n-3} + a_4 y_{n-4} = \epsilon_n$$

where $\epsilon_n$ was zero-mean, unit-variance, white noise. The prediction-error filter $A(z)$ was defined in terms of its four zeros

$$A(z) = 1 + a_1 z^{-1} + a_2 z^{-2} + a_3 z^{-3} + a_4 z^{-4}$$

$$= (1 - z_1 z^{-1})(1 - z_1^* z^{-1})(1 - z_2 z^{-1})(1 - z_2^* z^{-1})$$

where the zeros were chosen as

$$z_1 = 0.99 \exp(0.2\pi j) \qquad z_2 = 0.99 \exp(0.4\pi j)$$

This gives for the filter coefficients

$$a_1 = -2.2137, \quad a_2 = 2.9403, \quad a_3 = -2.1697, \quad a_4 = 0.9606$$

The exact spectrum is given by Eq. (5.10.1) with $E_4 = \sigma_\epsilon^2 = 1$. Since the two zeros $z_1$ and $z_2$ are near the unit circle, the spectrum will have two very sharp peaks at the normalized frequencies

$$\omega_1 = 0.2\pi, \quad \omega_2 = 0.4\pi \qquad [\text{radians/sample}]$$

Using the above difference equation and a realization of $\epsilon_n$, a sequence of length 20 of $y_n$ samples was generated (the filter was run for a while until its transients

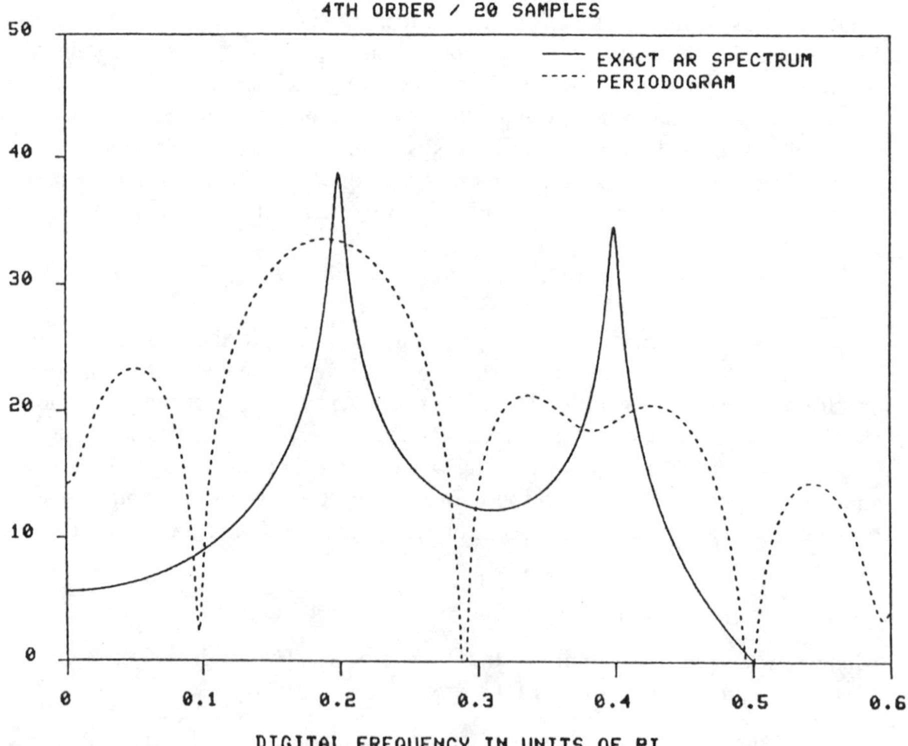


**Figure 5.13**  Ordinary Periodogram Based on 20 Samples

died out and stationarity of $y_n$ was reached). The *same* set of 20 samples was used to compute the ordinary periodogram spectrum and the autoregressive spectra using the Yule-Walker and Burg methods of extracting the model parameters. The resulting spectra are shown together with the exact spectrum in Figs. 5.13 through 5.15. Figures 5.16 through 5.18 show the spectra obtained by increasing the length of the data sequence $y_n$ to 100 samples. The lack of sufficient resolution of both the periodogram and the Yule-Walker spectrum estimates for the shorter data record is attributed to the *windowing* of the sequence $y_n$. But as the length increases the effects of windowing become less and less pronounced and both methods improve. Burg's method is remarkable in that it works very well even on the basis of very short data records. The Burg spectral estimate is sometimes called the "maximum entropy" spectral estimate. The connection to entropy concepts is discussed in the above references.

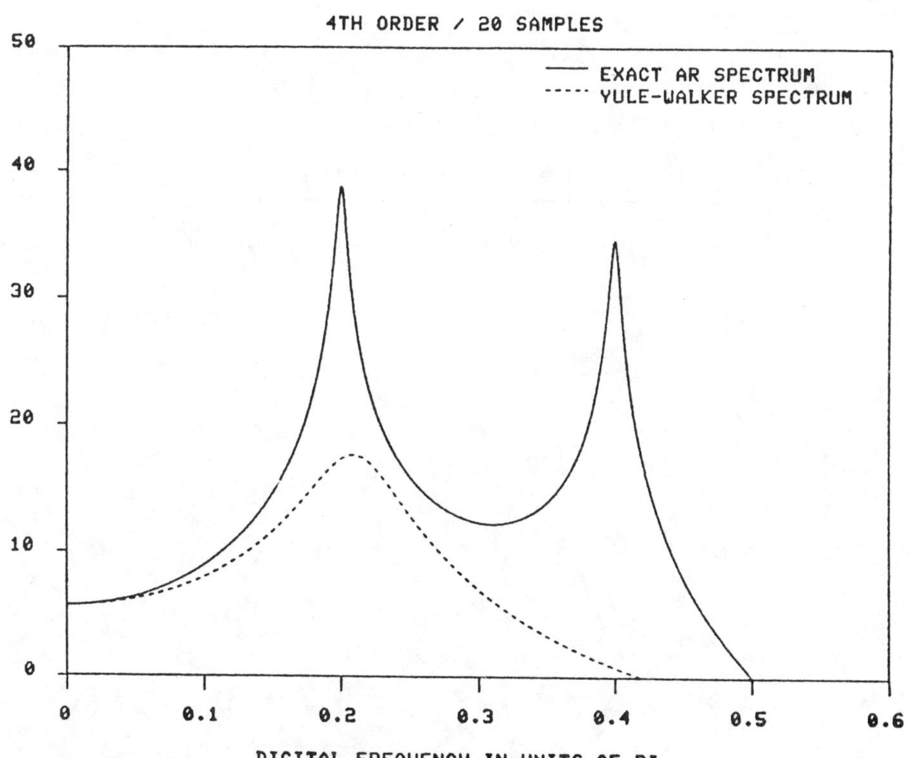


**Figure 5.14**   Yule-Walker Spectrum Based on 20 Samples

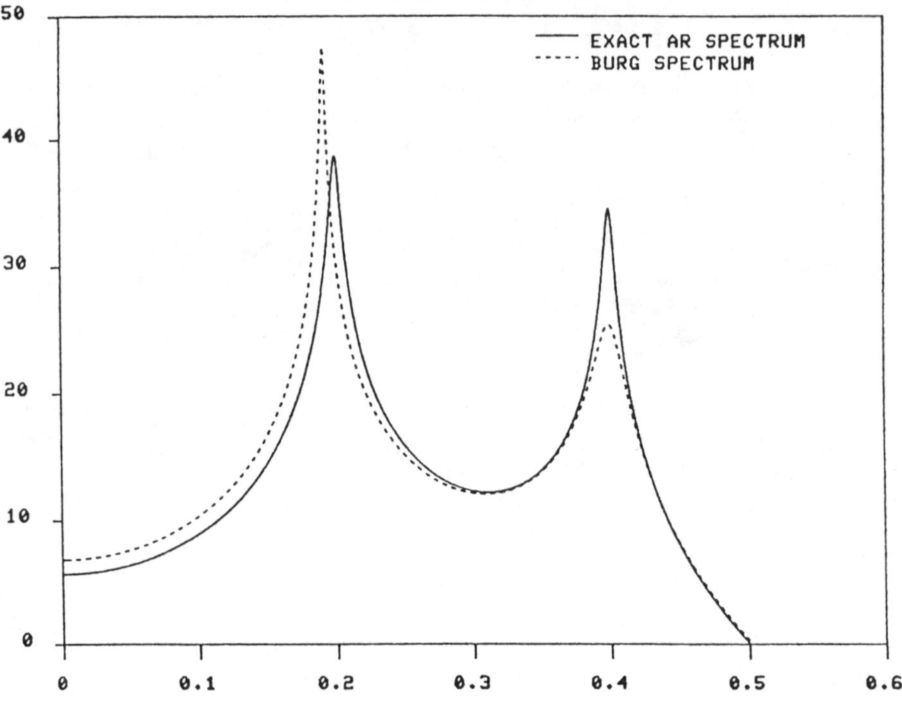


**Figure 5.15**  Burg Spectrum Based on 20 Samples

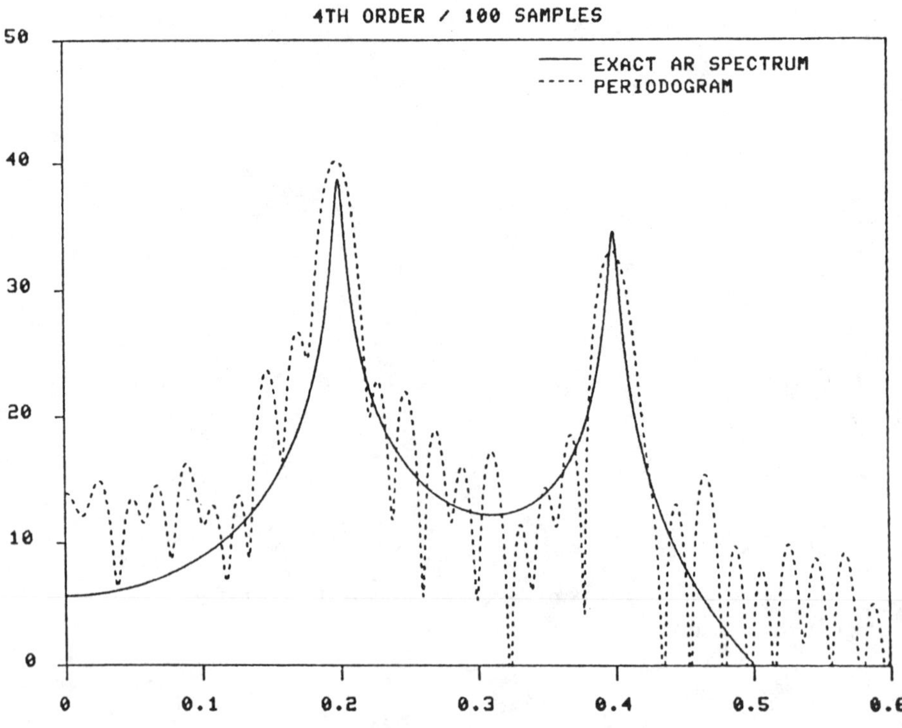


**Figure 5.16**  Ordinary Periodogram Based on 100 Samples

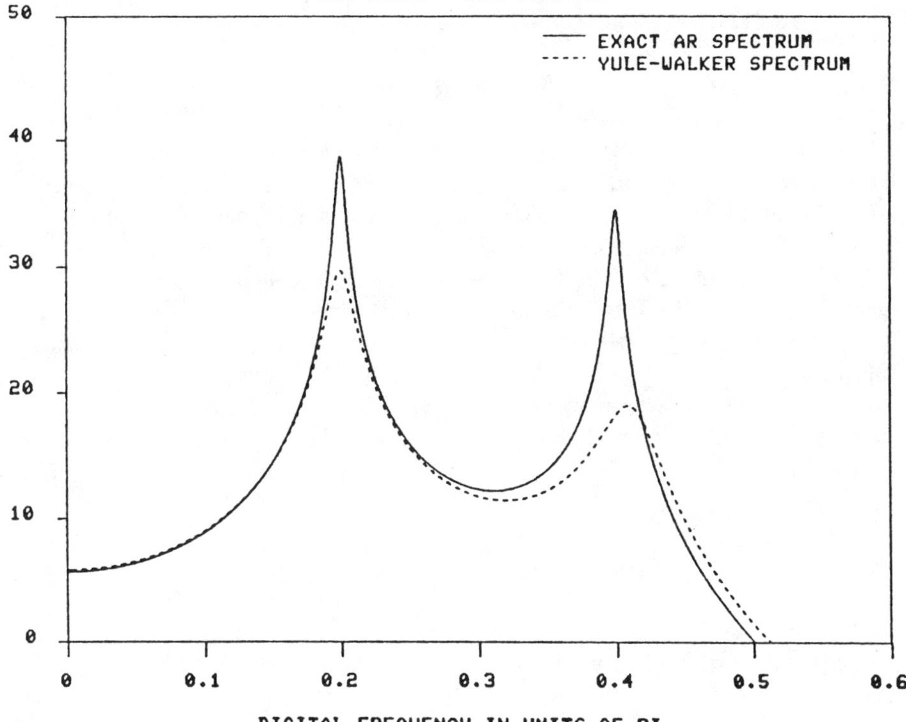

**Figure 5.17**    Yule-Walker Spectrum Based on 100 Samples

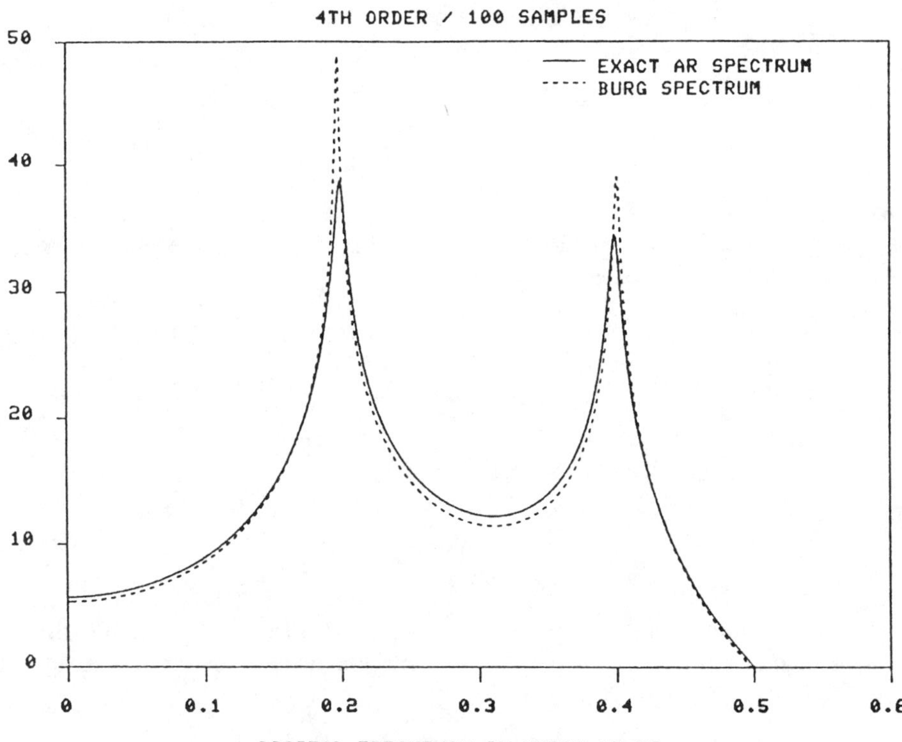

**Figure 5.18**    Burg Spectrum Based on 100 Samples

## 5.11 *Spectral Analysis of Sinusoids in Noise*

One of the most important signal processing problems is the estimation of the frequencies and amplitudes of sinusoidal signals buried in additive noise [10,11,33,35,59,61,64–72]. In addition to its practical importance, this problem has served as the testing ground for all spectrum estimation techniques, new or old. In this section we discuss four approaches to this problem: (1) the classical method, based on the Fourier transform of the windowed autocorrelation; (2) the maximum entropy method, based on the autoregressive modeling of the spectrum; (3) the maximum likelihood, or minimum energy, method; and (4) Pisarenko's method of harmonic retrieval which offers the highest resolution.

Consider a signal consisting of $L$ complex sinusoids with random phases in additive noise:

$$y_n = v_n + \sum_{i=1}^{L} A_i \exp(j\omega_i n + j\phi_i) \qquad (5.11.1)$$

where the phases $\phi_i$ are uniformly distributed and independent of each other, and $v_n$ is zero-mean white noise of variance $\sigma_v^2$, assumed to be independent of the phases $\phi_i$:

$$E[v_n^* v_m] = \sigma_v^2 \delta_{nm}, \qquad E[\phi_i v_n] = 0 \qquad (5.11.2)$$

Under these assumptions, the autocorrelation of $y_n$ is easily found to be

$$R(k) = E[y_{n+k} y_n^*] = \sigma_v^2 \delta(k) + \sum_{i=1}^{L} P_i \exp(j\omega_i k) \qquad (5.11.3)$$

where $P_i$ denotes the power level of the $i$th sinusoid; that is, $P_i = |A_i|^2$. The basic problem is to extract the set of frequencies $\{\omega_1, \omega_2, \ldots, \omega_L\}$ and powers $\{P_1, P_2, \ldots, P_L\}$ by appropriate processing of a segment of signal samples $y_n$. The theoretical power spectrum is a line spectrum superimposed on a white-noise spectrum:

$$S(\omega) = \sigma_v^2 + \sum_{i=1}^{L} P_i \, 2\pi \delta(\omega - \omega_i), \qquad -\pi \leq \omega \leq \pi \qquad (5.11.4)$$

which is obtained by Fourier transforming (5.11.3):

$$S(\omega) = \sum_{k=-\infty}^{\infty} R(k) \exp(-j\omega k) \qquad (5.11.5)$$

Given a finite set of autocorrelation lags $\{R(0), R(1), \ldots, R(M)\}$, the classical spectrum analysis method consists of windowing these lags by an appropriate window and then computing the sum (5.11.5), truncated to $-M \leq k \leq M$. We will use the triangular or Bartlett window which corresponds to the mean value of the ordinary periodogram spectrum [73]. This window is defined by

$$w_B(k) = \begin{cases} \dfrac{1}{M+1}(M+1-|k|), & \text{if } -M \leq k \leq M \\ 0 & \text{otherwise} \end{cases}$$

Replacing $R(k)$ by $w_B(k)R(k)$ in Eq. (5.11.5), we obtain the classical Bartlett spectrum estimate:

$$\hat{S}_B(\omega) = \sum_{k=-M}^{M} w_B(k)R(k)\exp(-j\omega k) \tag{5.11.6}$$

We choose the Bartlett window because this expression can be written in a compact matrix form by introducing the $(M+1)$-dimensional phase vector

$$\mathbf{s}_\omega = \begin{bmatrix} 1 \\ e^{j\omega} \\ e^{2j\omega} \\ \vdots \\ e^{j\omega M} \end{bmatrix}$$

and the $(M+1) \times (M+1)$ autocorrelation matrix $R$, defined as

$$R(k,m) = R(k-m) = \sigma_v^2 \delta(k-m) + \sum_{i=1}^{L} P_i \exp[j\omega_i(k-m)]$$

for $0 \leq k, m \leq M$. Ignoring the $1/(M+1)$ scale factor arising from the definition of the Bartlett window, we may write Eq. (5.11.6) as

$$\hat{S}_B(\omega) = \mathbf{s}_\omega^\dagger R \mathbf{s}_\omega \qquad \text{(classical Bartlett spectrum)} \tag{5.11.7}$$

The autocorrelation matrix $R$ of the sinusoids can also be written in terms of the phasing vectors as

$$R = \sigma_v^2 I + \sum_{i=1}^{L} P_i \mathbf{s}_{\omega_i} \mathbf{s}_{\omega_i}^\dagger \tag{5.11.8}$$

where $I$ is the $(M+1) \times (M+1)$ identity matrix. It can be written even more compactly by introducing the $L \times L$ diagonal *power matrix*, and the $(M+1) \times L$ *sinusoid matrix*

$$P = \mathrm{diag}\{P_1, P_2, \ldots, P_L\}, \qquad S = [\mathbf{s}_{\omega_1}, \mathbf{s}_{\omega_2}, \ldots, \mathbf{s}_{\omega_L}]$$

Then, Eq. (5.11.8) becomes

$$R = \sigma_v^2 I + SPS^\dagger \tag{5.11.9}$$

Inserting Eq. (5.11.8) into Eq. (5.11.7) we find

$$\hat{S}_B(\omega) = \sigma_v^2 \mathbf{s}_\omega^\dagger \mathbf{s}_\omega + \sum_{i=1}^{L} P_i \mathbf{s}_\omega^\dagger \mathbf{s}_{\omega_i} \mathbf{s}_{\omega_i}^\dagger \mathbf{s}_\omega$$

Defining the function

$$W(\omega) = \sum_{m=0}^{M} \exp(-j\omega m) = \frac{1 - e^{-j\omega(M+1)}}{1 - e^{-j\omega}}$$

$$= \frac{\sin\left[\dfrac{\omega(M+1)}{2}\right]}{\sin\left(\dfrac{\omega}{2}\right)} \, e^{-j\omega M/2} \quad (5.11.10)$$

we note that

$$\mathbf{s}_\omega^\dagger \mathbf{s}_{\omega_i} = W(\omega - \omega_i) \qquad \text{and} \qquad \mathbf{s}_\omega^\dagger \mathbf{s}_\omega = W(0) = M + 1$$

Then, in this notation the Bartlett spectrum (5.11.7) becomes

$$\hat{S}_B(\omega) = \sigma_v^2(M + 1) + \sum_{i=1}^{L} P_i |W(\omega - \omega_i)|^2 \quad (5.11.11)$$

The effect of $W(\omega - \omega_i)$ is to smear each spectral line $\delta(\omega - \omega_i)$ of the true spectrum. If the frequencies $\omega_i$ are too close to each other the smeared peaks will tend to overlap with a resulting loss of resolution. The function $W(\omega)$ is the Fourier transform of the *rectangular* window and it is depicted below:

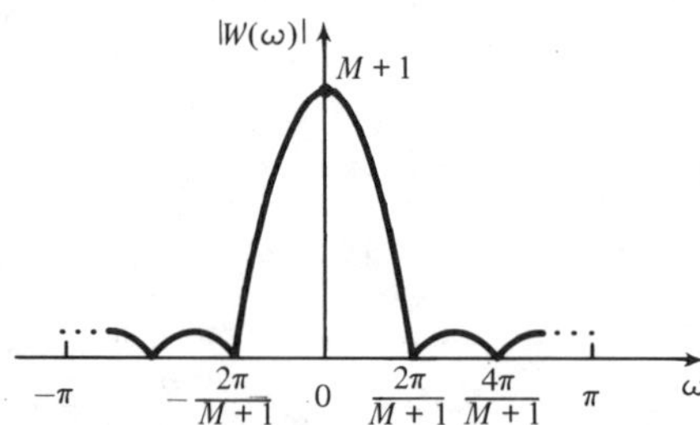


It has an effective resolution width $\Delta\omega = 2\pi/(M + 1)$. For fairly large $M$s, the first side lobe is about 13 dB down from the main lobe. As $M$ increases, the main lobe becomes higher and thinner, resembling more and more a delta function, which improves the frequency resolution capability of this estimate.

Next, we derive a closed form expression [68,72] for the AR, or maximum entropy, spectral estimate. It is given by Eq. (5.10.1) and is obtained by fitting an order-$M$ autoregressive model to the autocorrelation lags $\{R(0), R(1), \ldots, R(M)\}$. This can be done for any desired value of $M$. Autoregressive spectrum estimates generally work well in modeling ''peaked'' or resonant spectra; therefore, it is expected that they will work in this case, too. However, it should be kept in mind that AR models are not really appropriate for such sinusoidal signals. Indeed, AR models are characterized by all-pole stable filters that always result in autocorrelation functions $R(k)$ which decay exponentially with the lag $k$; whereas Eq. (5.11.3) is persistent in $k$ and never decays. As a rule, AR modeling of sinusoidal spectra works very well as long as the signal to noise ratios (SNRs) are fairly high. Pisarenko's method, to be discussed later, provides unbiased frequency estimates *regardless* of the SNRs.

The LPC model parameters for the AR spectrum estimate (5.10.1) are obtained by minimizing the mean-squared prediction error:

$$\mathcal{E} = E[e_n^* e_n] = \mathbf{a}^\dagger R \mathbf{a} = \min, \quad e_n = \sum_{m=0}^{M} a_m y_{n-m} \qquad (5.11.12)$$

where $\mathbf{a} = [1, a_1, a_2, \ldots, a_M]^T$ is the prediction-error filter and $R$ the autocorrelation matrix (5.11.9). The minimization of $\mathcal{E}$ must be subject to the linear constraint that the first entry of $\mathbf{a}$ be unity. This constraint can be expressed in vector form

$$a_0 = \mathbf{u}_0^\dagger \mathbf{a} = 1 \qquad (5.11.13)$$

where $\mathbf{u}_0 = [1, 0, 0, \ldots, 0]^T$ is the unit vector consisting of one followed by $M$ zeros. Incorporating this constraint with a Lagrange multiplier, we solve the minimization problem

$$\mathcal{E} = \mathbf{a}^\dagger R \mathbf{a} + \mu(1 - \mathbf{u}_0^\dagger \mathbf{a}) = \min$$

Differentiating with respect to $\mathbf{a}$ we obtain the normal equations

$$R\mathbf{a} = \mu \mathbf{u}_0$$

To fix the Lagrange multiplier, multiply from the left by $\mathbf{a}^\dagger$ and use Eq. (5.11.13) to get $\mathbf{a}^\dagger R \mathbf{a} = \mu \mathbf{a}^\dagger \mathbf{u}_0$, or $\mathcal{E} = \mu$. Thus, $\mu$ is the minimized value of $\mathcal{E}$ which we denote by $E$; in summary, we have

$$R\mathbf{a} = E\mathbf{u}_0 \quad \text{and} \quad \mathbf{a} = E R^{-1} \mathbf{u}_0 \qquad (5.11.14)$$

Acting from the left by $\mathbf{u}_0^T$ we also find $1 = E\mathbf{u}_0^T R^{-1} \mathbf{u}_0$, or

$$E^{-1} = \mathbf{u}_0^T R^{-1} \mathbf{u}_0 = (R^{-1})_{00} \qquad (5.11.15)$$

which is, of course, the same as Eq. (5.7.18). The special structure of $R$ allows the computation of $\mathbf{a}$ and the AR spectrum in closed form. Applying the matrix inversion lemma to Eq. (5.11.9), we find the inverse of $R$,

$$R^{-1} = \frac{1}{\sigma_v^2} (I + SDS^\dagger) \qquad (5.11.16)$$

where $D$ is a $L \times L$ matrix given by

$$D = -[\sigma_v^2 P^{-1} + S^\dagger S]^{-1} \qquad (5.11.17)$$

Equation (5.11.16) can also be derived directly by making such an Ansatz for $R^{-1}$ and then fixing $D$. The quantity $\sigma_v^2 P^{-1}$ in $D$ is a matrix of noise to signal ratios. Inserting Eq. (5.11.16) into Eq. (5.11.14), we find for $\mathbf{a}$

$$\mathbf{a} = ER^{-1}\mathbf{u}_0 = \frac{E}{\sigma_v^2} [\mathbf{u}_0 + SDS^{\dagger}\mathbf{u}_0] = \frac{E}{\sigma_v^2} [\mathbf{u}_0 + S\mathbf{d}]$$

where we used the fact that $\mathbf{s}_{\omega_i}^{\dagger}\mathbf{u}_0 = 1$, which implies that

$$S^{\dagger}\mathbf{u}_0 = \begin{bmatrix} \mathbf{s}_{\omega_1}^{\dagger} \\ \mathbf{s}_{\omega_2}^{\dagger} \\ \vdots \\ \mathbf{s}_{\omega_L}^{\dagger} \end{bmatrix} \mathbf{u}_0 = \begin{bmatrix} 1 \\ 1 \\ \vdots \\ 1 \end{bmatrix} = \mathbf{v}; \qquad \text{i.e., a column of } L \text{ ones}$$

and defined

$$\mathbf{d} = \begin{bmatrix} d_1 \\ d_2 \\ \vdots \\ d_L \end{bmatrix} = D\mathbf{v} \qquad \text{or} \qquad d_i = \sum_{j=1}^{L} D_{ij}$$

Using Eq. (5.11.15) we have also

$$E^{-1} = \mathbf{u}_0^T R^{-1}\mathbf{u}_0 = \frac{1}{\sigma_v^2} \mathbf{u}_0^T[I + SDS^{\dagger}]\mathbf{u}_0$$

$$= \frac{1}{\sigma_v^2} [1 + \mathbf{v}^T D\mathbf{v}]$$

$$= \frac{1}{\sigma_v^2} [1 + \mathbf{v}^T\mathbf{d}] = \frac{1}{\sigma_v^2} \left[ 1 + \sum_{i=1}^{L} d_i \right]$$

and, therefore,

$$E = \sigma_v^2 \left[ 1 + \sum_{i=1}^{L} d_i \right]^{-1} \tag{5.11.18}$$

We finally find for the prediction-error filter

$$\mathbf{a} = \frac{\mathbf{u}_0 + S\mathbf{d}}{1 + \mathbf{v}^T\mathbf{d}} = \frac{\mathbf{u}_0 + \sum_{i=1}^{L} d_i\mathbf{s}_{\omega_i}}{1 + \sum_{i=1}^{L} d_i} \tag{5.11.19}$$

The frequency response $A(\omega)$ of the filter is obtained by dotting the phasing vector $\mathbf{s}_\omega$ into $\mathbf{a}$:

$$A(\omega) = \sum_{m=0}^{M} a_m \exp(-j\omega m) = \mathbf{s}_\omega^\dagger \mathbf{a} = \frac{1 + \sum_{i=1}^{L} d_i \, \mathbf{s}_\omega^\dagger \mathbf{s}_{\omega_i}}{1 + \sum_{i=1}^{L} d_i}$$

and using the result $\mathbf{s}_\omega^\dagger \mathbf{s}_{\omega_i} = W(\omega - \omega_i)$, we finally find

$$A(\omega) = \frac{1 + \sum_{i=1}^{L} d_i W(\omega - \omega_i)}{1 + \sum_{i=1}^{L} d_i} \tag{5.11.20}$$

and for the AR, or maximum entropy, spectrum estimate:

$$\hat{S}_{AR}(\omega) = \frac{E}{|A(\omega)|^2} = \sigma_v^2 \, \frac{1 + \sum_{i=1}^{L} d_i}{\left| 1 + \sum_{i=1}^{L} d_i W(\omega - \omega_i) \right|^2} \tag{5.11.21}$$

The frequency dependence is shown explicitly. Note, that the matrix $S^\dagger S$, appearing in the definition of $D$, can also be expressed in terms of $W(\omega)$. Indeed, the $ij$th element of $S^\dagger S$ is

$$(S^\dagger S)_{ij} = \mathbf{s}_{\omega_i}^\dagger \mathbf{s}_{\omega_j} = W(\omega_i - \omega_j)$$

One interesting consequence of Eq. (5.11.21) is that in the limit of very weak noise $\sigma_v^2 \rightarrow 0$, it vanishes. In this limit the mean-squared prediction error (5.11.18) vanishes. This is to be expected, since in this case the noise term $v_n$ is absent from the sum (5.11.1), rendering $y_n$ a deterministic signal; that is, one that can be predicted from a few past values with zero prediction error. To avoid such behavior when $\sigma_v^2$ is small, the factor $E$ is sometimes dropped altogether from the spectral estimate resulting in the "pseudo-spectrum"

$$\hat{S}_{AR}(\omega) = \frac{1}{|A(\omega)|^2} \qquad \text{(AR spectrum estimate)} \tag{5.11.22}$$

This expression will exhibit fairly sharp peaks at the sinusoid frequencies, but the magnitude of these peaks will no longer be representative of the power levels $P_i$. This expression can only be used to extract the frequencies $\omega_i$. Up to a scale factor, Eq. (5.11.22) can also be written in the form

$$\hat{S}_{AR}(\omega) = \frac{1}{|\mathbf{s}_\omega^\dagger R^{-1} \mathbf{u}_0|^2}$$

*Example 5.11.1:*  To see the effect of the SNR on the sharpness of the peaks in the AR spectrum, consider the case $M = L = 1$. Then,

$$S^\dagger S = \mathbf{s}_{\omega_1}^\dagger \mathbf{s}_{\omega_1} = [1, e^{-j\omega_1}] \begin{bmatrix} 1 \\ e^{j\omega_1} \end{bmatrix} = M + 1 = 2$$

$$D = -[\sigma_v^2 P_1^{-1} + 2]^{-1}$$

$$\mathbf{a} = \frac{\mathbf{u}_0 + d_1 \mathbf{s}_{\omega_1}}{1 + d_1} = \begin{bmatrix} 1 \\ \dfrac{d_1}{1 + d_1} e^{j\omega_1} \end{bmatrix}$$

Using $d_1 = D$, we find

$$\mathbf{a} = \begin{bmatrix} 1 \\ \dfrac{-P_1}{P_1 + \sigma_v^2} e^{j\omega_1} \end{bmatrix}, \qquad A(z) = 1 + a_1 z^{-1}$$

The prediction-error filter has a zero at

$$z_1 = -a_1 = \frac{P_1}{P_1 + \sigma_v^2} \exp(j\omega_1)$$

The zero $z_1$ is inside the unit circle, as it should. The lower the SNR $= P_1/\sigma_v^2$, the more inside it lies, resulting in a more smeared peak about $\omega_1$. As the SNR increases, the zero moves closer to the unit circle at the right frequency $\omega_1$, resulting in a very sharp peak in the spectrum (5.11.22).

More generally, in the case of multiple sinusoids, if the SNRs are high the spectrum (5.11.22) will exhibit sharp peaks at the desired sinusoid frequencies. The mechanism by which this happens can be seen qualitatively from Eq. (5.11.20) as follows: The matrix $S^\dagger S$ in $D$ introduces cross-coupling among the various frequencies $\omega_i$. However, if these frequencies are well separated from each other (by more than $2\pi/(M + 1)$), then the off-diagonal elements of $S^\dagger S$, namely $W(\omega_i - \omega_j)$, will be small, and for the purpose of this argument may be taken to be zero. This makes the matrix $S^\dagger S$ diagonal. Since $W(0) = M + 1$, it follows that $S^\dagger S = (M + 1)I$, and $D$ will become diagonal with diagonal elements

$$d_i = D_{ii} = -[\sigma_v^2 P_i^{-1} + M + 1]^{-1} = -\frac{P_i}{\sigma_v^2 + (M + 1)P_i}$$

Evaluating $A(\omega)$ at $\omega_i$ and keeping only the $i$th contribution in the sum we find, approximately,

$$A(\omega_i) \equiv \frac{1 + d_i W(0)}{1 + \displaystyle\sum_{j=1}^{L} d_j} = \frac{1}{1 + \displaystyle\sum_{j=1}^{L} d_i} \; \frac{1}{1 + (M + 1)\left(\dfrac{P_i}{\sigma_v^2}\right)}$$

which shows that if the SNRs $P_i/\sigma_v^2$ are high, $A(\omega_i)$ will be very small, resulting in large spectral peaks in Eq. (5.11.22). The resolvability properties of the AR estimate improve *both* when the *SNRs increase* and when the *order M increases*. The mutual interaction of the various sinusoid components cannot be ignored altogether. One effect of this interaction is *biasing* in the estimates of the frequencies; that is, even if two nearby peaks are clearly separated, the peaks may not occur exactly at the desired sinusoid frequencies, but may be slightly shifted. The degree of bias depends on the *relative separation* of the peaks and on *the SNRs*. With the above qualifications in mind, we can state that the LPC approach to this problem is one of the most successful ones.

The maximum likelihood (ML), or minimum energy, spectral estimator is given by the expression [64]

$$\hat{S}_{ML}(\omega) = \frac{1}{\mathbf{s}_\omega^\dagger R^{-1} \mathbf{s}_\omega} \qquad \text{(ML spectrum estimate)} \qquad (5.11.23)$$

It can be justified by envisioning a bank of narrowband filters, each designed to allow a sinewave through at the filter's center frequency and to attenuate all other frequency components. Thus, the narrowband filter with center frequency $\omega$ is required to let this frequency go through unchanged, that is,

$$A(\omega) = \mathbf{s}_\omega^\dagger \mathbf{a} = 1$$

while at the same time it is required to minimize the output power

$$\mathbf{a}^\dagger R \mathbf{a} = \min$$

The solution of this minimization problem subject to the above constraint is readily found to be

$$\mathbf{a} = \frac{R^{-1} \mathbf{s}_\omega}{\mathbf{s}_\omega^\dagger R^{-1} \mathbf{s}_\omega}$$

which gives for the minimized output power at this frequency

$$\mathbf{a}^\dagger R \mathbf{a} = \frac{1}{\mathbf{s}_\omega^\dagger R^{-1} \mathbf{s}_\omega}$$

Using Eq. (5.11.16), we find

$$\mathbf{s}_\omega^\dagger R^{-1}\mathbf{s}_\omega = \frac{1}{\sigma_v^2}\left[\mathbf{s}_\omega^\dagger\mathbf{s}_\omega + \sum_{i,j=1}^{L} D_{ij}\mathbf{s}_\omega^\dagger\mathbf{s}_{\omega_i}\mathbf{s}_{\omega_j}^\dagger\mathbf{s}_\omega\right]$$

$$= \frac{1}{\sigma_v^2}\left[(M + 1) + \sum_{i,j=1}^{L} D_{ij}W(\omega - \omega_i)W(\omega - \omega_j)^*\right]$$

and the theoretical ML spectrum becomes in this case

$$\hat{S}_{ML}(\omega) = \frac{\sigma_v^2}{(M + 1) + \sum_{i,j=1}^{L} D_{ij}W(\omega - \omega_i)W(\omega - \omega_j)^*} \tag{5.11.24}$$

*Example 5.11.2:*   Determine the matrix $D$ and vector $\mathbf{d}$ for the case of $L = 2$ and arbitrary $M$. The matrix $S^\dagger S$ is in this case

$$S^\dagger S = \begin{bmatrix} W(0) & W(\omega_1 - \omega_2) \\ W(\omega_2 - \omega_1) & W(0) \end{bmatrix} \equiv \begin{bmatrix} M + 1 & W_{12} \\ W_{12}^* & M + 1 \end{bmatrix}$$

so that $D$ becomes

$$D = -\begin{bmatrix} \sigma_v^2 P_1^{-1} + M + 1 & W_{12} \\ W_{12}^* & \sigma_v^2 P_2^{-1} + M + 1 \end{bmatrix}^{-1}$$

$$= \frac{1}{|W_{12}|^2 - (\sigma_v^2 P_1^{-1} + M + 1)(\sigma_v^2 P_2^{-1} + M + 1)}$$

$$\times \begin{bmatrix} \sigma_v^2 P_2^{-1} + M + 1 & -W_{12} \\ -W_{12}^* & \sigma_v^2 P_1^{-1} + M + 1 \end{bmatrix}$$

and

$$\mathbf{d} = D\mathbf{v} = D\begin{bmatrix} 1 \\ 1 \end{bmatrix} = \frac{1}{|W_{12}|^2 - (\sigma_v^2 P_1^{-1} + M + 1)(\sigma_v^2 P_2^{-1} + M + 1)}$$

$$\times \begin{bmatrix} \sigma_v^2 P_2^{-1} + M + 1 - W_{12} \\ \sigma_v^2 P_1^{-1} + M + 1 - W_{12}^* \end{bmatrix}$$

Using the results of Example 5.11.2, we have carried out a computation illustrating the three spectral estimates: Figure 5.19 shows the theoretical autoregressive, Bartlett, and maximum likelihood spectral estimates given by Eqs. (5.11.11), (5.11.22), and (5.11.24), respectively, for two sinusoids of frequencies

$$\omega_1 = 0.4\pi, \qquad \omega_2 = 0.6\pi$$

and equal powers SNR $= 10\log_{10}(P_1/\sigma_v^2) = 6$ dB, and $M = 6$. To facilitate the comparison, all three spectra have been *normalized* to 0 dB at the frequency $\omega_1$ of the first sinusoid. It is seen that the length $M = 6$ is too short for the

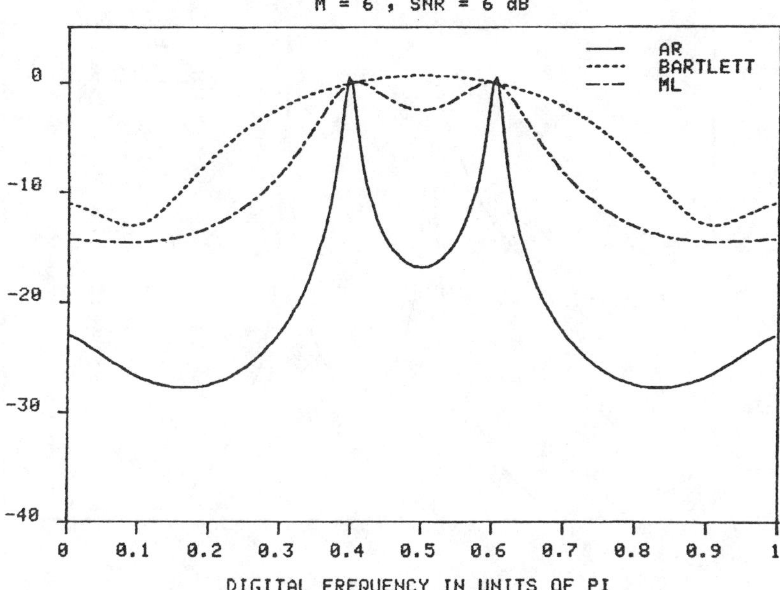



**Figure 5.19**  Maximum Entropy, Bartlett, and Maximum Likelihood Spectrum Estimates of Two Sinusoids in Noise

Bartlett spectrum to resolve the two peaks. The AR spectrum is the best (however, close inspection of the graph will reveal a small *bias* in the frequency of the peaks, arising from the mutual interaction of the two sinewaves). The effect of increasing the SNR is shown in Fig. 5.20, where the SNR has been changed to SNR = 12 dB. It is seen that the AR spectral peaks become narrower, thus increasing their resolvability. To show the effect of increasing the length $M$, we kept SNR = 6 dB, and increased $M$ to $M = 12$ and $M = 18$. The resulting spectra are shown in Figs. 5.21 and 5.22. It is seen that all spectra tend to become better. The interplay between resolution, order, SNR, and bias has been studied in [66,68,71].

The main motivation behind the definition (5.11.22) for the pseudospectrum was to obtain an expression that exhibits very sharp spectral peaks at the sinusoid frequencies $\omega_i$. Infinite resolution can, in principle, be achieved if we can find a polynomial $A(z)$ that has zeros *on the unit circle* at the desired frequency angles; namely,

$$z_i = \exp(j\omega_i); \qquad i = 1,2,\ldots,L \qquad (5.11.25)$$

Pisarenko's method determines such a polynomial on the basis of the autocorrelation matrix $R$. The desired conditions on the polynomial are

$$A(z_i) = A(\omega_i) = 0; \qquad i = 1,2,\ldots,L \qquad (5.11.26)$$

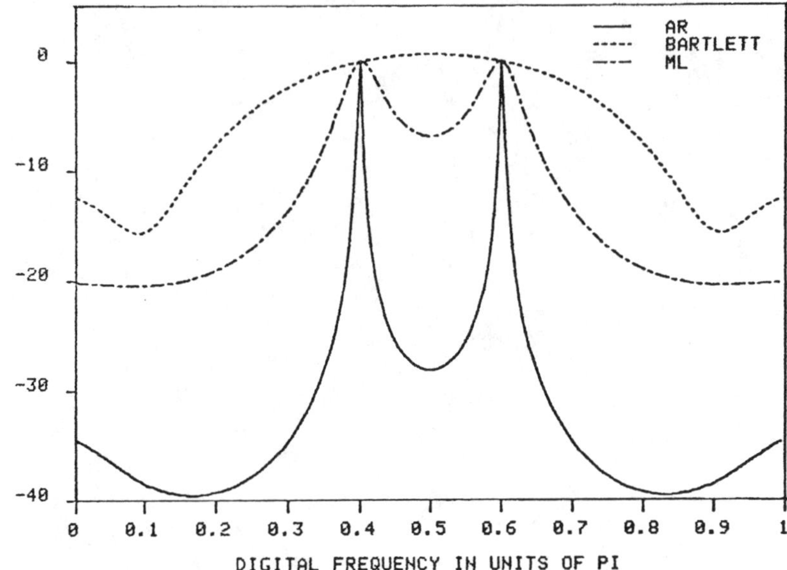

**Figure 5.20**  Effect of Increasing SNR on the Resolution Capability of the Three Spectrum Estimates

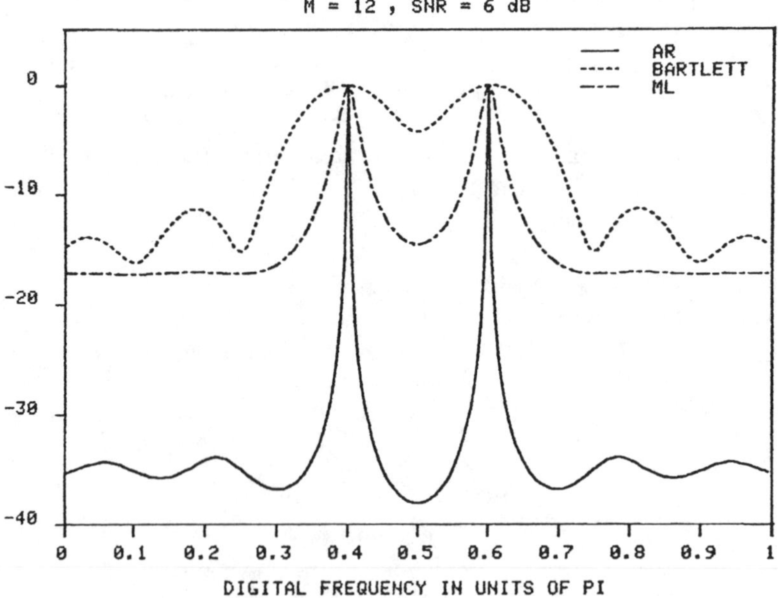

**Figure 5.21**  Effect of Increasing the Order on the Resolution Capability

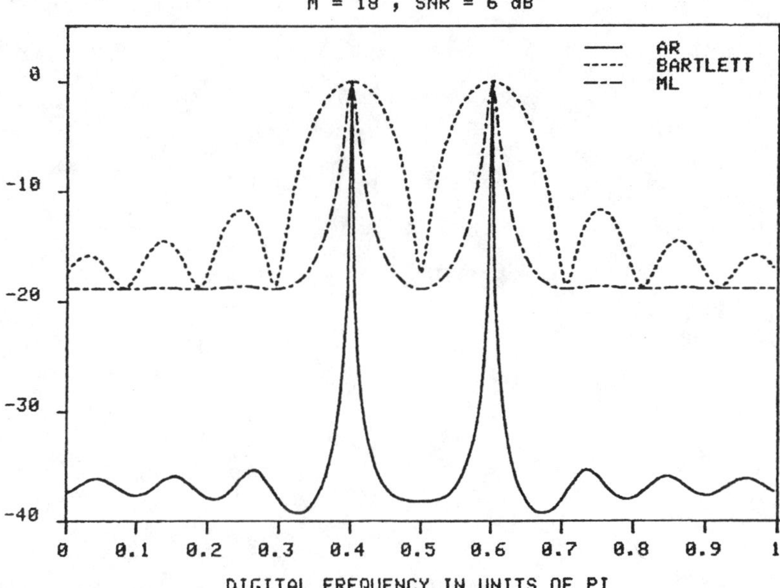



**Figure 5.22**  Further Increase of the Order $M$

where we slightly abuse the notation and write $A(e^{j\omega}) = A(\omega)$. To satisfy these conditions, the degree $M$ of the polynomial $A(z)$ must necessarily be $M \geq L$; then, the remaining $M - L$ zeros of $A(z)$ could be arbitrary. Let $\mathbf{a}$ be the vector of coefficients of $A(z)$, so that

$$\mathbf{a} = \begin{bmatrix} a_0 \\ a_1 \\ \vdots \\ a_M \end{bmatrix} \qquad A(z) = a_0 + a_1 z^{-1} + \cdots + a_M z^{-M}$$

Noting that $A(\omega) = \mathbf{s}_\omega^\dagger \mathbf{a}$, Eqs. (5.11.26) may be combined into one vectorial equation

$$S^\dagger \mathbf{a} = \begin{bmatrix} A(\omega_1) \\ A(\omega_2) \\ \vdots \\ A(\omega_L) \end{bmatrix} = 0 \qquad (5.11.27)$$

But then, Eq. (5.11.9) implies that

$$R\mathbf{a} = \sigma_v^2 \mathbf{a} + SPS^\dagger \mathbf{a} = \sigma_v^2 \mathbf{a}$$

or that $\sigma_v^2$ must be an *eigenvalue* of $R$ with $\mathbf{a}$ the corresponding *eigenvector*:

$$R\mathbf{a} = \sigma_v^2\mathbf{a} \tag{5.11.28}$$

The quantity $\sigma_v^2$ is actually the *smallest* eigenvalue of $R$. To see this, consider any other eigenvector $\mathbf{a}$ of $R$, and normalize it to unit norm

$$R\mathbf{a} = \lambda\mathbf{a}, \quad \text{with } \mathbf{a}^\dagger\mathbf{a} = 1 \tag{5.11.29}$$

Then, (5.11.9) implies that

$$\lambda = \lambda\mathbf{a}^\dagger\mathbf{a} = \mathbf{a}^\dagger R\mathbf{a} = \sigma_v^2\mathbf{a}^\dagger\mathbf{a} + \mathbf{a}^\dagger SPS^\dagger\mathbf{a}$$

$$= \sigma_v^2 + [A(\omega_1)^*, A(\omega_2)^*, \ldots, A(\omega_L)^*]\, P \begin{bmatrix} A(\omega_1) \\ A(\omega_2) \\ \vdots \\ A(\omega_L) \end{bmatrix}$$

$$= \sigma_v^2 + \sum_{i=1}^{L} P_i|A(\omega_i)|^2$$

which shows that any $\lambda$ is equal to $\sigma_v^2$ shifted by a nonnegative amount. If the eigenvector satisfies the conditions (5.11.26), then the shift in $\lambda$ vanishes. Thus, the desired polynomial $A(z)$ can be found by solving the *eigenvalue problem* (5.11.29) and selecting the eigenvector belonging to the *minimum eigenvalue*. This is Pisarenko's method [67]. As a byproduct of the procedure, the noise power level $\sigma_v^2$ is also determined, which in turn allows the determination of the power matrix $P$, as follows: Writing Eq. (5.11.9) as

$$R - \sigma_v^2 I = SPS^\dagger$$

and acting by $S^\dagger$ and $S$ from the left and right, we obtain

$$P = U^\dagger(R - \sigma_v^2 I)U \quad \text{where } U = S(S^\dagger S)^{-1} \tag{5.11.30}$$

Since there is a freedom in selecting the *remaining* $M - L$ zeros of the polynomial $A(z)$, it follows that there are $(M - L) + 1$ eigenvectors all belonging to the minimum eigenvalue $\sigma_v^2$. Thus, the $(M + 1)$-dimensional eigenvalue problem (5.11.29) has: (a) $M + 1 - L$ *degenerate eigenvalues* equal to $\sigma_v^2$, and (b) $L$ additional eigenvalues which are strictly greater than $\sigma_v^2$. The $(M + 1 - L)$-dimensional subspace spanned by the degenerate eigenvectors belonging to $\sigma_v^2$ is called the *noise* subspace. The $L$-dimensional subspace spanned by the eigenvectors belonging to the remaining $L$ eigenvalues is called the *signal* subspace. Since the signal subspace is orthogonal to the noise subspace, and the $L$ linearly independent signal vectors $\mathbf{s}_{\omega_i}$; $i = 1, 2, \ldots, L$, are also orthogonal to the noise subspace, it follows that the signal subspace is *spanned* by the $\mathbf{s}_{\omega_i}$s. In the special case when $L = M$, there is no degeneracy in the minimum eigenvalue, and there is a *unique* minimum eigenvector. In this case, all $M = L$ zeros of $A(z)$ lie on the unit circle at the desired angles $\omega_i$.

*Example 5.11.3:*  Consider the case $L = M = 2$. The matrix $R$ is written explicitly as

$$R = \sigma_v^2 I + P_1 \mathbf{s}_{\omega_1} \mathbf{s}_{\omega_1}^\dagger + P_2 \mathbf{s}_{\omega_2} \mathbf{s}_{\omega_2}^\dagger$$

$$= \begin{bmatrix} \sigma_v^2 + P_1 + P_2 & P_1 e^{-j\omega_1} + P_2 e^{-j\omega_2} & P_1 e^{-2j\omega_1} + P_2 e^{-2j\omega_2} \\ P_1 e^{j\omega_1} + P_2 e^{j\omega_2} & \sigma_v^2 + P_1 + P_2 & P_1 e^{-j\omega_1} + P_2 e^{-j\omega_2} \\ P_1 e^{2j\omega_1} + P_2 e^{2j\omega_2} & P_1 e^{j\omega_1} + P_2 e^{j\omega_2} & \sigma_v^2 + P_1 + P_2 \end{bmatrix}$$

It is easily verified that the (unnormalized) vector

$$\mathbf{a} = \begin{bmatrix} 1 \\ -(e^{j\omega_1} + e^{j\omega_2}) \\ e^{j\omega_1} e^{j\omega_2} \end{bmatrix}$$

is an eigenvector of $R$ belonging to $\lambda = \sigma_v^2$. In this case, the polynomial $A(z)$ is

$$A(z) = a_0 + a_1 z^{-1} + a_2 z^{-2} = 1 - (e^{j\omega_1} + e^{j\omega_2}) z^{-1} + e^{j\omega_1} e^{j\omega_2} z^{-2}$$

$$= (1 - e^{j\omega_1} z^{-1})(1 - e^{j\omega_2} z^{-1})$$

exhibiting the two desired zeros at the sinusoid frequencies.

*Example 5.11.4:*  Consider the case $M = 2$, $L = 1$. The matrix $R$ is

$$R = \sigma_v^2 I + P_1 \mathbf{s}_{\omega_1} \mathbf{s}_{\omega_1}^\dagger = \begin{bmatrix} \sigma_v^2 + P_1 & P_1 e^{-j\omega_1} & P_1 e^{-2j\omega_1} \\ P_1 e^{j\omega_1} & \sigma_v^2 + P_1 & P_1 e^{-j\omega_1} \\ P_1 e^{2j\omega_1} & P_1 e^{j\omega_1} & \sigma_v^2 + P_1 \end{bmatrix}$$

It is easily verified that the three eigenvectors of $R$ are

$$\mathbf{e}_1 = \begin{bmatrix} 1 \\ -e^{j\omega_1} \\ 0 \end{bmatrix}, \qquad \mathbf{e}_2 = \begin{bmatrix} 0 \\ 1 \\ -e^{j\omega_1} \end{bmatrix}, \qquad \mathbf{e}_3 = \begin{bmatrix} 1 \\ e^{j\omega_1} \\ e^{2j\omega_1} \end{bmatrix} = \mathbf{s}_{\omega_1}$$

belonging to the eigenvalues

$$\lambda = \sigma_v^2, \qquad \lambda = \sigma_v^2, \qquad \lambda = \sigma_v^2 + 3P_1$$

The first two eigenvectors span the noise subspace and the third the signal subspace. *Any* linear combination of the noise eigenvectors also belongs to $\lambda = \sigma_v^2$. For example, if we take

$$\mathbf{a} = \begin{bmatrix} a_0 \\ a_1 \\ a_2 \end{bmatrix} = \begin{bmatrix} 1 \\ -e^{j\omega_1} \\ 0 \end{bmatrix} - \rho \begin{bmatrix} 0 \\ 1 \\ -e^{j\omega_1} \end{bmatrix} = \begin{bmatrix} 1 \\ -(\rho + e^{j\omega_1}) \\ \rho e^{j\omega_1} \end{bmatrix}$$

the corresponding polynomial is

$$A(z) = 1 - (e^{j\omega_1} + \rho)\, z^{-1} + e^{j\omega_1}\rho z^{-2} = (1 - e^{j\omega_1}z^{-1})\,(1 - \rho z^{-1})$$

showing one desired zero at $z_1 = e^{j\omega_1}$ and a *spurious* zero.

The Pisarenko method can also be understood in terms of a *minimization criterion* of the type (5.11.12), as follows: For any set of coefficients **a**, define the output signal

$$e_n = a_0 y_n + a_1 y_{n-1} + \cdots + a_M y_{n-M}$$

Then, the mean output power is expressed as

$$\mathcal{E} = E[e_n^* e_n] = \mathbf{a}^\dagger R \mathbf{a} = \sigma_v^2\, \mathbf{a}^\dagger \mathbf{a} + \sum_{i=1}^{L} P_i |A(\omega_i)|^2$$

Imposing the quadratic constraint

$$\mathbf{a}^\dagger \mathbf{a} = 1 \tag{5.11.31}$$

we obtain

$$\mathcal{E} = E[e_n^* e_n] = \mathbf{a}^\dagger R \mathbf{a} = \sigma_v^2 + \sum_{i=1}^{L} P_i |A(\omega_i)|^2 \tag{5.11.32}$$

It is evident that the *minimum* of this expression is obtained when conditions (5.11.26) are satisfied. Thus, an *equivalent* formulation of the Pisarenko method is to *minimize the performance index* (5.11.32) subject to the *quadratic constraint* (5.11.31). The AR and the Pisarenko spectrum estimation techniques differ only in the *type* of constraint imposed on the filter weights **a**.

In summary, we have discussed the theoretical aspects of four methods of estimating the frequencies of sinusoids in noise. In practice, an *estimate* of the correlation matrix $R$ can be obtained in terms of the sample autocorrelations from a block of data values $y_n$. The AR and Pisarenko methods can also be implemented *adaptively*. The adaptive approach is based on the minimization criteria (5.11.12) and (5.11.32) and will be discussed in Chapter 6, where also some simulations will be presented.

## 5.12 *Superresolution Array Processing*

One of the main signal processing functions of sonar, radar, or seismic arrays of sensors is to detect the presence of one or more radiating point-sources. This is a problem of spectral analysis, and it is the *spatial frequency* analog of the problem of extracting sinusoids in noise discussed in the previous section. The same spectral analysis techniques can be applied to this problem. All methods aim at producing a high-resolution estimate of the *spatial* frequency power spectrum of the signal field incident on the array of sensors. The *directions* of point-source emitters can be extracted by identifying the sharpest peaks in this spectrum. In this section, we discuss conventional (Bartlett) *beamforming*, as well as the *maximum-likelihood*, *linear prediction*, and *eigenvector* based methods, all of which are of current interest [74–91].

Consider a *linear array* of $M + 1$ sensors *equally-spaced* at distances $d$, and a plane wave incident on the array at an angle $\theta_1$ with respect to the array normal, as shown below.

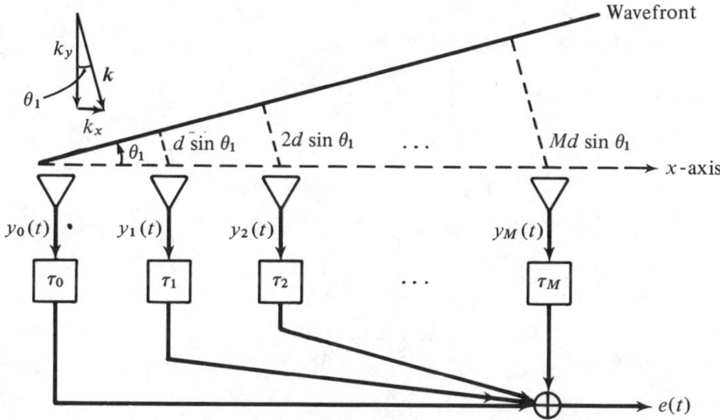


The conventional *beamformer* introduces appropriate delays at the outputs of each sensor to compensate the propagation delays of the wavefront reaching the array. The output of the beamformer (the ''beam'') is the sum

$$e(t) = \sum_{m=0}^{M} y_m(t - \tau_m) \tag{5.12.1}$$

where $y_m(t)$; $m = 0,1, \ldots ,M$ is the signal at the $m$th sensor. To reach sensor 1, the wavefront must travel an extra distance $d \sin\theta_1$; to reach sensor 2 it must travel distance $2 d \sin \theta_1$, and so on. Thus, it reaches these sensors with a propagation delay of $d \sin \theta_1/c$, $2 d \sin \theta_1/c$, and so on. The last sensor is reached with a delay of $M d \sin \theta_1/c$ seconds. Thus, to time-align the first and the last sensor, the output of the first sensor must be delayed by $\tau_0 = M d \sin \theta_1/c$, and similarly, the $m$th sensor is time-aligned with the last one, with a delay of

$$\tau_m = (M - m)d \sin\theta_1/c; \qquad m = 0,1, \ldots ,M \tag{5.12.2}$$

In this case, all terms in the sum (5.12.1) are equal to the value measured at the last sensor—that is, $y_m(t - \tau_m) = y_M(t)$—and the output of the beamformer is $e(t) = (M + 1) y_M(t)$, thus enhancing the received signal by a factor of $M + 1$ and hence its power by a factor $(M + 1)^2$. The concept of beamforming is the same as that of *signal averaging* discussed in Example (2.3.5). If there is additive noise present, it will contribute incoherently to the output power—that is, by a factor of $(M + 1)$, whereas the signal power is enhanced by

$(M + 1)^2$. Thus, the *gain* in the signal to noise ratio at the output of the array (the array gain) is a factor of $M + 1$.

In the frequency domain, the above delay-and-sum operation becomes equivalent to linear weighting. Fourier transforming Eq. (5.12.1) we have

$$e(\omega) = \sum_{m=0}^{M} y_m(\omega) e^{-j\omega\tau_m}$$

which can be written compactly as

$$e = \mathbf{a}^T \mathbf{y} \tag{5.12.3}$$

where $\mathbf{a}$ and $\mathbf{y}$ are the $(M + 1)$-vectors of weights and sensor outputs

$$\mathbf{a} = \begin{bmatrix} e^{-j\omega\tau_0} \\ e^{-j\omega\tau_1} \\ \vdots \\ e^{-j\omega\tau_M} \end{bmatrix}, \qquad \mathbf{y} = \begin{bmatrix} y_0(\omega) \\ y_1(\omega) \\ \vdots \\ y_M(\omega) \end{bmatrix}$$

From now on, we will concentrate on *narrow-band* arrays operating at a given frequency $\omega$ and the dependence on $\omega$ will not be shown explicitly. This assumes that the signals from all the sensors have been subjected to narrow-band prefiltering that leaves only the narrow operating frequency band. The beamformer now acts as a linear combiner, as shown in Fig. 5.23. A plane wave at the operating frequency $\omega$, of amplitude $A_1$, and incident at the above angle $\theta_1$, will have a value at the space-time point $(t,\mathbf{r})$ given by

$$A_1 e^{j\omega t - j\mathbf{k}\cdot\mathbf{r}}$$

Dropping the sinusoidal $t$-dependence and evaluating this expression on the $x$-axis, we have

$$A_1 e^{-jk_x x}$$

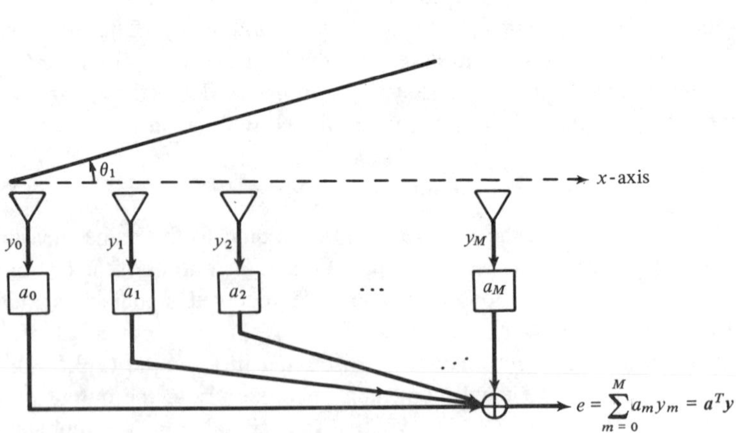


**Figure 5.23** Beamforming

where $k_x$ is the $x$-component of the wavenumber $\mathbf{k}$

$$k_x = \frac{\omega}{c} \sin \theta_1$$

The value of this field at the $m$th sensor, $x_m = md$, is then

$$A_1 \, e^{-jmk_1}$$

where $k_1$ denotes the *normalized* wavenumber

$$k_1 = k_x d = \frac{\omega d}{c} \sin \theta_1 = \frac{2\pi d}{\lambda} \sin \theta_1 \tag{5.12.4}$$

This is the analog of the digital frequency. To avoid aliasing effects arising from the spatial sampling process, the spatial sampling frequency $1/d$ must be greater than or equal to twice the spatial frequency of the wave; namely, $1/\lambda$. Thus, we must have $1/d \geq 2(1/\lambda)$, or $d \leq \lambda/2$. Since $\sin \theta_1$ has magnitude less than one, the sampling condition forces $k_1$ to lie within the Nyquist interval

$$-\pi \leq k_1 \leq \pi$$

In this case the correspondence between $k_1$ and $\theta_1$ is unique. For any angle $\theta$ and corresponding normalized wavenumber $k$, we introduce the phasing, or *steering vector*,

$$\mathbf{s}_k = \begin{bmatrix} 1 \\ e^{jk} \\ e^{2jk} \\ \vdots \\ e^{Mjk} \end{bmatrix}, \qquad k = \frac{2\pi d}{\lambda} \sin \theta \tag{5.12.5}$$

In this notation, the plane wave measured at the sensors is represented by the vector

$$\mathbf{y} = A_1 \, \mathbf{s}_{k_1}^* = A_1 \begin{bmatrix} 1 \\ e^{-jk_1} \\ e^{-2jk_1} \\ \vdots \\ e^{-Mjk_1} \end{bmatrix}$$

The steering vector of array weights $\mathbf{a}$, steered towards an arbitrary direction $\theta$, is also expressed in terms of the phasing vector $\mathbf{s}_k$; we have

$$a_m = e^{-j\omega\tau_m} = e^{-j\omega(M-m)(d\sin\theta/c)} = e^{-jMk} \, e^{+jmk}$$

or, ignoring the overall common phase $e^{-jMk}$, we have

$$\mathbf{a} = \mathbf{s}_k \qquad \text{(steering vector towards } k = \frac{2\pi d}{\lambda} \sin \theta) \tag{5.12.6}$$

The output of the beamformer, steered towards $\theta$, is

$$e = \mathbf{a}^T\mathbf{y} = \mathbf{s}_k^T\mathbf{y} = A_1\mathbf{s}_k^T\mathbf{s}_{k_1}^* = A_1\mathbf{s}_{k_1}^\dagger\mathbf{s}_k = A_1W(k - k_1)^*$$

where $W(\cdot)$ was defined in Section 5.11. The mean output power of the beamformer steered towards $k$ is

$$S(k) = E[e^*e] = \mathbf{a}^\dagger E[\mathbf{y}^*\mathbf{y}^T]\mathbf{a} = \mathbf{a}^\dagger R\mathbf{a} = \mathbf{s}_k^\dagger R\mathbf{s}_k$$

Using $\mathbf{y} = A_1\mathbf{s}_{k_1}^*$, we find $R = E[\mathbf{y}^*\mathbf{y}^T] = P_1\mathbf{s}_{k_1}\mathbf{s}_{k_1}^\dagger$, where $P_1 = E[|A_1|^2]$; and

$$S(k) = \mathbf{s}_k^\dagger R\mathbf{s}_k = P_1\mathbf{s}_k^\dagger\mathbf{s}_{k_1}\mathbf{s}_{k_1}^\dagger\mathbf{s}_k$$

$$= P_1|W(k - k_1)|^2$$

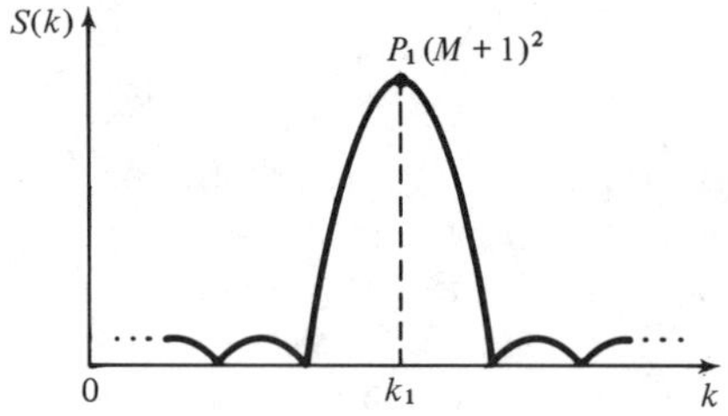

If the beam is steered on target, that is, if $\theta = \theta_1$ or $k = k_1$, then $S(k_1) = P_1(M + 1)^2$, and the output power is enhanced. The response pattern of the array has the same shape as $W(k)$, and therefore its resolution capability is limited to the width $\Delta k = 2\pi/(M + 1)$ of the main lobe of the smearing function $W(k)$. Using $\Delta k = (2\pi d/\lambda)\Delta\theta$ we find the basic angular resolution to be $\Delta\theta = [\lambda/(M + 1)d]$, or $\Delta\theta = \lambda/D$, where $D = (M + 1)d$ is the effective *aperture* of the array. This is the classic Rayleigh limit on the resolving power of an optical system with aperture $D$.

Next, we consider the problem of resolving the directions of arrival of *multiple* plane waves incident on an array in the presence of background noise. We assume $L$ planes waves incident on an array of $M + 1$ sensors from angles $\theta_i$; $i = 1,2, \ldots ,L$. The incident field is sampled at the sensors giving rise to a series of "snapshots." At the $n$th snapshot time instant, the field received at the $m$th sensor is of the form [75]

$$y_m(n) = v_m(n) + \sum_{i=1}^{L} A_i(n)e^{-jmk_i}, \qquad m = 0,1, \ldots ,M \qquad (5.12.7)$$

where $A_i(n)$ is the amplitude of the $i$th wave (it would be a constant independent of time if we had exact sinusoidal dependence at the operating frequency), and $k_i$ are the normalized wavenumbers related to the angles of arrival by

$$k_i = \frac{2\pi d}{\lambda}\sin\theta_i; \; i = 1,2, \ldots ,L \qquad (5.12.8)$$

and $v_m(n)$ is the background noise, which is assumed to be *spatially incoherent*, and also uncorrelated with the signal amplitudes $A_i(n)$; that is,

$$E[v_m(n)^*v_k(n)] = \sigma_v^2\delta_{mk}; \qquad E[v_m(n)^*A_i(n)] = 0 \qquad (5.12.9)$$

Eq. (5.12.7) can be written in vector form as

$$\mathbf{y}(n) = \mathbf{v}(n) + \sum_{i=1}^{L} A_i(n)\mathbf{s}_{k_i}^* \qquad (5.12.10)$$

The autocorrelation matrix of the signal field sensed by the array is

$$R = E[\mathbf{y}(n)^*\mathbf{y}(n)^T] = \sigma_v^2 I + \sum_{i,j=1}^{L} \mathbf{s}_{k_i} P_{ij} \mathbf{s}_{k_j}^\dagger \qquad (5.12.11)$$

where $I$ is the $(M + 1) \times (M + 1)$ unit matrix, and $P_{ij}$ is the amplitude correlation matrix

$$P_{ij} = E[A_i(n)^* A_j(n)] \qquad 1 \le i,j \le L \qquad (5.12.12)$$

If the sources are uncorrelated with respect to each other, the power matrix $P_{ij}$ is diagonal. Introducing the $(M + 1) \times L$ signal matrix

$$S = [\mathbf{s}_{k_1}, \mathbf{s}_{k_2}, \ldots, \mathbf{s}_{k_L}]$$

we write Eq. (5.12.11) as

$$R = \sigma_v^2 I + SPS^\dagger \qquad (5.12.13)$$

which is the same as Eq. (5.11.9) of the previous section. Therefore, the analytical expressions of the various spectral estimators can be transferred to this problem as well. We summarize the various spectrum estimators below:

$$\hat{S}_B(k) = \mathbf{s}_k^\dagger R \mathbf{s}_k \qquad \text{(conventional Bartlett beamformer)}$$

$$\hat{S}_{AR}(k) = \frac{1}{|\mathbf{s}_k^\dagger R^{-1}\mathbf{u}_0|^2} \qquad \text{(LP spectral estimate)}$$

$$\hat{S}_{ML}(k) = \frac{1}{\mathbf{s}_k^\dagger R^{-1}\mathbf{s}_k} \qquad \text{(ML beamformer)}$$

For example, for uncorrelated sources $P_{ij} = P_i \delta_{ij}$, the Bartlett spatial spectrum will be

$$\hat{S}_B(k) = \mathbf{s}_k^\dagger R \mathbf{s}_k = \sigma_v^2(M + 1) + \sum_{i=1}^{L} P_i |W(k - k_i)|^2$$

which gives rise to peaks at the desired wavenumbers $k_i$ from which the angles $\theta_i$ can be extracted. When the beam is steered towards the $i$th plane wave, the measured power at the output of the beamformer will be

$$\hat{S}_B(k_i) = \sigma_v^2(M + 1) + P_i(M + 1)^2 + \sum_{j \ne i} P_j |W(k_i - k_j)|^2$$

Ignoring the third term for the moment, we observe the basic *improvement* in the SNR offered by beamforming:

$$\frac{P_i(M + 1)^2}{\sigma_v^2(M + 1)} = \frac{P_i}{\sigma_v^2}(M + 1)$$

If the sources are too close to each other [closer than the beamwidth of $W(k)$], the resolution ability of the beamformer worsens. In such cases, the alternative spectral estimates offer better resolution, with the LP estimate typically having a better performance. The resolution capability of both the ML and the LP estimates improves with higher SNR, whereas that of the conventional beamformer does not.

The Pisarenko method can also be applied here. As discussed in the previous section, the $(M + 1)$-dimensional eigenvalue problem $R\mathbf{a} = \lambda\mathbf{a}$ has an $L$-dimensional *signal subspace* with eigenvalues greater than $\sigma_v^2$, and a $(M + 1 - L)$-dimensional *noise subspace* spanned by the degenerate eigenvectors belonging to the minimum eigenvalue of $\sigma_v^2$. *Any* vector $\mathbf{a}$ in the *noise subspace* will have at least $L$ zeros at the desired wavenumber frequencies $k_i$; that is, the polynomial

$$A(z) = a_0 + a_1 z^{-1} + \cdots + a_M z^{-M}$$

will have $L$ zeros at

$$z_i = \exp(jk_i); \qquad i = 1, 2, \ldots, L$$

and $(M - L)$ other spurious zeros. This can be seen as follows: If $R\mathbf{a} = \sigma_v^2\,\mathbf{a}$, then Eq. (5.12.13) implies

$$(\sigma_v^2 I + SPS^\dagger)\mathbf{a} = \sigma_v^2\mathbf{a} \qquad \text{or} \qquad SPS^\dagger\mathbf{a} = 0$$

dotting with $\mathbf{a}^\dagger$ we find $\mathbf{a}^\dagger SPS^\dagger\mathbf{a} = 0$, and since $P$ is assumed to be strictly positive definite, it follows that $S^\dagger\mathbf{a} = 0$, or

$$S^\dagger\mathbf{a} = \begin{bmatrix} A(k_1) \\ A(k_2) \\ \vdots \\ A(k_L) \end{bmatrix} = 0$$

In practice, *estimates* of the covariance matrix $R$ are used. For example, if the sensor outputs are recorded over $N$ snapshots—that is, $\mathbf{y}(n)$; $n = 0, 1, \ldots, N - 1$, then, the covariance matrix $R$ may be estimated by replacing the ensemble average of Eq. (5.12.11) with the *time-average*:

$$\hat{R} = \frac{1}{N}\sum_{n=0}^{N-1} \mathbf{y}(n)^* \, \mathbf{y}(n)^T = \text{empirical } R$$

Since the empirical $R$ will not be of the exact theoretical form of Eq. (5.12.11), the degeneracy of the noise subspace will be lifted somewhat. The degree to which this happens depends on how much the empirical $R$ differs from the theoretical $R$. One can still use the *minimum eigenvector* **a** to define the polynomial $A(z)$ and from it an approximate Pisarenko spectral estimator

$$\hat{S}_P(k) = \frac{1}{|A(z)|^2} \qquad \text{where } z = e^{jk}$$

which will have sharp and possibly biased peaks at the desired wavenumber frequencies.

*Example 5.12.1:* Consider the case $L = M = 1$, defined by the theoretical autocorrelation matrix

$$R = \sigma_v^2 I + P_1 \mathbf{s}_{k_1} \mathbf{s}_{k_1}^\dagger = \begin{bmatrix} \sigma_v^2 + P_1 & P_1 e^{-jk_1} \\ P_1 e^{jk_1} & \sigma_v^2 + P_1 \end{bmatrix}$$

Its eigenvectors are

$$\mathbf{e}_1 = \begin{bmatrix} 1 \\ -e^{jk_1} \end{bmatrix}, \qquad \mathbf{e}_2 = \mathbf{s}_{k_1} = \begin{bmatrix} 1 \\ e^{jk_1} \end{bmatrix}$$

belonging to the eigenvalues $\lambda_1 = \sigma_v^2$ and $\lambda_2 = \sigma_v^2 + 2P_1$, respectively. Selecting as the array vector

$$\mathbf{a} = \mathbf{e}_1 = \begin{bmatrix} 1 \\ -e^{jk_1} \end{bmatrix}$$

we obtain a polynomial with a zero at the desired location:

$$A(z) = 1 + a_1 z^{-1} = 1 - e^{jk_1} z^{-1}$$

Now, suppose that the analysis is based on an empirical autocorrelation matrix $\hat{R}$ which differs from the theoretical one by a small amount:

$$\hat{R} = R + \Delta R$$

Using standard first order perturbation theory, we find the correction to the minimum eigenvalue $\lambda_1$ and eigenvector $\mathbf{e}_1$

$$\hat{\lambda}_1 = \lambda_1 + \Delta\lambda_1, \qquad \hat{\mathbf{e}}_1 = \mathbf{e}_1 + \Delta c \mathbf{e}_2$$

where the first order correction terms are

$$\Delta\lambda_1 = \frac{\mathbf{e}_1^\dagger \Delta R \mathbf{e}_1}{\mathbf{e}_1^\dagger \mathbf{e}_1} \qquad \text{and} \qquad \Delta c = \frac{\mathbf{e}_2^\dagger \Delta R \mathbf{e}_1}{(\lambda_1 - \lambda_2)\mathbf{e}_2^\dagger \mathbf{e}_2}$$

The change induced in the zero of the eigenpolynomial is found as follows

$$\hat{\mathbf{a}} = \hat{\mathbf{e}}_1 = \begin{bmatrix} 1 \\ -e^{jk_1} \end{bmatrix} + \Delta c \begin{bmatrix} 1 \\ e^{jk_1} \end{bmatrix} = \begin{bmatrix} 1 + \Delta c \\ -(1 - \Delta c)e^{jk_1} \end{bmatrix}$$

so that

$$\hat{A}(z) = (1 + \Delta c) - (1 - \Delta c)e^{jk_1}z^{-1}$$

and the zero is now at $z_1 = e^{jk_1}\dfrac{1 - \Delta c}{1 + \Delta c} = e^{jk_1}(1 - 2\Delta c)$ to first order in

$\Delta c$. Since $\Delta c$ is generally complex, the factor $(1 - 2\Delta c)$ will cause both a change (bias) in the phase of the zero $e^{jk_1}$, and will move it off the unit circle reducing the resolution. Another way to see this is to compute the value of the polynomial steered on target; that is,

$$\hat{A}(k_1) = \mathbf{s}_{k_1}^\dagger\hat{\mathbf{a}} = \mathbf{s}_{k_1}^\dagger(\mathbf{e}_1 + \Delta c\mathbf{e}_2) = \Delta c\mathbf{s}_{k_1}^\dagger\mathbf{e}_2 = 2\Delta c$$

which is small but not zero.

There are a number of alternative eigenvector-based methods, which utilize *all* the noise subspace eigenvectors, not just the minimum one. Let $\mathbf{e}_{Ni}$; $i = 1,2, \ldots ,N$, where $N = M + 1 - L$, be the degenerate eigenvectors (belonging to $\sigma_v^2$) of the theoretical autocorrelation matrix $R$ of Eq. (5.12.13). And let $\mathbf{e}_{Si}$; $i = 1,2, \ldots ,L$ be the eigenvectors spanning the signal subspace of $R$. Assume these eigenvectors have been *orthonormalized*. Define the eigenvector matrices

$$E_N = [\mathbf{e}_{N1},\mathbf{e}_{N2}, \ldots ,\mathbf{e}_{NN}], \qquad E_S = [\mathbf{e}_{S1},\mathbf{e}_{S2}, \ldots ,\mathbf{e}_{SL}]$$

Since every noise eigenvector has nulls at $k_i$, so will $E_N$; that is, the quantity

$$\mathbf{s}_k^\dagger E_N$$

steered towards any of the $L$ directions, $k = k_i$, vanishes. Thus, a general class of spectral estimators can be defined by

$$\hat{S}_{EV}(k) = \frac{1}{\mathbf{s}_k^\dagger E_N D E_N^\dagger \mathbf{s}_k} \qquad \text{(eigenvector spectral estimator)}$$

where $D$ is any positive semi-definite matrix. This expression will have infinite peaks whenever it is steered toward the plane waves. In practice, the eigenvectors $E_N$ extracted from the empirical covariance matrix $R$ will only be approximately orthogonal to the signal directions, and the estimator will produce sharp but, finite (and possibly biased) peaks. The choice $D = I_N$ corresponds to Schmidt's [85] method, and the choice $D = \Lambda_N^{-1}$ to Johnson and DeGraaf's [88] method, where $I_N$ is the unit matrix of the noise subspace and $\Lambda_N$ is the diagonal matrix of the noise eigenvalues (which are slightly different from $\sigma_v^2$ due to the lifting of the degeneracy). This expression also encompasses Pisarenko's single eigen-vector method.

To see the effect of inaccuracies in the knowledge of $R$ let us carry out a first order perturbation analysis of the noise eigenvectors. The eigenvalue problem based on the theoretical $R$ is

$$RE_N = \sigma_v^2 E_N, \qquad RE_S = E_S \Lambda_S$$

where the first equation expresses the degeneracy of the noise eigenvectors, and $\Lambda_S$ is the diagonal matrix containing all the signal subspace eigenvalues. Let $\hat{R} = R + \Delta R$ represent a small perturbation of $R$. The perturbation $\Delta R$ lifts the degeneracy of $\sigma_v^2$ and the new set of noise eigenvalues will be

$$\hat{\Lambda}_N = \sigma_v^2 I_N + \Delta\Lambda_N, \qquad \Delta\Lambda_N = \mathrm{diag}[\Delta\lambda_1, \Delta\lambda_2, \ldots, \Delta\lambda_N]$$

where $\Delta\Lambda_N$ is a diagonal matrix. These represent the solution to the eigenvalue problem

$$\hat{R}\hat{E}_N = \hat{E}_N \hat{\Lambda}_N$$

where $\hat{E}_N$ is the new set of noise eigenvectors. Using standard degenerate first order perturbation theory we find

$$\hat{E}_N = E_N C + E_S \Delta C$$

where the matrices $C$, $\Delta C$, and $\Delta\Lambda_N$ are found as follows: The corrections to the eigenvalues are found from the secular equation

$$\det[\Delta\lambda I_N - E_N^\dagger \Delta R E_N] = 0$$

and the matrix $C$ is determined as the solution to

$$E_N^\dagger \Delta R E_N C = C\Delta\Lambda_N$$

normalized so that $C^\dagger C = I_N$. Then, $\Delta C$ is obtained by

$$\Delta C = -(\Lambda_S - \sigma_v^2 I_L)^{-1} E_S^\dagger \Delta R E_N C$$

The point of this analysis is to see the effect of the perturbation on the spectral estimator. We have

$$\mathbf{s}_k^\dagger \hat{E}_N = \mathbf{s}_k^\dagger E_N C + \mathbf{s}_k^\dagger E_s \Delta C$$

Thus, when steered towards any of the plane waves, we have

$$\mathbf{s}_{k_i}^\dagger \hat{E}_N = \mathbf{s}_{k_i}^\dagger E_S \Delta C$$

which is small, thus resulting in sharp spectral peaks in the estimator.

The high resolution properties of the Pisarenko and the other eigenvector methods depend directly on the assumption that the background noise field is *spatially incoherent*, resulting in the special structure of the autocorrelation matrix $R$. When the noise is spatially coherent, a different eigen-analysis must be carried out. Suppose the covariance matrix of the noise field $\mathbf{v}$ is

$$E[\mathbf{v}^*\mathbf{v}^T] = \sigma_v^2 Q$$

where $Q$ reflects the spatial coherence of $\mathbf{v}$. Then the covariance matrix of Eq. (5.12.13) is replaced by

$$R = \sigma_v^2 Q + SPS^\dagger$$

The relevant eigenvalue problem is now the *generalized* eigenvalue problem

$$R\mathbf{a} = \lambda Q\mathbf{a}$$

Consider any such generalized eigenvector $\mathbf{a}$, and assume it is normalized such that

$$\mathbf{a}^\dagger Q\mathbf{a} = 1$$

Then, the corresponding eigenvalue is expressed as

$$\lambda = \lambda\,\mathbf{a}^\dagger Q\mathbf{a} = \mathbf{a}^\dagger R\mathbf{a} = \sigma_v^2\mathbf{a}^\dagger Q\mathbf{a} + \mathbf{a}^\dagger SPS^\dagger\mathbf{a}$$

$$= \sigma_v^2 + \sum_{i=1}^{L} P_i|A(k_i)|^2$$

which shows that the minimum eigenvalue is $\sigma_v^2$ and it is attained whenever $A(k_i) = 0;\ i = 1,2,\ \ldots\ ,L$. Therefore, the eigenpolynomial $A(z)$ can be used to determine the wavenumbers $k_i$. Thus, the procedure is to solve the generalized eigenvalue problem and select the minimum eigenvector. This eigenvalue problem is also equivalent to the minimization problem

$$\mathcal{E} = \mathbf{a}^\dagger R\mathbf{a} = \text{min}, \qquad \text{subject to } \mathbf{a}^\dagger Q\mathbf{a} = 1$$

The other versions of the eigenvector methods are defined in a similar manner. The practical implementation of the method requires knowledge of the noise background covariance matrix $Q$, which is not always possible to obtain.

## 5.13 *Dynamic Predictive Deconvolution— Waves in Layered Media*

The analysis and synthesis lattice filters, implemented via the Levinson recursion, were obtained within the context of linear prediction. Here, we would like to point out the remarkable fact that the same analysis and synthesis lattice structures also occur naturally in the problem of *wave propagation* in *layered media* [6,7,9,19,21,34,38,92–100]. This is perhaps the reason behind the great success of linear prediction methods in speech and seismic signal processing. In fact, historically many linear prediction techniques were originally developed within the context of these two application areas.

In speech, the vocal tract is modeled as an acoustic tube of varying cross-sectional area. It can be approximated by the *piece-wise* constant area approximation shown below:

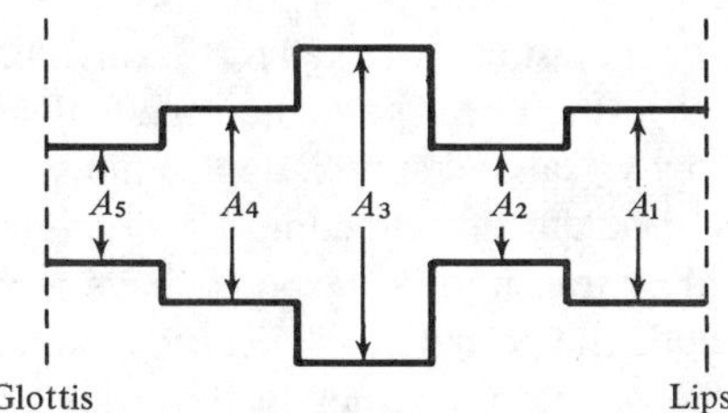


The acoustic impedance of a sound wave varies inversely with the tube area

$$Z = \frac{\rho c}{A}$$

where $\rho$, $c$, and $A$ are the air density, speed of sound, and tube area, respectively. Therefore, as the sound wave propagates from the glottis to the lips, it will suffer reflections every time it encounters an interface; that is, every time it enters a tube segment of different diameter. Multiple reflections will be set up within each segment and the tube will *reverberate* in a complicated manner depending on the number of segments and the diameter of each segment. By measuring the speech wave that eventually comes out of the lips, it is possible to remove, or deconvolve, the reverberatory effects of the tube and, in the process, extract the *tube parameters*, such as the areas of the segments or, equivalently, the reflection coefficients at the interfaces. During speech, the configuration of the vocal tract tube changes continuously. But being a mechanical system, it does so fairly slowly, and for short periods of time (of the order of 20–30 msec) it may be assumed to maintain a fixed configuration. From each such short segment of speech, a set of configuration parameters (e.g., reflection coefficients) may be extracted. This set may be used to synthesize the speech segment.

The seismic problem is somewhat different. Here it is not the transmitted wave that is experimentally accessible, but rather the overall reflected wave:

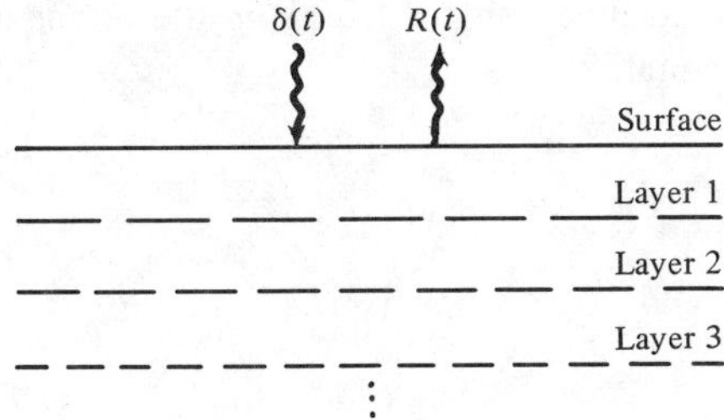

An impulsive input to the earth, such as a dynamite explosion near the surface, will set up seismic elastic waves propagating downwards. As the various earth layers are encountered, reflections will take place. Eventually each layer will be reverberating and an overall reflected wave will be measured at the surface. On the basis of this reflected wave, the layered structure (i.e., reflection coefficients, impedances, etc.) must be extracted by deconvolution techniques. These are essentially identical to the linear prediction methods.

In addition to geophysical and speech applications, this wave problem and the associated *inverse* problem of extracting the structure of the medium from the observed (reflected or transmitted) response have a number of other applications. Examples include the probing of dielectric materials by electromagnetic waves, the study of the optical properties of thin films, the probing of tissues by ultrasound, and the design of broadband terminations of transmission lines.

The mathematical analysis of such wave propagation problems has been done more or less independently in each of these application areas, and is well known dating back to the time of Stokes.

In this type of wave propagation problem there are always *two* associated propagating field quantities, the ratio of which is constant and equal to the corresponding *characteristic impedance* of the propagation medium. Examples of these include the electric and magnetic fields in the case of EM waves, the air pressure and particle volume velocity for sound waves, the stress and particle displacement for seismic waves, and the voltage and current waves in the case of TEM transmission lines.

As a concrete example, we have chosen to present in some detail the case of EM waves propagating in lossless dielectrics. The simplest and most basic scattering problem arises when there is a single interface separating two semi-infinite dielectrics of characteristic impedances $Z$ and $Z'$, as shown



where $\mathcal{E}_+$ and $\mathcal{E}_-$ are the right and left moving electric fields in medium $Z$, and $\mathcal{E}'_+$ and $\mathcal{E}'_-$ are those in medium $Z'$. *Matching* the boundary conditions at this interface gives two equations

$$\mathcal{E}_+ + \mathcal{E}_- = \mathcal{E}'_+ + \mathcal{E}'_- \qquad \text{(continuity of electric field)}$$

$$\frac{1}{Z}(\mathcal{E}_+ - \mathcal{E}_-) = \frac{1}{Z'}(\mathcal{E}'_+ - \mathcal{E}'_-) \qquad \text{(continuity of magnetic field)}$$

Introducing the reflection and transmission coefficients

$$\rho = \frac{Z' - Z}{Z' + Z}, \qquad \tau = 1 + \rho, \qquad \rho' = -\rho,$$

$$\tau' = 1 + \rho' = 1 - \rho \qquad (5.13.1)$$

the above equations can be written in a transmission matrix form

$$\begin{bmatrix} \mathcal{E}_+ \\ \mathcal{E}_- \end{bmatrix} = \frac{1}{\tau} \begin{bmatrix} 1 & \rho \\ \rho & 1 \end{bmatrix} \begin{bmatrix} \mathcal{E}'_+ \\ \mathcal{E}'_- \end{bmatrix} \qquad (5.13.2)$$

The flow of energy carried by these waves is given by the Poynting vector

$$P = \frac{1}{2} Re \left[ (\mathcal{E}_+ + \mathcal{E}_-)^* \frac{1}{Z} (\mathcal{E}_+ - \mathcal{E}_-) \right] = \frac{1}{Z} (\mathcal{E}_+^* \mathcal{E}_+ - \mathcal{E}_-^* \mathcal{E}_-) \qquad (5.13.3)$$

One consequence of the above matching conditions is that the total energy flow to the right is preserved *across* the interface: that is,

$$\frac{1}{Z} (\mathcal{E}_+^* \mathcal{E}_+ - \mathcal{E}_-^* \mathcal{E}_-) = \frac{1}{Z'} (\mathcal{E}_+'^* \mathcal{E}_+' - \mathcal{E}_-'^* \mathcal{E}_-') \qquad (5.13.4)$$

It proves convenient to absorb the factors $1/Z$ and $1/Z'$ into the definitions for the fields by *renormalizing* them as follows:

$$\begin{bmatrix} E_+ \\ E_- \end{bmatrix} = \frac{1}{\sqrt{Z}} \begin{bmatrix} \mathcal{E}_+ \\ \mathcal{E}_- \end{bmatrix}, \qquad \begin{bmatrix} E'_+ \\ E'_- \end{bmatrix} = \frac{1}{\sqrt{Z'}} \begin{bmatrix} \mathcal{E}'_+ \\ \mathcal{E}'_- \end{bmatrix}$$

Then, Eq. (5.13.4) reads

$$E_+^* E_+ - E_-^* E_- = E_+'^* E_+' - E_-'^* E_-' \qquad (5.13.5)$$

and the matching equations (5.13.2) can be written in the normalized form

$$\begin{bmatrix} E_+ \\ E_- \end{bmatrix} = \frac{1}{t} \begin{bmatrix} 1 & \rho \\ \rho & 1 \end{bmatrix} \begin{bmatrix} E'_+ \\ E'_- \end{bmatrix}, \qquad t = \sqrt{1 - \rho^2} = \sqrt{\tau \tau'} \qquad (5.13.6)$$

They may also be written in a *scattering matrix* form that relates the *outgoing* fields to the *incoming* fields, as follows:

$$\begin{bmatrix} E'_+ \\ E_- \end{bmatrix} = \begin{bmatrix} t & \rho' \\ \rho & t \end{bmatrix} \begin{bmatrix} E_+ \\ E'_- \end{bmatrix} = S \begin{bmatrix} E_+ \\ E'_- \end{bmatrix} \qquad (5.13.7)$$

This is the most elementary scattering matrix of all, and $\rho$ and $t$ are the most elementary reflection and transmission responses. From these, the reflection and transmission response of more complicated structures can be built up. In the more general case, we have a dielectric structure consisting of $M$ slabs stacked together as shown in Fig. 5.24.

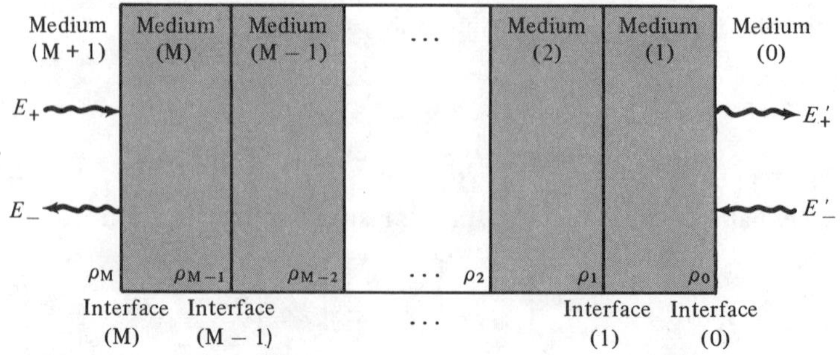



**Figure 5.24**  Layered Structure

The media to the left and right in the figure are assumed to be semi-infinite. The *reflection* and *transmission responses* (from the *left,* or from the *right*) of the structure are defined as the responses of the structure to an impulse (incident from the left, or from the right) as shown in Fig. 5.25.

The corresponding scattering matrix is defined as

$$S = \begin{bmatrix} T & R' \\ R & T' \end{bmatrix}$$

and by linear superposition, the relationship between arbitrary incoming and outgoing waves is

The *inverse scattering problem* that we pose is how to extract the detailed properties of the layered structure, such as the reflection coefficients $\rho_0, \rho_1, \ldots, \rho_M$, from the knowledge of the scattering matrix $S$; that is, from observations of the reflection response $R$ or the transmission response $T$.

Without loss of generality, we may assume the $M$ slabs have *equal travel*

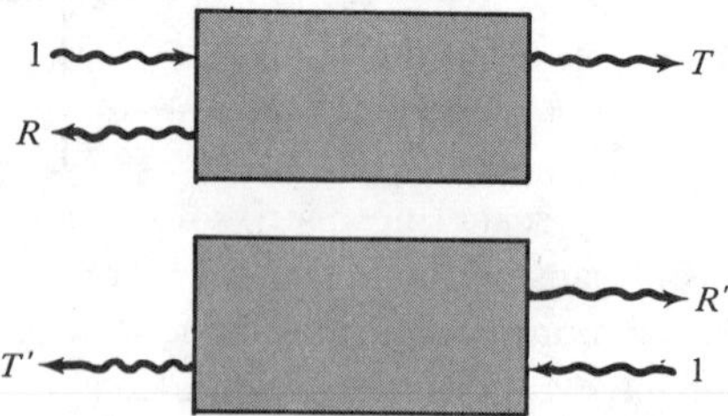



**Figure 5.25**  Reflection and Transmission Responses

*time*. We denote the common one-way travel time by $T_1$ and the two-way travel time by $T_2 = 2T_1$. As an impulse $\delta(t)$ is incident from the left on interface $M$, there will be immediately a reflected wave and a transmitted wave into medium $M$. When the latter reaches interface $M - 1$, part of it will be transmitted into medium $M - 1$, and part will be reflected back towards interface $M$ where it will be partially rereflected towards $M - 1$ and partially transmitted to the left into medium $M + 1$, thus contributing towards the overall reflection response. Since the wave had to travel to interface $M - 1$ and back, this latter contribution will occur at time $T_2$. Similarly, another wave will return back to interface $M$ due to reflection from the second interface $M - 2$; this wave will return $2T_2$ seconds later and will add to the contribution from the zig-zag path within medium $M$ which is also returning at $2T_2$, and so on. The timing diagram below shows all the possible return paths up to time $t = 3T_2$, during which the original impulse can only travel as far as interface $M - 3$:

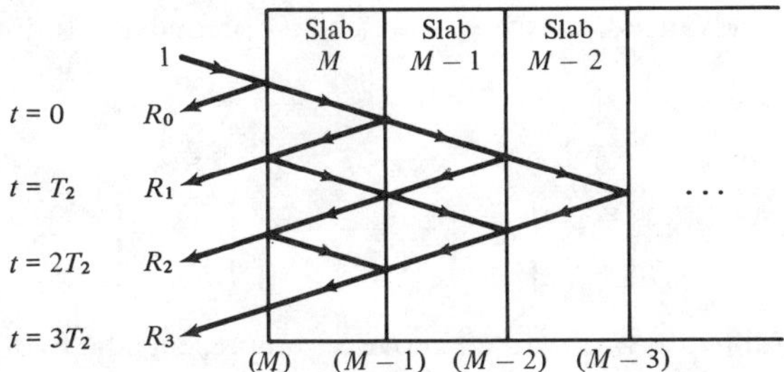


When we add the contributions of all the returned waves we see that the reflection response will be a linear superposition of returned impulses

$$R(t) = \sum_{k=0}^{\infty} R_k \delta(t - kT_2)$$

It has a Fourier transform expressible more conveniently as the $z$-transform

$$R(z) = \sum_{k=0}^{\infty} R_k z^{-k} \qquad \text{where } z = e^{j\omega T_2}$$

Observe that $R$ is *periodic in frequency* $\omega$ with period $2\pi/T_2$, which plays a role analogous to the sampling frequency. Therefore, it is enough to specify $R$ within the *Nyquist interval* $[-\pi/T_2, \pi/T_2]$.

Next, we develop the *lattice recursions* that facilitate the solution of the direct and the inverse scattering problems. Consider the $m$th slab and let $E_m^{\pm}$ be the right/left moving waves incident on the left side of the $m$th interface. To relate them to the same quantities $E_{m-1}^{\pm}$ incident on the left side of the $(m - 1)$st interface, first we use the *matching* equations to "pass" to the other side of the

$m$th interface and into the $m$th slab, and then we *propagate* these quantities to reach the left side of the $(m - 1)$st interface. This is shown below.

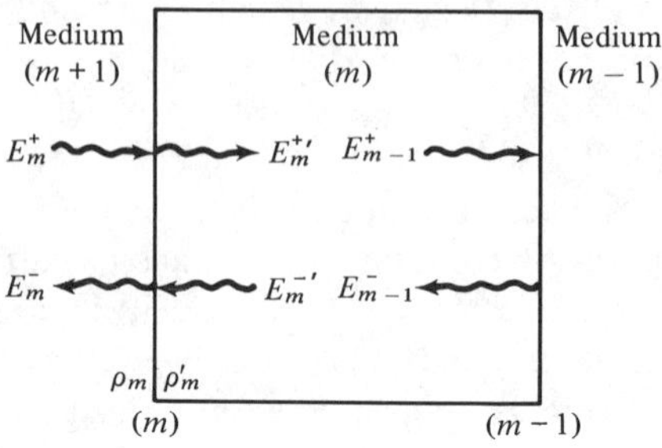


$$\begin{bmatrix} E_m^+ \\ E_m^- \end{bmatrix} = \frac{1}{t_m} \begin{bmatrix} 1 & \rho_m \\ \rho_m & 1 \end{bmatrix} \begin{bmatrix} E_m^{+\prime} \\ E_m^{-\prime} \end{bmatrix}, \qquad t_m = (1 - \rho_m^2)^{1/2} \qquad (5.13.8)$$

Since the left-moving wave $E_m^{-\prime}$ is the delayed replica of $E_{m-1}^-$ by $T_1$ seconds, and $E_m^{+\prime}$ is the advanced replica of $E_{m-1}^+$ by $T_1$ seconds, it follows that

$$E_m^{+\prime} = z^{1/2} E_{m-1}^+, \qquad E_m^{-\prime} = z^{-1/2} E_{m-1}^-$$

or, in matrix form,

$$\begin{bmatrix} E_m^{+\prime} \\ E_m^{-\prime} \end{bmatrix} = \begin{bmatrix} z^{1/2} & 0 \\ 0 & z^{-1/2} \end{bmatrix} \begin{bmatrix} E_{m-1}^+ \\ E_{m-1}^- \end{bmatrix} \qquad (5.13.9)$$

where the variable $z^{-1}$ was defined above and represents the two-way travel time delay, while $z^{-1/2}$ represents the one-way travel time delay. Combining the matching and propagation equations (5.13.8) and (5.13.9), we obtain the desired relationship between $E_m^\pm$ and $E_{m-1}^\pm$:

$$\begin{bmatrix} E_m^+ \\ E_m^- \end{bmatrix} = \frac{z^{1/2}}{t_m} \begin{bmatrix} 1 & \rho_m z^{-1} \\ \rho_m & z^{-1} \end{bmatrix} \begin{bmatrix} E_{m-1}^+ \\ E_{m-1}^- \end{bmatrix} \qquad (5.13.10)$$

Or, introducing some convenient vector notation

$$E_m(z) = \psi_m(z) E_{m-1}(z) \qquad (5.13.11)$$

where

$$E_m(z) = \begin{bmatrix} E_m^+(z) \\ E_m^-(z) \end{bmatrix}, \qquad \psi_m(z) = \frac{z^{1/2}}{t_m} \begin{bmatrix} 1 & \rho_m z^{-1} \\ \rho_m & z^{-1} \end{bmatrix} \qquad (5.13.12)$$

The "match-and-propagate" transition matrix $\psi_m(z)$ has two interesting properties; namely, defining $\bar{\psi}_m(z) = \psi_m(z^{-1})$

$$\bar{\psi}_m(z)^T J_3 \psi(z) = \psi_m(z^{-1})^T J_3 \psi_m(z) = J_3, \qquad J_3 = \begin{bmatrix} 1 & 0 \\ 0 & -1 \end{bmatrix} \qquad (5.13.13)$$

$$\bar{\psi}_m(z) = \psi_m(z^{-1}) = J_1 \psi_m(z) J_1, \qquad J_1 = \begin{bmatrix} 0 & 1 \\ 1 & 0 \end{bmatrix} \qquad (5.13.14)$$

From Eq. (5.3.13) we have

$$\bar{E}_m^+ E_m^+ - \bar{E}_m^- E_m^- = \bar{E}_m^T J_3 E_m = \bar{E}_{m-1}^T \bar{\psi}_m^T J_3 \psi_m E_{m-1} = \bar{E}_{m-1}^T J_3 E_{m-1}$$
$$= \bar{E}_{m-1}^+ E_{m-1}^+ - \bar{E}_{m-1}^- E_{m-1}^- \qquad (5.13.15)$$

which is equivalent to *energy conservation*, according to Eq. (5.13.5). The second property, Eq. (5.13.14), expresses *time-reversal invariance* and allows the construction of a *second*, linearly independent, solution of the recursive equations (5.13.11):

$$\hat{E}_m = J_1 \bar{E}_m = \begin{bmatrix} \bar{E}_m^- \\ \bar{E}_m^+ \end{bmatrix} = J_1 \bar{\psi}_m \bar{E}_{m-1} = J_1 \bar{\psi}_m J_1 J_1 \bar{E}_{m-1} = \psi_m \hat{E}_{m-1} \qquad (5.13.16)$$

The recursions (5.13.11) may be iterated now down to $m = 0$. By an additional boundary match, we may pass to the right side of interface $m = 0$:

$$E_m = \psi_m \psi_{m-1} \cdots \psi_1 E_0 = \psi_m \psi_{m-1} \cdots \psi_1 \psi_0 E_0'$$

where we defined $\psi_0$ by

$$\psi_0 = \frac{1}{t_0} \begin{bmatrix} 1 & \rho_0 \\ \rho_0 & 1 \end{bmatrix}$$

or, more explicitly,

$$\begin{bmatrix} E_m^+ \\ E_m^- \end{bmatrix} = \frac{z^{m/2}}{t_m t_{m-1} \cdots t_1 t_0} \begin{bmatrix} 1 & \rho_m z^{-1} \\ \rho_m & z^{-1} \end{bmatrix}$$
$$\cdots \begin{bmatrix} 1 & \rho_1 z^{-1} \\ \rho_1 & z^{-1} \end{bmatrix} \begin{bmatrix} 1 & \rho_0 \\ \rho_0 & 1 \end{bmatrix} \begin{bmatrix} E_0^{+\prime} \\ E_0^{-\prime} \end{bmatrix} \qquad (5.13.17)$$

To deal with this product of matrices, we define

$$\begin{bmatrix} A_m & C_m \\ B_m & D_m \end{bmatrix} = \begin{bmatrix} 1 & \rho_m z^{-1} \\ \rho_m & z^{-1} \end{bmatrix} \cdots \begin{bmatrix} 1 & \rho_1 z^{-1} \\ \rho_1 & z^{-1} \end{bmatrix} \begin{bmatrix} 1 & \rho_0 \\ \rho_0 & 1 \end{bmatrix} \qquad (5.13.18)$$

where $A_m, B_m, C_m, D_m$ are *polynomials* of degree $m$ in the variable $z^{-1}$. The energy conservation and time-reversal invariance properties of the $\psi_m$ matrices imply similar properties for these polynomials. Writing Eq. (5.13.18) in terms of the $\psi_m$s, we have

$$\begin{bmatrix} A_m & C_m \\ B_m & D_m \end{bmatrix} = z^{-m/2} \sigma_m \psi_m \psi_{m-1} \cdots \psi_1 \psi_0$$

where we defined the quantity

$$\sigma_m = t_m t_{m-1} \cdots t_1 t_0 = \prod_{i=0}^{m} (1 - \rho_i^2)^{1/2} \qquad (5.13.19)$$

The property (5.13.13) implies the same for the above product of matrices; that is,

$$\begin{bmatrix} \bar{A}_m & \bar{C}_m \\ \bar{B}_m & \bar{D}_m \end{bmatrix}^T \begin{bmatrix} 1 & 0 \\ 0 & -1 \end{bmatrix} \begin{bmatrix} A_m & C_m \\ B_m & D_m \end{bmatrix} = \begin{bmatrix} 1 & 0 \\ 0 & -1 \end{bmatrix} \sigma_m^2$$

which implies that the quantity

$$\bar{A}_m(z)A_m(z) - \bar{B}_m(z)B_m(z) = \sigma_m^2 \tag{5.13.20}$$

is independent of $z$. The property (5.13.14) implies that $C_m$ and $D_m$ are the reverse polynomials $B_m^R$ and $A_m^R$, respectively; in fact

$$\begin{bmatrix} A_m^R & C_m^R \\ B_m^R & D_m^R \end{bmatrix} = z^{-m} \begin{bmatrix} \bar{A}_m & \bar{C}_m \\ \bar{B}_m & \bar{D}_m \end{bmatrix} = z^{-m} z^{m/2} \sigma_m \bar{\psi}_m \cdots \bar{\psi}_0$$

$$= z^{-m/2} \sigma_m J_1 (\psi_m \cdots \psi_0) J_1 = J_1 \begin{bmatrix} A_m & C_m \\ B_m & D_m \end{bmatrix} J_1$$

$$= \begin{bmatrix} 0 & 1 \\ 1 & 0 \end{bmatrix} \begin{bmatrix} A_m & C_m \\ B_m & D_m \end{bmatrix} \begin{bmatrix} 0 & 1 \\ 1 & 0 \end{bmatrix} = \begin{bmatrix} D_m & B_m \\ C_m & A_m \end{bmatrix} \tag{5.13.21}$$

from which it follows that $C_m(z) = B_m^R(z)$ and $D_m(z) = A_m^R(z)$. The definition (5.13.18) also implies the recursion

$$\begin{bmatrix} A_m & B_m^R \\ B_m & A_m^R \end{bmatrix} = \begin{bmatrix} 1 & \rho_m z^{-1} \\ \rho_m & z^{-1} \end{bmatrix} \begin{bmatrix} A_{m-1} & B_{m-1}^R \\ B_{m-1} & A_{m-1}^R \end{bmatrix}$$

Therefore each column of the *ABCD* matrix satisfies the same recursion.

To summarize, we have

$$\begin{bmatrix} A_m(z) & B_m^R(z) \\ B_m(z) & A_m^R(z) \end{bmatrix} = \begin{bmatrix} 1 & \rho_m z^{-1} \\ \rho_m & z^{-1} \end{bmatrix} \cdots \begin{bmatrix} 1 & \rho_1 z^{-1} \\ \rho_1 & z^{-1} \end{bmatrix} \begin{bmatrix} 1 & \rho_0 \\ \rho_0 & 1 \end{bmatrix} \tag{5.13.22}$$

with the *lattice recursion*

$$\begin{bmatrix} A_m(z) \\ B_m(z) \end{bmatrix} = \begin{bmatrix} 1 & \rho_m z^{-1} \\ \rho_m & z^{-1} \end{bmatrix} \begin{bmatrix} A_{m-1}(z) \\ B_{m-1}(z) \end{bmatrix} \tag{5.13.23}$$

and the property (5.13.20). The lattice recursion may be *initialized* at $m = 0$ by

$$A_0(z) = 1, \quad B_0(z) = \rho_0, \quad \text{or} \quad \begin{bmatrix} A_0(z) & B_0^R(z) \\ B_0(z) & A_0^R(z) \end{bmatrix} = \begin{bmatrix} 1 & \rho_0 \\ \rho_0 & 1 \end{bmatrix} \tag{5.13.24}$$

Furthermore, it follows from the lattice recursion (5.13.23) that the reflection coefficients $\rho_m$ always appear in the *first* and *last* coefficients of the polynomials $A_m(z)$ and $B_m(z)$, as follows:

$$b_m(0) = \rho_m, \quad a_m(0) = 1, \quad b_m(m) = \rho_0, \quad a_m(m) = \rho_0 \rho_m \tag{5.13.25}$$

Eq. (5.13.17) for the field components reads now

$$\begin{bmatrix} E_m^+ \\ E_m^- \end{bmatrix} = \frac{z^{m/2}}{\sigma_m} \begin{bmatrix} A_m & B_m^R \\ B_m & A_m^R \end{bmatrix} \begin{bmatrix} E_0^{+\,\prime} \\ E_0^{-\,\prime} \end{bmatrix}$$

Setting $m = M$, we find the relationship between the fields incident on the dielectric slab structure from the left to those incident from the right:

$$\begin{bmatrix} E_M^+ \\ E_M^- \end{bmatrix} = \frac{z^{M/2}}{\sigma_M} \begin{bmatrix} A_M & B_M^R \\ B_M & A_M^R \end{bmatrix} \begin{bmatrix} E_0^{+\,\prime} \\ E_0^{-\,\prime} \end{bmatrix} \tag{5.13.26}$$

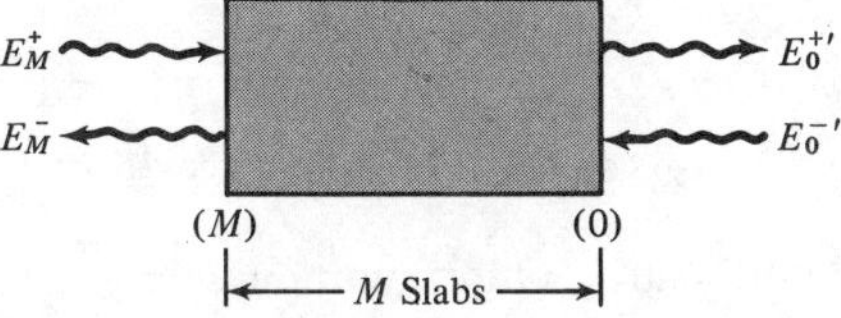


All the multiple reflections and reverberatory effects of the structure are buried in the transition matrix

$$\begin{bmatrix} A_M & B_M^R \\ B_M & A_M^R \end{bmatrix}$$

In reference to Fig. 5.25, the reflection and transmission responses $R$, $T$, $R'$, $T'$ of the structure can be obtained from Eq. (5.13.26) by noting that

$$\begin{bmatrix} 1 \\ R \end{bmatrix} = \frac{z^{M/2}}{\sigma_M} \begin{bmatrix} A_M & B_M^R \\ B_M & A_M^R \end{bmatrix} \begin{bmatrix} T \\ 0 \end{bmatrix}, \qquad \begin{bmatrix} 0 \\ T' \end{bmatrix} = \frac{z^{M/2}}{\sigma_M} \begin{bmatrix} A_M & B_M^R \\ B_M & A_M^R \end{bmatrix} \begin{bmatrix} R' \\ 1 \end{bmatrix}$$

which may be combined into one equation

$$\begin{bmatrix} 1 & 0 \\ R & T' \end{bmatrix} = \frac{z^{M/2}}{\sigma_M} \begin{bmatrix} A_M & B_M^R \\ B_M & A_M^R \end{bmatrix} \begin{bmatrix} T & R' \\ 0 & 1 \end{bmatrix}$$

that can be written as follows:

$$\frac{z^{M/2}}{\sigma_M} \begin{bmatrix} A_M & B_M^R \\ B_M & A_M^R \end{bmatrix} = \begin{bmatrix} 1 & 0 \\ R & T' \end{bmatrix} \begin{bmatrix} T & R' \\ 0 & 1 \end{bmatrix}^{-1}$$

$$= \begin{bmatrix} 1 & 0 \\ R & 1 \end{bmatrix} \begin{bmatrix} T^{-1} & 0 \\ 0 & T' \end{bmatrix} \begin{bmatrix} 1 & -R' \\ 0 & 1 \end{bmatrix}$$

Solving these for the reflection and transmission responses, we find

$$R(z) = \frac{B_M(z)}{A_M(z)} \qquad T(z) = \frac{\sigma_M z^{-M/2}}{A_M(z)}$$

$$R'(z) = -\frac{B_M^R(z)}{A_M(z)} \qquad T'(z) = \frac{\sigma_M z^{-M/2}}{A_M(z)} \tag{5.13.27}$$

Note that $T(z) = T'(z)$. Since on physical grounds the transmission response $T(z)$ must be a *stable* and *causal* $z$-transform, it follows that necessarily the polynomial $A_M(z)$ must be a *minimal-phase polynomial*. The overall delay factor $z^{-M/2}$ in $T(z)$ is of no consequence. It just means that before anything can be transmitted through the structure, it must traverse all $M$ slabs, each with a travel time delay of $T_1$ seconds; that is, with overall delay of $MT_1$ seconds.

The equations (5.13.27) for the scattering responses $R$, $T$, $R'$, $T'$ imply the *unitarity* of the scattering matrix $S$

$$S = \begin{bmatrix} T & R' \\ R & T' \end{bmatrix}$$

that is,

$$\bar{S}(z)^T S(z) = S(z^{-1})^T S(z) = I \qquad (5.13.28)$$

where $I$ is the $2 \times 2$ unit matrix. On the unit circle $z = e^{j\omega}$ the scattering matrix becomes a *unitary* matrix: $S(\omega)^\dagger S(\omega) = I$. Component-wise, Eq. (5.13.28) becomes

$$\bar{T}T + \bar{R}R = \bar{T}'T' + \bar{R}'R' = 1, \qquad \bar{T}R' + \bar{R}T' = 0 \qquad (5.13.29)$$

Robinson and Treitel's *dynamic predictive deconvolution method* [19] of solving the *inverse scattering problem* is based on the above unitarity equation. In the inverse problem, it is required to extract the set of reflection coefficients from measurements of either the reflection response $R$ or the transmission response $T$. In speech processing it is the transmission response that is available. In geophysical applications, or in studying the reflectivity properties of thin films, it is the reflection response that is available. The problem of designing terminations of transmission lines also falls in the latter category. In this case, an appropriate termination is desired that must have a specified reflection response $R(z)$; e.g., to be reflectionless over a wide band of frequencies about some operating frequency.

The solution of both types of problems follows the same steps. First, from the knowledge of the reflection response $R(z)$, or the transmission response $T(z)$, the *spectral function* of the structure is defined

$$\Phi(z) = 1 - R(z)\bar{R}(z) = T(z)\bar{T}(z) = \frac{\sigma_M^2}{A_M(z)A_M(z^{-1})} \qquad (5.13.30)$$

This is recognized as the *power spectrum* of the *transmission* response, and it is of the autoregressive type. Thus, linear prediction methods can be used in the solution.

In the time domain, the autocorrelation lags $\phi(k)$ of the spectral function are obtained from the sample autocorrelations of the reflection sequence, or the transmission sequence

$$\phi(k) = \delta(k) - C(k) = D(k) \qquad (5.13.31)$$

where $C(k)$ and $D(k)$ are the sample autocorrelations

$$C(k) = \sum_n R(n + k)R(n), \qquad D(k) = \sum_n T(n + k)T(n) \qquad (5.13.32)$$

In practice, only a *finite record* of the reflection (or transmission) sequence will be available, say $\{R(0),R(1), \ldots ,R(N)\}$. Then, an approximation to $C(k)$ must be used, as follows:

$$C(k) = \sum_{n=0}^{N-k} R(n + k)R(n) \qquad (5.13.33)$$

The polynomial $A_M(z)$ may be recovered from the knowledge of the first $M$ lags of the spectral function; that is, $\{\phi(0),\phi(1), \ldots ,\phi(M)\}$. The determining equations for the *coefficients* of $A_M(z)$ are precisely the *normal* equations of linear prediction. In the present context, they may be derived directly by noting that $\Phi(z)$ is a stable spectral density and is already factored into its minimal-phase factors, in Eq. (5.13.30). Thus, writing

$$\Phi(z)A_M(z) = \frac{\sigma_M^2}{A_M(z^{-1})}$$

it follows that the right-hand side is expandable in positive powers of $z$; the negative powers of $z$ in the left-hand side must be set equal to zero. This gives the normal equations:

$$\begin{bmatrix} \phi(0) & \phi(1) & \phi(2) & \cdots & \phi(M) \\ \phi(1) & \phi(0) & \phi(1) & \cdots & \phi(M-1) \\ \phi(2) & \phi(1) & \phi(0) & \cdots & \phi(M-2) \\ \vdots & \vdots & \vdots & & \vdots \\ \phi(M) & \phi(M-1) & \phi(M-2) & \cdots & \phi(0) \end{bmatrix} \begin{bmatrix} 1 \\ a_M(1) \\ a_M(2) \\ \vdots \\ a_M(M) \end{bmatrix} = \begin{bmatrix} \sigma_M^2 \\ 0 \\ 0 \\ \vdots \\ 0 \end{bmatrix}$$

$$(5.13.34)$$

which can be solved efficiently using Levinson's algorithm.

Having obtained $A_M(z)$ and noting that $B_M(z) = A_M(z)R(z)$, the coefficients of the polynomial $B_M(z)$ may be recovered by *convolution*:

$$b_M(n) = \sum_{m=0}^{n} a_M(n - m)R(m), \qquad n = 0,1, \ldots ,M \qquad (5.13.35)$$

Having obtained both $A_M(z)$ and $B_M(z)$ and noting that $\rho_M = b_M(0)$, the lattice recursion (5.13.23) may be inverted to recover the polynomials $A_{M-1}(z)$ and $B_{M-1}(z)$, as well as the next reflection coefficient $\rho_{M-1} = b_{M-1}(0)$, and so on. The inverse of the lattice recursion matrix is

$$\begin{bmatrix} 1 & \rho_m z^{-1} \\ \rho_m & z^{-1} \end{bmatrix}^{-1} = \frac{1}{1 - \rho_m^2} \begin{bmatrix} 1 & -\rho_m \\ -\rho_m z & z \end{bmatrix}$$

Therefore, the *backward recursion* becomes

$$\begin{bmatrix} A_{m-1}(z) \\ B_{m-1}(z) \end{bmatrix} = \frac{1}{1 - \rho_m^2} \begin{bmatrix} 1 & -\rho_m \\ -\rho_m z & z \end{bmatrix} \begin{bmatrix} A_m(z) \\ B_m(z) \end{bmatrix}, \text{ with } \rho_m = b_m(0) \quad (5.13.36)$$

In this manner, all the reflection coefficients $\rho_0, \rho_1, \ldots, \rho_M$ can be extracted. The computational algorithm is summarized as follows:

1. Measure $R(0), R(1), \ldots, R(N)$.
2. Select a reasonable value for the number of slabs $M$.
3. Compute the $M + 1$ sample autocorrelation lags $C(0), C(1), \ldots, C(M)$ of the reflection response $R(n)$, using Eq. (5.13.33).
4. Compute $\phi(k) = \delta(k) - C(k);\ k = 0, 1, \ldots, M$.
5. Using Levinson's algorithm, solve the normal equations (5.13.34) for the coefficients of $A_M(z)$.
6. Convolve $A_M(z)$ with $R(z)$ to find $B_M(z)$.
7. Compute $\rho_M = b_M(0)$, and iterate the backward recursion (5.13.36) from $m = M$ down to $m = 0$.

The subroutine DPD (see appendix) is an implementation of the dynamic predictive deconvolution procedure. The inputs to the subroutine are $N + 1$ samples of the reflection response $\{R(0), R(1), \ldots, R(N)\}$ and the number of layers $M$. The outputs are the lattice polynomials $A_i(z)$ and $B_i(z)$, for $i = 0, 1, \ldots, M$, arranged in the two *lower-triangular* matrices $A$ and $B$ whose *rows* hold the coefficients of these polynomials; that is, $A(i, j) = a_i(j)$, or

$$A(i, z) = \sum_{j=0}^{i} A(i, j) z^{-j}; \qquad i = 0, 1, \ldots, M$$

and similarly for $B(i, z)$. The subroutine invokes the routine LEV to solve the normal equations (5.13.34). The *forward scattering problem* is implemented by the subroutine SCATTER, whose inputs are the set of reflection coefficients $\{\rho_0, \rho_1, \ldots, \rho_M\}$ and whose outputs are the lattice polynomials $A(i, z)$ and $B(i, z)$, for $i = 0, 1, \ldots, M$, as well as a prespecified number $N + 1$ of reflection response samples $R(0), R(1), \ldots, R(N)$. It utilizes the forward lattice recursion (5.13.23) to obtain the lattice polynomials, and then computes the reflection response samples by taking the inverse $z$-transform of Eq. (5.13.27).

Next, we present a number of deconvolution examples simulated by means of the routines SCATTER and DPD. In each case, we specified the five reflection coefficients of a structure consisting of four layers. Using SCATTER we generated the exact lattice polynomials whose coefficients are arranged in the matrices $A$ and $B$, and also generated 16 samples of the reflection response $R(n)$; $n = 0, 1, \ldots, 15$. These 16 samples were sent through the DPD routine to extract the lattice polynomials $A$ and $B$. The first graph of each example displays the reflection response samples and the exact and extracted polynomials. Note that the *first column* of the matrix $B$ is the vector of *reflection coefficients,*

|  $K$  |  $R(K)$  |
|-------|----------|
|  0  |  0.5000  |
|  1  |  0.3750  |
|  2  |  0.1875  |
|  3  |  0.0234  |
|  4  |  $-0.0586$  |
|  5  |  $-0.1743$  |
|  6  |  0.1677  |
|  7  |  0.0265  |
|  8  |  $-0.0601$  |
|  9  |  $-0.0259$  |
|  10  |  0.0238  |
|  11  |  0.0314  |
|  12  |  $-0.0225$  |
|  13  |  $-0.0153$  |
|  14  |  0.0109  |
|  15  |  0.0097  |

Exact $A(i,j)$

| | | | | |
|---|---|---|---|---|
| 1.0000 | 0.0000 | 0.0000 | 0.0000 | 0.0000 |
| 1.0000 | 0.2500 | 0.0000 | 0.0000 | 0.0000 |
| 1.0000 | 0.5000 | 0.2500 | 0.0000 | 0.0000 |
| 1.0000 | 0.7500 | 0.5625 | 0.2500 | 0.0000 |
| 1.0000 | 1.0000 | 0.9375 | 0.6250 | 0.2500 |

Extracted $A(i,j)$

| | | | | |
|---|---|---|---|---|
| 1.0000 | 0.0000 | 0.0000 | 0.0000 | 0.0000 |
| 1.0000 | 0.2509 | 0.0000 | 0.0000 | 0.0000 |
| 1.0000 | 0.5009 | 0.2510 | 0.0000 | 0.0000 |
| 1.0000 | 0.7509 | 0.5638 | 0.2508 | 0.0000 |
| 1.0000 | 1.0009 | 0.9390 | 0.6263 | 0.2504 |

Exact $B(i,j)$

| | | | | |
|---|---|---|---|---|
| 0.5000 | 0.0000 | 0.0000 | 0.0000 | 0.0000 |
| 0.5000 | 0.5000 | 0.0000 | 0.0000 | 0.0000 |
| 0.5000 | 0.6250 | 0.5000 | 0.0000 | 0.0000 |
| 0.5000 | 0.7500 | 0.7500 | 0.5000 | 0.0000 |
| 0.5000 | 0.8750 | 1.0313 | 0.8750 | 0.5000 |

Extracted $B(i,j)$

| | | | | |
|---|---|---|---|---|
| 0.5010 | 0.0000 | 0.0000 | 0.0000 | 0.0000 |
| 0.5000 | 0.5010 | 0.0000 | 0.0000 | 0.0000 |
| 0.5000 | 0.6255 | 0.5010 | 0.0000 | 0.0000 |
| 0.5000 | 0.7505 | 0.7510 | 0.5010 | 0.0000 |
| 0.5000 | 0.8755 | 1.0323 | 0.8764 | 0.5010 |

**Figure 5.26**  Reflection Response and Lattice Polynomials

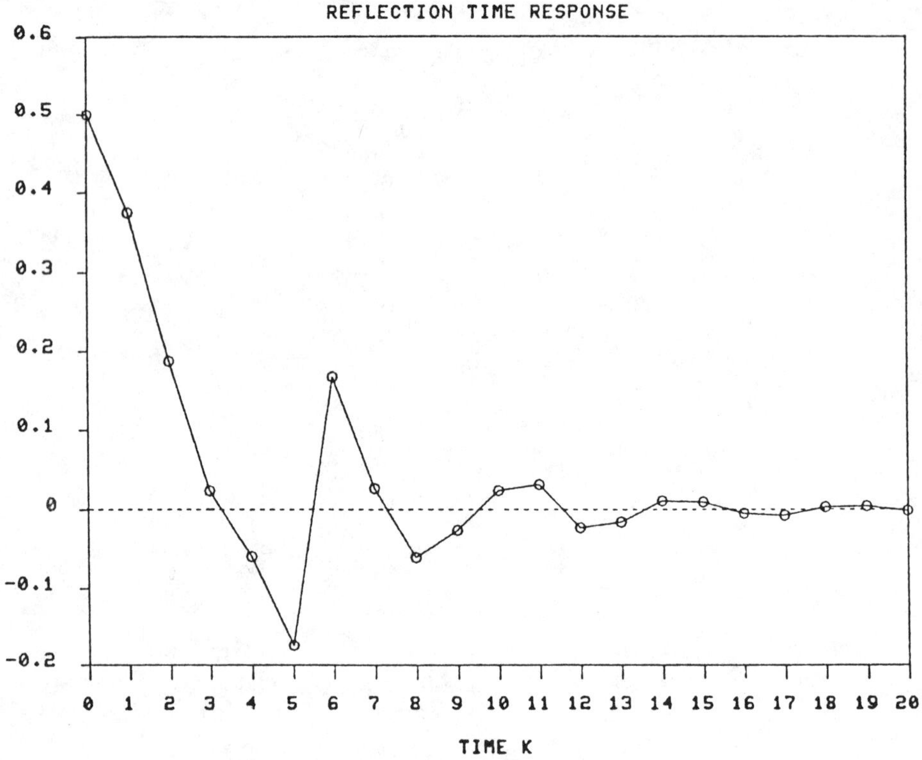

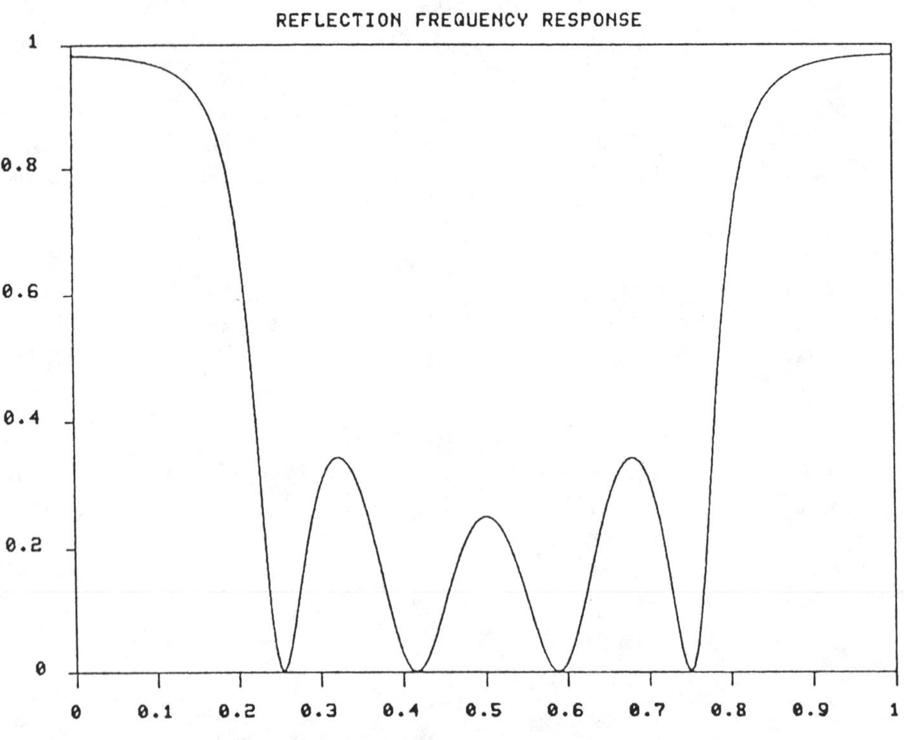



**Figure 5.27**   Reflection Response in the Time Domain



**Figure 5.28**   Reflection Response in the Frequency Domain

| $K$ | $R(K)$ |
|---|---|
| 0 | 0.3000 |
| 1 | 0.3640 |
| 2 | 0.3385 |
| 3 | 0.0664 |
| 4 | $-0.0468$ |
| 5 | $-0.1309$ |
| 6 | 0.0594 |
| 7 | 0.0373 |
| 8 | $-0.0146$ |
| 9 | $-0.0148$ |
| 10 | 0.0014 |
| 11 | 0.0075 |
| 12 | $-0.0001$ |
| 13 | $-0.0029$ |
| 14 | $-0.0003$ |
| 15 | 0.0010 |

Exact $A(i,j)$

| | | | | |
|---|---|---|---|---|
| 1.0000 | 0.0000 | 0.0000 | 0.0000 | 0.0000 |
| 1.0000 | 0.1200 | 0.0000 | 0.0000 | 0.0000 |
| 1.0000 | 0.3200 | 0.1500 | 0.0000 | 0.0000 |
| 1.0000 | 0.5200 | 0.3340 | 0.1200 | 0.0000 |
| 1.0000 | 0.6400 | 0.5224 | 0.2760 | 0.0900 |

Extracted $A(i,j)$

| | | | | |
|---|---|---|---|---|
| 1.0000 | 0.0000 | 0.0000 | 0.0000 | 0.0000 |
| 1.0000 | 0.1200 | 0.0000 | 0.0000 | 0.0000 |
| 1.0000 | 0.3200 | 0.1500 | 0.0000 | 0.0000 |
| 1.0000 | 0.5200 | 0.3340 | 0.1200 | 0.0000 |
| 1.0000 | 0.6400 | 0.5224 | 0.2760 | 0.0900 |

Exact $B(i,j)$

| | | | | |
|---|---|---|---|---|
| 0.3000 | 0.0000 | 0.0000 | 0.0000 | 0.0000 |
| 0.4000 | 0.3000 | 0.0000 | 0.0000 | 0.0000 |
| 0.5000 | 0.4600 | 0.3000 | 0.0000 | 0.0000 |
| 0.4000 | 0.6280 | 0.5200 | 0.3000 | 0.0000 |
| 0.3000 | 0.5560 | 0.7282 | 0.5560 | 0.3000 |

Extracted $B(i,j)$

| | | | | |
|---|---|---|---|---|
| 0.3000 | 0.0000 | 0.0000 | 0.0000 | 0.0000 |
| 0.4000 | 0.3000 | 0.0000 | 0.0000 | 0.0000 |
| 0.5000 | 0.4600 | 0.3000 | 0.0000 | 0.0000 |
| 0.4000 | 0.6280 | 0.5200 | 0.3000 | 0.0000 |
| 0.3000 | 0.5560 | 0.7282 | 0.5560 | 0.3000 |

**Figure 5.29**  Reflection Response and Lattice Polynomials

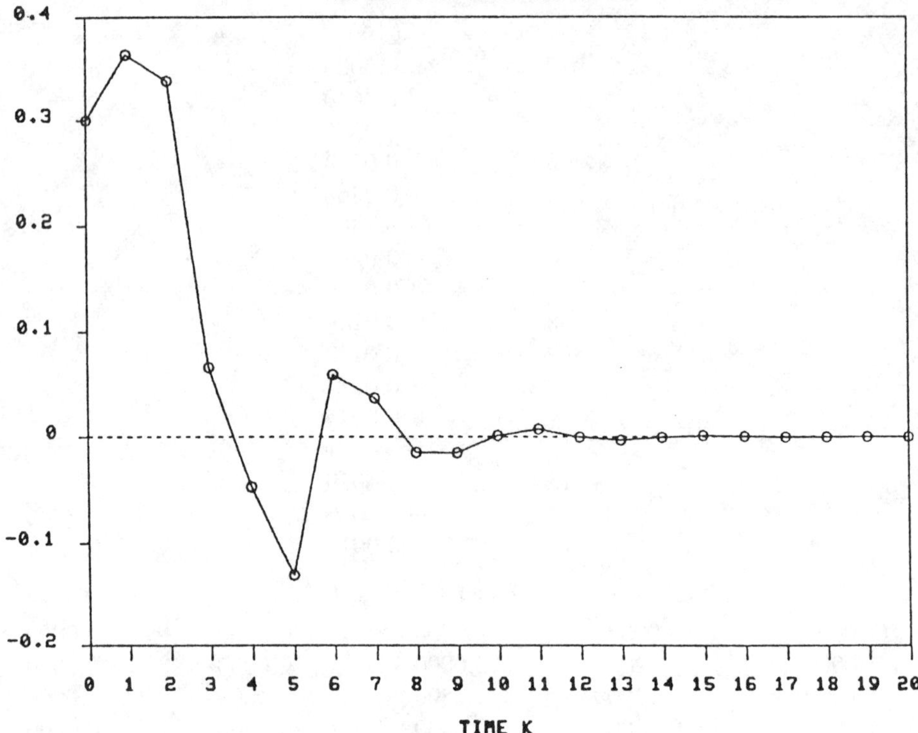

**Figure 5.30**   Reflection Response in the Time Domain

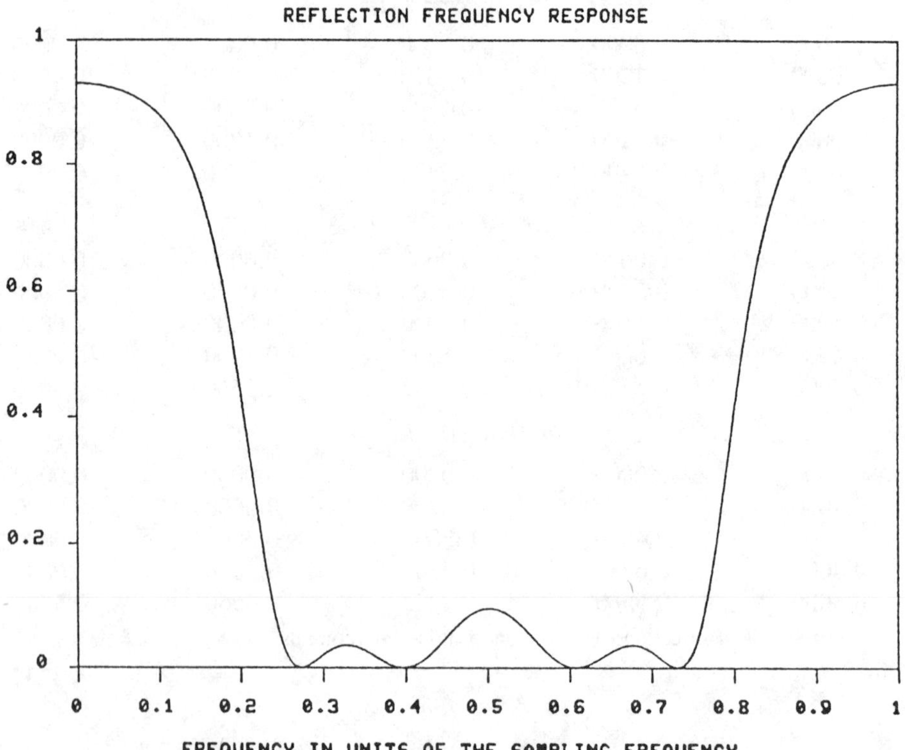

**Figure 5.31**   Reflection Response in the Frequency Domain

| $K$ | $R(K)$ |
|---|---|
| 0 | 0.1000 |
| 1 | 0.1980 |
| 2 | 0.2812 |
| 3 | 0.1445 |
| 4 | 0.0388 |
| 5 | −0.0346 |
| 6 | −0.0072 |
| 7 | 0.0017 |
| 8 | 0.0015 |
| 9 | 0.0002 |
| 10 | −0.0002 |
| 11 | −0.0001 |
| 12 | 0.0000 |
| 13 | 0.0000 |
| 14 | 0.0000 |
| 15 | 0.0000 |

Exact $A(i, j)$

| | | | | |
|---|---|---|---|---|
| 1.0000 | 0.0000 | 0.0000 | 0.0000 | 0.0000 |
| 1.0000 | 0.0200 | 0.0000 | 0.0000 | 0.0000 |
| 1.0000 | 0.0800 | 0.0300 | 0.0000 | 0.0000 |
| 1.0000 | 0.1400 | 0.0712 | 0.0200 | 0.0000 |
| 1.0000 | 0.1600 | 0.1028 | 0.0412 | 0.0100 |

Extracted $A(i, j)$

| | | | | |
|---|---|---|---|---|
| 1.0000 | 0.0000 | 0.0000 | 0.0000 | 0.0000 |
| 1.0000 | 0.0200 | 0.0000 | 0.0000 | 0.0000 |
| 1.0000 | 0.0800 | 0.0300 | 0.0000 | 0.0000 |
| 1.0000 | 0.1400 | 0.0712 | 0.0200 | 0.0000 |
| 1.0000 | 0.1600 | 0.1028 | 0.0412 | 0.0100 |

Exact $B(i, j)$

| | | | | |
|---|---|---|---|---|
| 0.1000 | 0.0000 | 0.0000 | 0.0000 | 0.0000 |
| 0.2000 | 0.1000 | 0.0000 | 0.0000 | 0.0000 |
| 0.3000 | 0.2060 | 0.1000 | 0.0000 | 0.0000 |
| 0.2000 | 0.3160 | 0.2120 | 0.1000 | 0.0000 |
| 0.1000 | 0.2140 | 0.3231 | 0.2140 | 0.1000 |

Extracted $B(i, j)$

| | | | | |
|---|---|---|---|---|
| 0.1000 | 0.0000 | 0.0000 | 0.0000 | 0.0000 |
| 0.2000 | 0.1000 | 0.0000 | 0.0000 | 0.0000 |
| 0.3000 | 0.2060 | 0.1000 | 0.0000 | 0.0000 |
| 0.2000 | 0.3160 | 0.2120 | 0.1000 | 0.0000 |
| 0.1000 | 0.2140 | 0.3231 | 0.2140 | 0.1000 |

**Figure 5.32**   Reflection Response and Lattice Polynomials

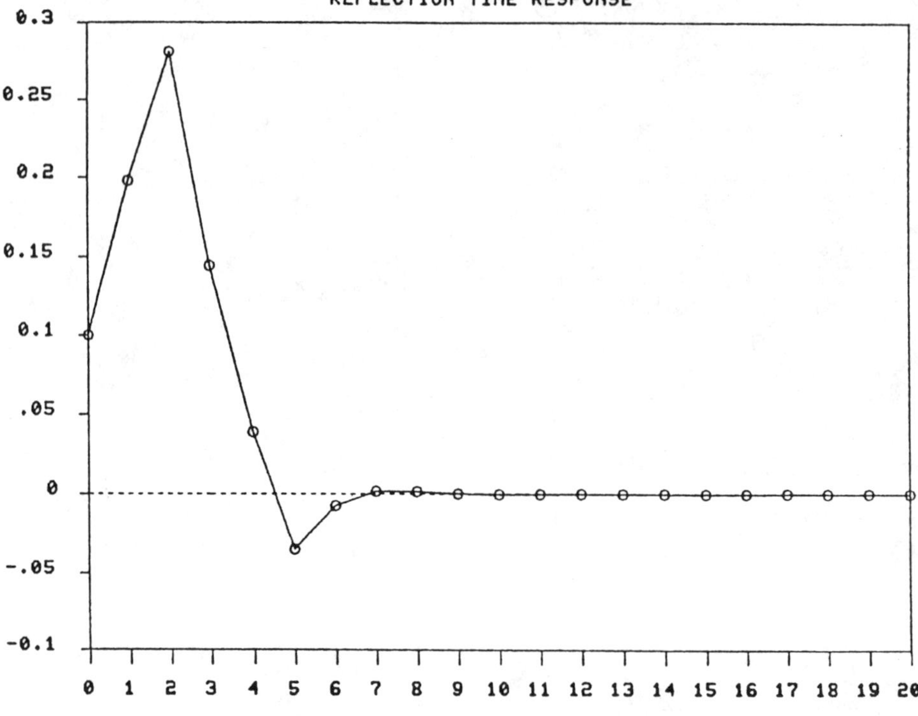

**Figure 5.33**    Reflection Response in the Time Domain

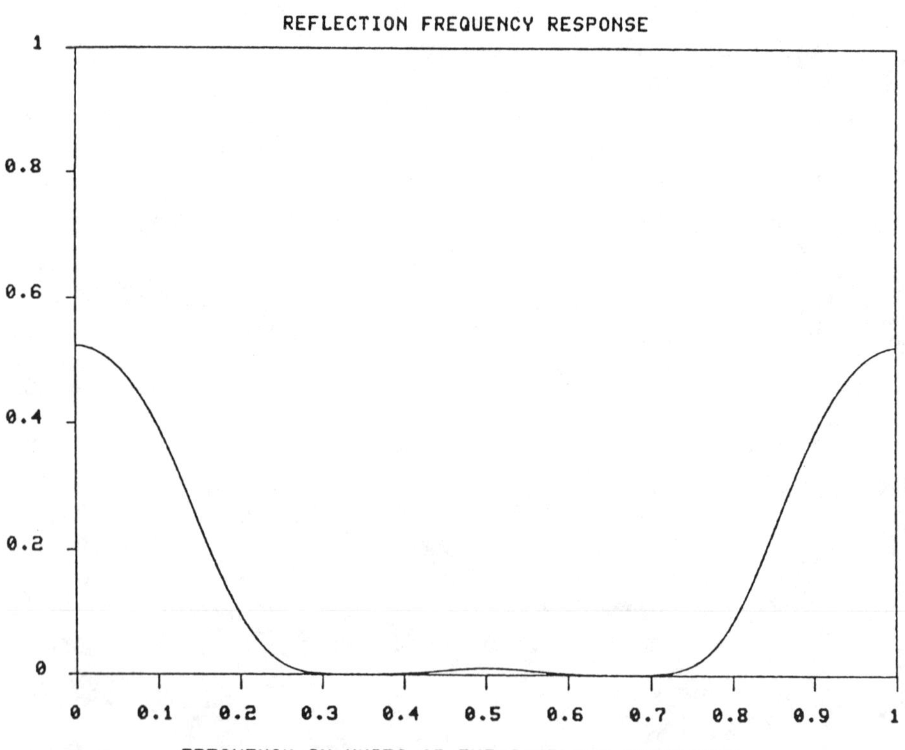

**Figure 5.34**    Reflection Response in the Frequency Domain

| $K$ | $R(K)$ |
|---|---|
| 0 | 0.5000 |
| 1 | $-0.3750$ |
| 2 | 0.1875 |
| 3 | $-0.0234$ |
| 4 | $-0.0586$ |
| 5 | 0.1743 |
| 6 | 0.1677 |
| 7 | $-0.0265$ |
| 8 | $-0.0601$ |
| 9 | 0.0259 |
| 10 | 0.0238 |
| 11 | $-0.0314$ |
| 12 | $-0.0225$ |
| 13 | 0.0153 |
| 14 | 0.0109 |
| 15 | $-0.0097$ |

Exact $A(i,j)$

| | | | | |
|---|---|---|---|---|
| 1.0000 | 0.0000 | 0.0000 | 0.0000 | 0.0000 |
| 1.0000 | $-0.2500$ | 0.0000 | 0.0000 | 0.0000 |
| 1.0000 | $-0.5000$ | 0.2500 | 0.0000 | 0.0000 |
| 1.0000 | $-0.7500$ | 0.5625 | $-0.2500$ | 0.0000 |
| 1.0000 | $-1.0000$ | 0.9375 | $-0.6250$ | 0.2500 |

Extracted $A(i,j)$

| | | | | |
|---|---|---|---|---|
| 1.0000 | 0.0000 | 0.0000 | 0.0000 | 0.0000 |
| 1.0000 | $-0.2509$ | 0.0000 | 0.0000 | 0.0000 |
| 1.0000 | $-0.5009$ | 0.2510 | 0.0000 | 0.0000 |
| 1.0000 | $-0.7509$ | 0.5638 | $-0.2508$ | 0.0000 |
| 1.0000 | $-1.0009$ | 0.9390 | $-0.6263$ | 0.2504 |

Exact $B(i,j)$

| | | | | |
|---|---|---|---|---|
| 0.5000 | 0.0000 | 0.0000 | 0.0000 | 0.0000 |
| $-0.5000$ | 0.5000 | 0.0000 | 0.0000 | 0.0000 |
| 0.5000 | $-0.6250$ | 0.5000 | 0.0000 | 0.0000 |
| $-0.5000$ | 0.7500 | $-0.7500$ | 0.5000 | 0.0000 |
| 0.5000 | $-0.8750$ | 1.0313 | $-0.8750$ | 0.5000 |

Extracted $B(i,j)$

| | | | | |
|---|---|---|---|---|
| 0.5010 | 0.0000 | 0.0000 | 0.0000 | 0.0000 |
| $-0.5000$ | 0.5010 | 0.0000 | 0.0000 | 0.0000 |
| 0.5000 | $-0.6255$ | 0.5010 | 0.0000 | 0.0000 |
| $-0.5000$ | 0.7505 | $-0.7510$ | 0.5010 | 0.0000 |
| 0.5000 | $-0.8755$ | 1.0323 | $-0.8764$ | 0.5010 |

**Figure 5.35**  Reflection Response and Lattice Polynomials

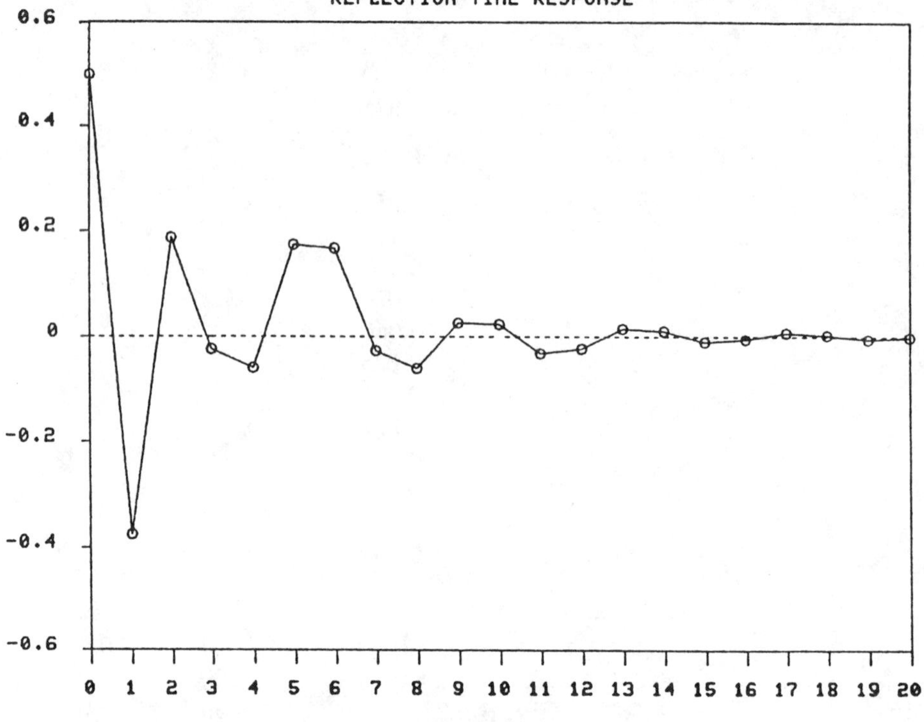

**Figure 5.36**  Reflection Response in the Time Domain

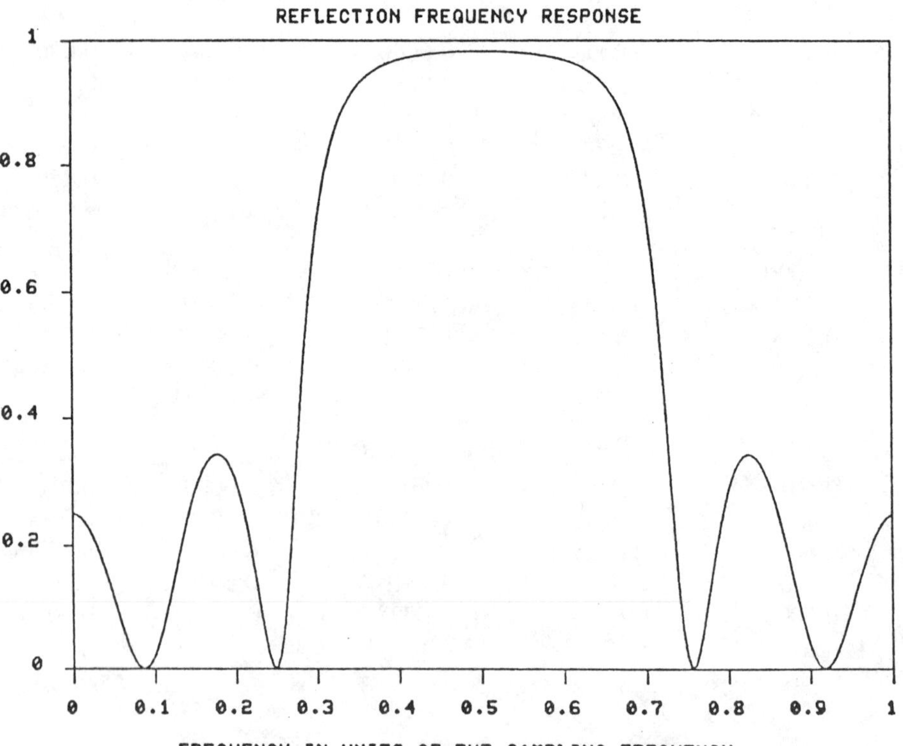

**Figure 5.37**  Reflection Response in the Frequency Domain

according to Eq. (5.13.25). The remaining two graphs of each example show the reflection response $R$ in the time domain and in the frequency domain. Note that the frequency response is plotted only over one Nyquist interval $[0, 2\pi/T_2]$, and it is symmetric about the Nyquist frequency $\pi/T_2$. Figures 5.26 through 5.28 correspond to the case of equal reflection coefficients $\{\rho_0, \rho_1, \rho_2, \rho_3, \rho_4\} = \{0.5, 0.5, 0.5, 0.5, 0.5\}$. In Figs. 5.29 through 5.31 the reflection coefficients have been tapered somewhat at the ends (windowed) and are $\{0.3, 0.4, 0.5, 0.4, 0.3\}$. Note the effect of tapering on the lobes of the reflection frequency response. Figures 5.32 through 5.34 correspond to the set of reflection coefficients $\{0.1, 0.2, 0.3, 0.2, 0.1\}$. Note the *broad band* of frequencies about the Nyquist frequency for which there is very little reflection. In contrast, the example in Figs. 5.35 through 5.37 exhibits *high reflectivity* over a broad band of frequencies about the Nyquist frequency. Its set of reflection coefficients is $\{0.5, -0.5, 0.5, -0.5, 0.5\}$.

In this section we have discussed the inverse problem of unraveling the structure of a medium from the knowledge of its reflection response. The connection of the dynamic predictive deconvolution method to the conventional inverse scattering methods based on the *Gelfand-Levitan-Marchenko* approach [101] has been discussed in [93,102,103]. The lattice recursions characteristic of the wave propagation problem were derived as a *direct* consequence of the *boundary conditions* at the interfaces between media, whereas the lattice recursions of linear prediction were a direct consequence of the Gram-Schmidt orthogonalization process and the *minimization* of the prediction-error performance index. Is there a deeper connection between these two problems [104–108]?

## 5.14 *Least-Squares Waveshaping and Spiking Filters*

In linear prediction, the three practical methods of estimating the prediction-error filter coefficients were all based on replacing the ensemble mean-squared minimization criterion by a *least-squares* criterion based on time averages. Similarly, the more general Wiener filtering problem may be recast in terms of such time averages. A practical formulation, which is analogous to the Yule-Walker or autocorrelation method, is as follows [19,20,34,109]: Given a record of available data

$$\boxed{y_0, y_1, \ldots, y_N}$$

find the *best* linear FIR filter of order $M$

$$\boxed{h_0, h_1, \ldots, h_M}$$

which reshapes $y_n$ into a desired signal $x_n$, specified in terms of the samples

$$\boxed{x_0, x_1, \ldots, x_{N+M}}$$

where for consistency of convolution we have assumed we know $N + M + 1$ samples of the desired signal. The actual *output* of the waveshaping filter will be

$$\hat{x}_n = \sum_{m=0}^{M} h_m y_{n-m}; \qquad 0 \leq n \leq N + M \qquad (5.14.1)$$

The *estimation error* will be

$$e_n = x_n - \hat{x}_n; \qquad 0 \leq n \leq N + M \qquad (5.14.2)$$

As the optimality criterion we choose the *least-squares* criterion

$$\mathcal{E} = \sum_{n=0}^{N+M} e_n^2 = \min \qquad (5.14.3)$$

The optimal filter weights $h_m$ are selected to minimize $\mathcal{E}$. It is convenient to recast the above in a compact matrix form. Define the $(N + M + 1) \times (M + 1)$ data matrix $Y$, the $(M + 1) \times 1$ vector of filter weights $\mathbf{h}$, the $(M + N + 1) \times 1$ vector of desired samples $\mathbf{x}$, and estimation errors $\mathbf{e}$, as follows:

$$Y = \begin{bmatrix} y_0 & 0 & 0 & \cdots & 0 \\ y_1 & y_0 & 0 & \cdots & 0 \\ y_2 & y_1 & y_0 & \cdots & 0 \\ \vdots & \vdots & \vdots & & \vdots \\ y_N & y_{N-1} & y_{N-2} & \cdots & y_{N-M} \\ 0 & y_N & y_{N-1} & \cdots & \vdots \\ 0 & 0 & y_N & \cdots & \vdots \\ \vdots & \vdots & \vdots & & \vdots \\ 0 & 0 & 0 & \cdots & y_N \end{bmatrix}, \quad \mathbf{h} = \begin{bmatrix} h_0 \\ h_1 \\ \vdots \\ h_M \end{bmatrix}, \quad \mathbf{x} = \begin{bmatrix} x_0 \\ x_1 \\ x_2 \\ \vdots \\ x_{N+M} \end{bmatrix} \qquad (5.14.4)$$

where the top of the matrix is labeled "$\leftarrow$ M zeros $\rightarrow$" and the lower-left is labeled "M zeros".

Equations (5.14.1) through (5.14.3) now become

$$\hat{\mathbf{x}} = Y\mathbf{h}, \qquad \mathbf{e} = \mathbf{x} - \hat{\mathbf{x}}, \qquad \mathcal{E} = \mathbf{e}^T \mathbf{e} \qquad (5.14.5)$$

Minimizing $\mathcal{E}$ with respect to the weight vector $\mathbf{h}$ results in the *orthogonality* equations

$$Y^T \mathbf{e} = Y^T (\mathbf{x} - Y\mathbf{h}) = 0 \qquad (5.14.6)$$

which are equivalent to the *normal* equations

$$Y^T Y \mathbf{h} = Y^T \mathbf{x} \qquad (5.14.7)$$

Solving for $\mathbf{h}$, we find

$$\mathbf{h} = (Y^T Y)^{-1} Y^T \mathbf{x} = R^{-1} \mathbf{r} \qquad (5.14.8)$$

where the quantities

$$R = Y^T Y, \qquad \mathbf{r} = Y^T \mathbf{x} \qquad (5.14.9)$$

may be recognized (see Section 1.9) as the $(M + 1) \times (M + 1)$ autocorrelation matrix formed by the sample autocorrelations $\hat{R}_{yy}(0), \hat{R}_{yy}(1), \ldots, \hat{R}_{yy}(M)$ of $y_n$, and as the $(M + 1) \times 1$ vector of sample cross-correlations $\hat{R}_{xy}(0), \hat{R}_{xy}(1), \ldots, \hat{R}_{xy}(M)$ between the desired and the available vectors $x_n$ and $y_n$. We have already used this expression for the weight vector $\mathbf{h}$ in the example of Section 5.8. Here we have justified it in terms of the least-squares criterion (5.14.3). The subroutine FIRW may be used to solve for the weights (5.14.8) and, if so desired, to give the corresponding lattice realization. The actual filter output $\hat{\mathbf{x}}$ is expressed as

$$\hat{\mathbf{x}} = Y\mathbf{h} = YR^{-1}Y^T \mathbf{x} = P\mathbf{x} \qquad (5.14.10)$$

where

$$P = YR^{-1}Y^T = Y(Y^T Y)^{-1}Y^T \qquad (5.14.11)$$

The error vector becomes $\mathbf{e} = (I - P)\mathbf{x}$. The "performance" matrix $P$ is a projection matrix, and thus, so is $(I - P)$. Then, the error squared becomes

$$\mathcal{E} = \mathbf{e}^T \mathbf{e} = \mathbf{x}^T (I - P)^2 \mathbf{x} = \mathbf{x}^T (I - P)\mathbf{x} \qquad (5.14.12)$$

The $(N + M + 1) \times (N + M + 1)$ matrix $P$ has trace equal to $M + 1$, as can be checked easily. Since its eigenvalues as a projection matrix are either 0 or 1, it follows that in order for the sum of all the eigenvalues (the trace) to be equal to $M + 1$, there must necessarily be $M + 1$ eigenvalues that are equal to 1, and $N$ eigenvalues equal to 0. Therefore the matrix $P$ has rank $M + 1$, and if the desired vector $\mathbf{x}$ is selected to be any of the $M + 1$ eigenvectors belonging to eigenvalue 1, the corresponding estimation error will be zero.

Among all possible waveshapes that may be chosen for the desired vector $\mathbf{x}$, of particular importance are the *spikes*, or impulses. In this case, $\mathbf{x}$ is a unit impulse, say at the origin; that is, $x_n = \delta_n$. The convolution $\hat{x}_n = h_n * y_n$ of the corresponding filter with $y_n$ is the best least-squares approximation to the unit impulse. In other words, $h_n$ is the best *least-squares inverse* filter to $y_n$ that attempts to reshape, or compress, $y_n$ into a unit impulse. Such least squares inverse filters are used extensively in *deconvolution* applications. More generally, the vector $\mathbf{x}$ may be chosen to be any one of the unit vectors

$$\mathbf{x} = \mathbf{u}_i = \begin{bmatrix} 0 \\ \vdots \\ 0 \\ 1 \\ 0 \\ \vdots \\ 0 \end{bmatrix} \leftarrow i\text{th slot}; \qquad i = 0, 1, \ldots, N + M \qquad (5.14.13)$$

which corresponds to a unit impulse occurring at the $i$th time instant instead of at the origin; that is, $x_n = \delta(n - i)$. The actual output from the spiking filter is given by

$$\hat{\mathbf{x}} = P\mathbf{x} = P\mathbf{u}_i = i\text{th column of } P \qquad (5.14.14)$$

Thus, the $i$th column of the matrix $P$ is the *output* of the *$i$th spiking filter* which attempts to compress $y_n$ into a spike with $i$ delays. The corresponding $i$th filter is $\mathbf{h} = R^{-1}Y^T\mathbf{u}_i$. Therefore, the *columns* of the matrix

$$H = R^{-1}Y^T \qquad (5.14.15)$$

are *all* the optimal *spiking filters*. The estimation error of the $i$th filter is

$$\mathscr{E}_i = \mathbf{u}_i^T(I - P)\mathbf{u}_i = 1 - P_{ii} \qquad (5.14.16)$$

where $P_{ii}$ is the $ii$th diagonal element of $P$. Since the delay $i$ may be positioned anywhere from $i = 0$ to $i = N + M$, there are $N + M + 1$ such spiking filters, each with error $\mathscr{E}_i$. Among these, there will be one that has the optimal delay $i$ which corresponds to the smallest of the $\mathscr{E}_i$s; or, equivalently, to the maximum of the diagonal elements $P_{ii}$. The design procedure for least-squares spiking filters for a given finite signal $y_n$; $n = 0,1, \ldots ,N$ is summarized as follows:

1. Compute $R = Y^TY$.
2. Compute the inverse $R^{-1}$ (preferably by the Levinson recursion).
3. Compute $H = R^{-1}Y^T = $ all the spiking filters.
4. Compute $P = YH = YR^{-1}Y^T = $ all spiking filter outputs.
5. Select that column $i$ of $P$ for which $P_{ii}$ is the largest.

If the Levinson-Cholesky algorithm is used to compute the inverse $R^{-1}$, this design procedure becomes fairly efficient. An implementation of the procedure is given by the subroutine SPIKE. The inputs to the subroutine are the $N + 1$ samples $y_0, y_1, \ldots ,y_N$, the desired order $M$ of the spiking filter, and a so-called "*prewhitening*" or Backus-Gilbert parameter EPSILON, which will be explained below. The outputs of the subroutine are the matrices $P$ and $H$.

To explain the role of the parameter $\epsilon$, let us go back to the waveshaping problem. When the data sequence $y_n$ to be reshaped into $x_n$ is inaccurately known—if, for example, it has been contaminated by white noise $v_n$—the least-squares minimization criterion (5.14.3) can be extended slightly to accomplish the double task of (1) producing the *best estimate* of $x_n$ and (2) reducing the noise *at the output* of the filter $h_n$ as much as possible. The input to the filter is the noisy sequence $y_n + v_n$, and its output is $h_n*y_n + y_n*v_n = \hat{x}_n + u_n$, where we set $u_n = h_n*v_n$. The term $u_n$ represents *the filtered noise*. The minimization criterion (5.14.3) may be replaced by

$$\mathscr{E} = \sum_n e_n^2 + \lambda E[u_n^2] \qquad (5.14.17)$$

where $\lambda$ is a positive parameter which can be chosen by the user. Large $\lambda$ emphasizes large *reduction* of the output noise, but this is done *at the expense of resolution*; that is, at the expense of obtaining a very good estimate. On the other hand, small $\lambda$ emphasizes higher resolution but with lesser noise reduction. This tradeoff between *resolution and noise reduction* is the basic property of this performance index. Assuming $v_n$ is white with variance $\sigma_v^2$, we have

$$E[u_u^2] = \sigma_v^2 \sum_{n=0}^{M} h_n^2 = \sigma_v^2 \mathbf{h}^T \mathbf{h}$$

Thus, Eq. (5.14.17) may be written as

$$\mathcal{E} = \mathbf{e}^T \mathbf{e} + \lambda \sigma_v^2 \mathbf{h}^T \mathbf{h} \tag{5.14.18}$$

Its minimization with respect to $\mathbf{h}$ gives the normal equations

$$(Y^T Y + \lambda \sigma_v^2 I)\mathbf{h} = Y^T \mathbf{x} \tag{5.14.19}$$

from which it is evident that the diagonal of $Y^T Y$ is shifted by an amount $\lambda \sigma_v^2$; that is

$$\hat{R}_{yy}(0) \to \hat{R}_{yy}(0) + \lambda \sigma_v^2 = (1 + \epsilon)\hat{R}_{yy}(0), \qquad \epsilon = \frac{\lambda \sigma_v^2}{\hat{R}_{yy}(0)}$$

In practice, $\epsilon$ may be taken to be a few percent or less. It is evident from Eq. (5.14.19) that one beneficial effect of the parameter $\epsilon$ is the *stabilization* of the inverse of the matrix $Y^T Y + \lambda \sigma_v^2$.

The main usage of spiking filters is in deconvolution problems [19,20,34,109–111], where the desired and the available signals $x_n$ and $y_n$ are related to each other by the convolutional relationship

$$y_n = f_n * x_n = \sum_m f_m x_{n-m} \tag{5.14.20}$$

where $f_n$ is a "blurring" function which is assumed to be approximately *known*. The basic deconvolution problem is to recover $x_n$ from $y_n$ if $f_n$ is known. For example, $y_n$ may represent the image of an object $x_n$ recorded through an optical system with a point-spread function $f_n$; or $y_n$ might represent the recorded seismic trace arising from the excitation of the layered earth by an impulsive waveform $f_n$ (the source wavelet) which is convolved with the reflection impulse response $x_n$ of the earth. (In the previous section $x_n$ was denoted by $R_n$.) If the effect of the source wavelet $f_n$ can be "deconvolved away," the resulting reflection sequence $x_n$ may be subjected to the dynamic predictive deconvolution procedure to unravel the earth structure. Or, $f_n$ may represent the impulse response of a channel, or a magnetic recording medium, which broadens and blurs (intersymbol interference) the desired message $x_n$. The least-squares inverse spiking filters offer a way to solve the deconvolution problem: Simply *design a least-squares* spiking filter $h_n$ corresponding to the blurring function $f_n$; that is, $h_n * f_n \approx \delta_n$

in the least squares sense. Then, filtering $y_n$ through $h_n$ will recover the desired signal $x_n$:

$$\hat{x}_n = h_n * y_n = (h_n * f_n) * x_n \cong \delta_n * x_n = x_n \qquad (5.14.21)$$

If the $i$th spiking filter is used, which compresses $f_n$ into an impulse with $i$ delays, $h_n * f_n \cong \delta(n - i)$, then the desired signal $x_n$ will be recovered with a delay of $i$ units of time.

This and all other approaches to deconvolution work well when the data $y_n$ are not noisy. In presence of noise, Eq. (5.14.20) becomes

$$y_n = f_n * x_n + v_n \qquad (5.14.22)$$

where $v_n$ may be assumed to be zero-mean white noise of variance $\sigma_v^2$. Even if the blurring function $f_n$ is known exactly and a good least-squares inverse filter $h_n$ can be designed, the presence of the noise term can distort the deconvolved signal beyond recognition. This may be explained as follows. Filtering $y_n$ through the inverse filter $h_n$ results in

$$h_n * y_n = (h_n * f_n) * x_n + h_n * v_n \cong x_n + u_n$$

where $u_n = h_n * v_n$ is the filtered noise. Its variance is

$$E[u_n^2] = \sigma_v^2 \mathbf{h}^T \mathbf{h} = \sigma_v^2 \sum_{m=0}^{M} h_m^2$$

which, depending on the particular shape of $h_n$, may be much larger than the original variance $\sigma_v^2$. This happens, for example, when $f_n$ consists mainly of low frequencies. For $h_n$ to compress $f_n$ into a spike with a high frequency content, the impulse response $h_n$ itself must be very spiky, which can result in values for $\mathbf{h}^T \mathbf{h}$ which are greater than one. To combat the effects of noise, the least-squares design criterion for $\mathbf{h}$ must be changed by adding to it a term $\lambda E[u_n^2]$ as was done in Eq. (5.14.17). The design criterion for $\mathbf{h}$ is then

$$\mathcal{E} = \sum_{n} (\delta_n - h_n * f_n)^2 + \lambda \sigma_v^2 \mathbf{h}^T \mathbf{h} = \min$$

which effectively amounts to changing the autocorrelation lag $\hat{R}_{ff}(0)$ to $(1 + \epsilon)\hat{R}_{ff}(0)$. The first term in this performance index tries to produce a good inverse filter; the second term tries to minimize the output power of the noise after filtering by the deconvolution filter $h_n$. Note that conceptually this index is somewhat different from that of Eq. (5.14.17), because now $v_n$ represents the noise in the data $y_n$, whereas there $v_n$ represented inaccuracies in the knowledge of the wavelet $f_n$. In this approach to deconvolution we are not attempting to determine the best least-squares estimate of the desired signal $x_n$, but rather the best least-squares inverse to the blurring function $f_n$. If the second order statistics of $x_n$ were known, we could, of course, determine the optimal (Wiener) estimates $\hat{x}_n$ of $x_n$. This is also done in many applications.

The performance of the spiking filters and their usage in deconvolution are illustrated by the following example: The blurring function $f_n$ to be spiked was chosen as

$$f_n = g(n - 25); n = 0,1,2, \ldots ,65$$

$$f_n = 0; \text{ otherwise}$$

where $g(k)$ was the "gaussian hat"

$$g(k) = \cos(0.15k) \exp(-0.004k^2)$$

The signal $x_n$ to be recovered, was taken to be the series of spikes

$$x_n = \sum_{i=0}^{9} a_i \delta(n - n_i)$$

where the amplitudes $a_i$ and delays $n_i$ were chosen as

$$a_i = 1, 0.8, 0.5, 0.95, 0.7, 0.5, 0.3, 0.9, 0.5, 0.85$$

$$n_i = 25, 50, 60, 70, 80, 90, 100, 120, 140, 160$$

for

$$i = 0, 1, 2, 3, 4, 5, 6, 7, 8, 9$$

Figure 5.38 shows the signal $f_n$ to be spiked. Since the gaussian hat is symmetric about the origin, we chose the spiking delay to be at $i = 25$. The order of the spiking filter $h_n$ was $M = 50$. Figure 5.39 shows the impulse response $h_n$ versus time. Note the spiky nature of $h_n$ which is required here because $f_n$ has a fairly low frequency content. Figure 5.40 shows the results of the convolution $h_n * f_n$, which is the best least-squares approximation to the impulse $\delta(n - 25)$. The "goodness" of the spiking filter is judged by the diagonal entries of the performance matrix $P$, according to Eq. (5.14.16). For the chosen delay $k = 25$, we find $P(25,25) = 0.97$. To get a better picture of the overall performance of the spiking filters, in Figure 5.41 we have plotted the diagonal elements $P(k,k)$ versus $k$. It is seen that the chosen delay $k = 25$ is nearly optimal. Figure 5.42 shows the composite signal $y_n$ obtained by convolving $f_n$ and $x_n$, according to Eq. (5.4.20). Figure 5.43 shows the deconvolved signal $x_n$ according to Eq. (5.14.21). The recovery of the amplitudes $a_i$ and delays $n_i$ of $x_n$ is very accurate. These results represent the idealistic case of *noise-free* data $y_n$ and *perfect knowledge* of the blurring function $f_n$. To study the sensitivity of the deconvolution technique to inaccuracies in the knowledge of the signal $f_n$, we have added a small high frequency perturbation on $f_n$ as follows:

$$f'_n = f_n + 0.05 \sin[1.5(n - 25)]$$

The approximate signal $f'_n$ is shown in Figure 5.44. The spiking filter was designed on the basis of $f'_n$ rather than $f_n$. The result of filtering the *same* composite signal $y_n$ through the corresponding inverse filter is shown in Figure 5.45. Both the delays and the amplitudes $a_i$ and $n_i$ are not well resolved, but

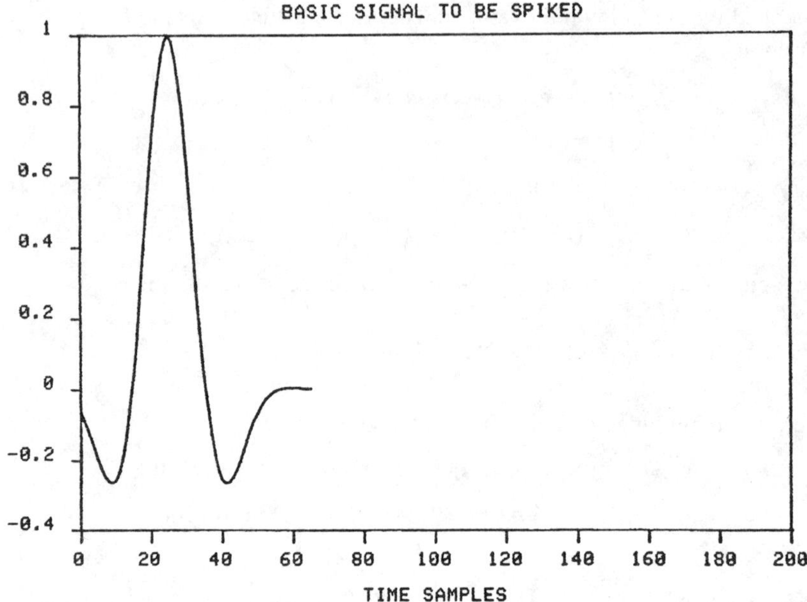


**Figure 5.38**    Signal $f_n$ To Be Spiked

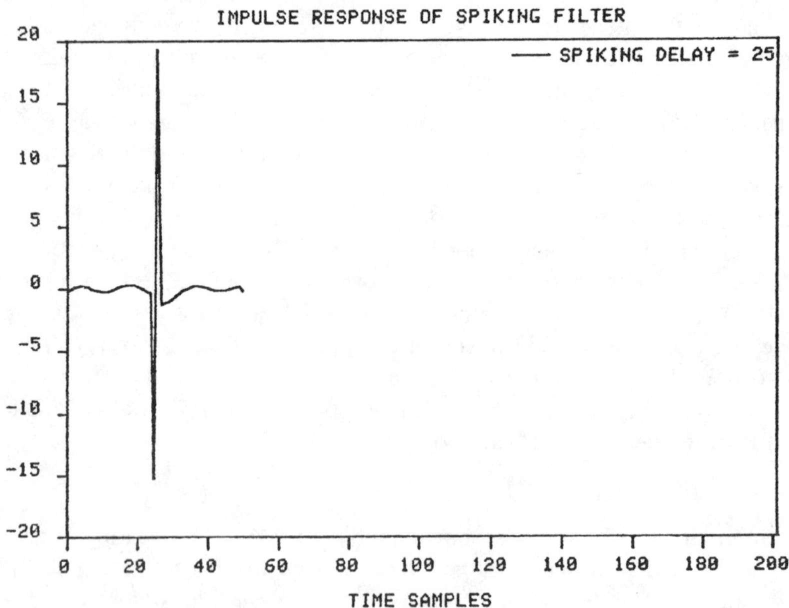


**Figure 5.39**    Impulse Response $h_n$ of Spiking Filter

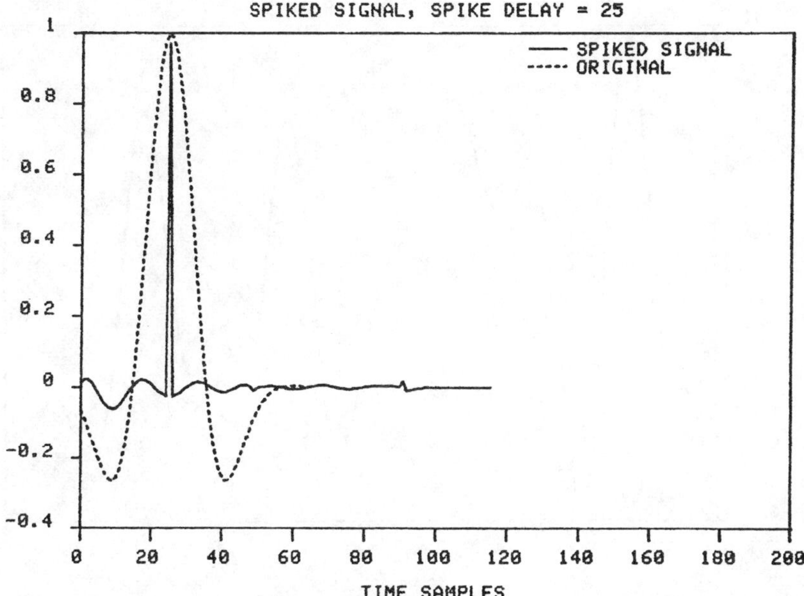



**Figure 5.40**   The Signal $h_n * f_n$

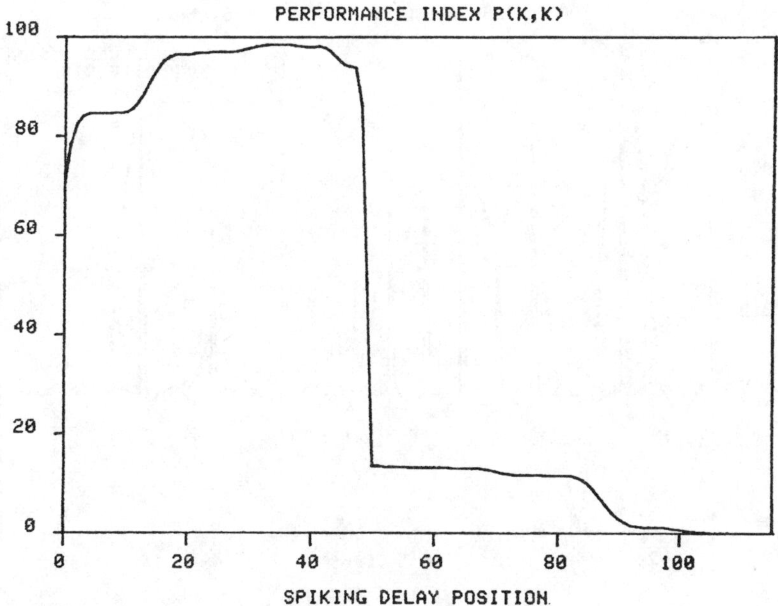



**Figure 5.41**   Performance of the Spiking Filters for $f_n$

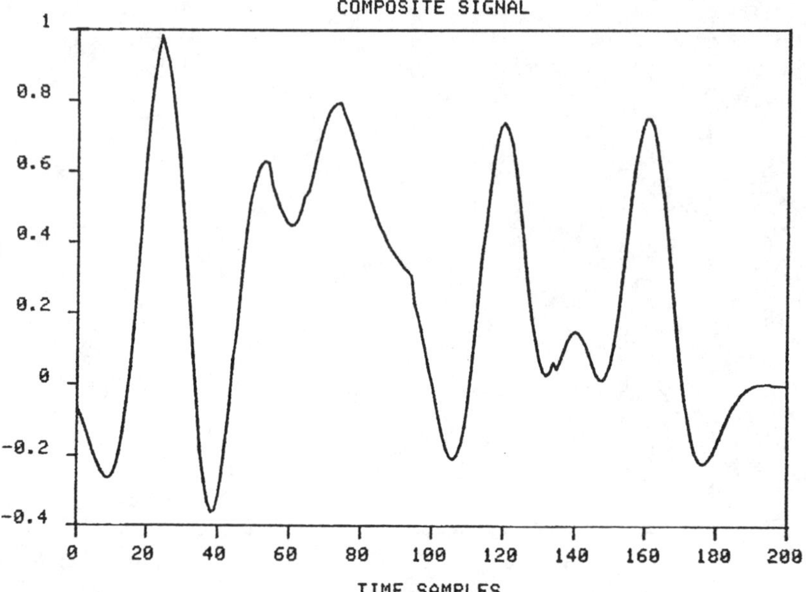

**Figure 5.42**   Composite Signal $y_n = f_n * x_n$

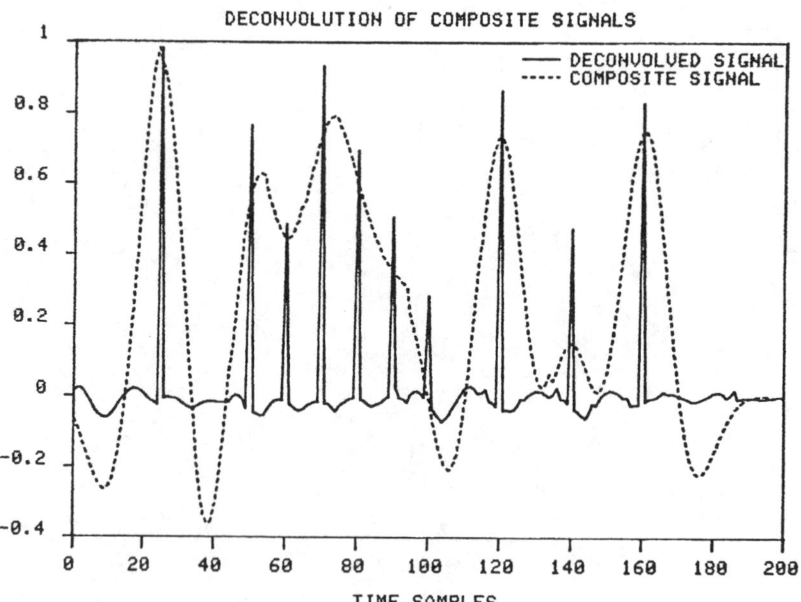

**Figure 5.43**   Deconvolved Signal $h_n * y_n$

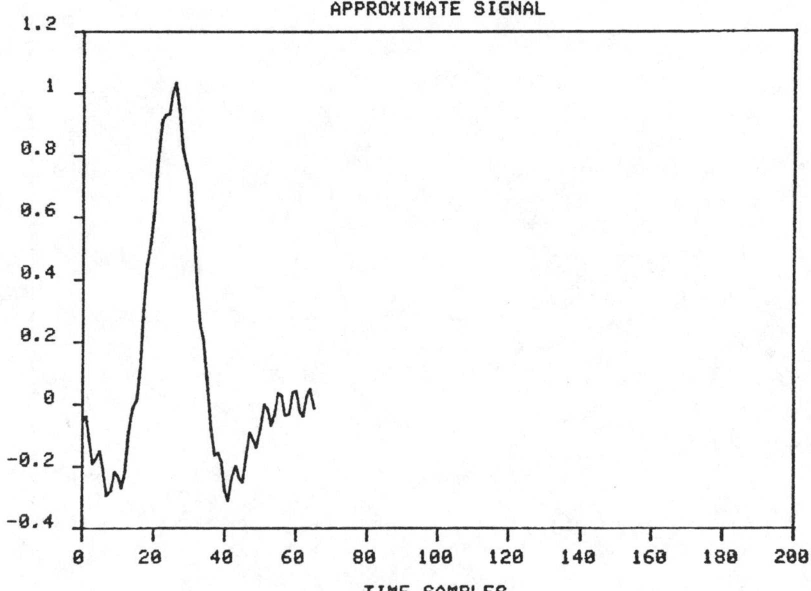


**Figure 5.44**  Approximately Known Signal $f'_n$

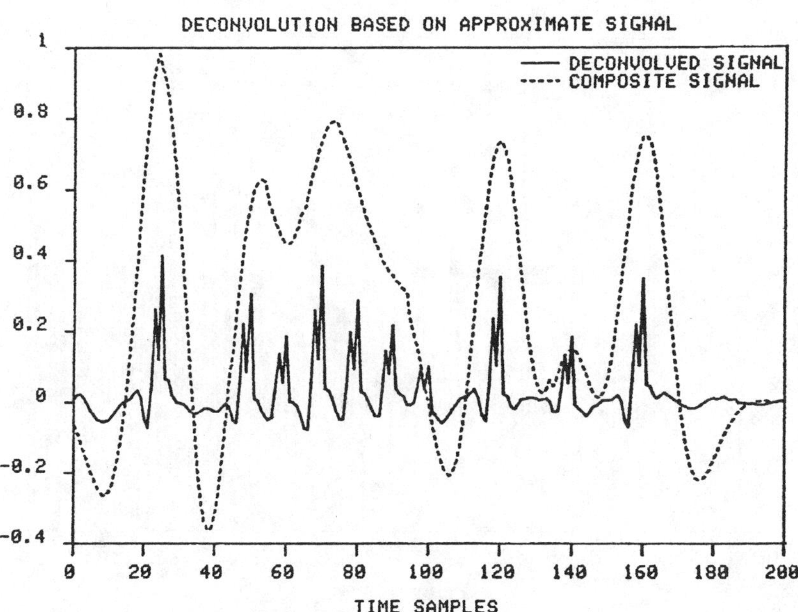


**Figure 5.45**  Deconvolved Signal $h'_n * y_n$, with $h'_n$ Designed on the Basis of $f'_n$

253

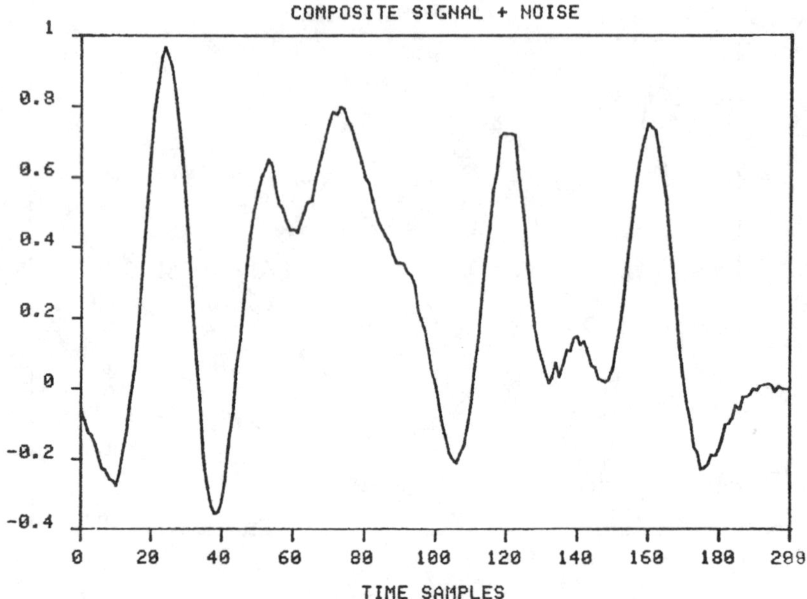



**Figure 5.46** Noisy Composite Signal $y_n = f_n * x_n + v_n$

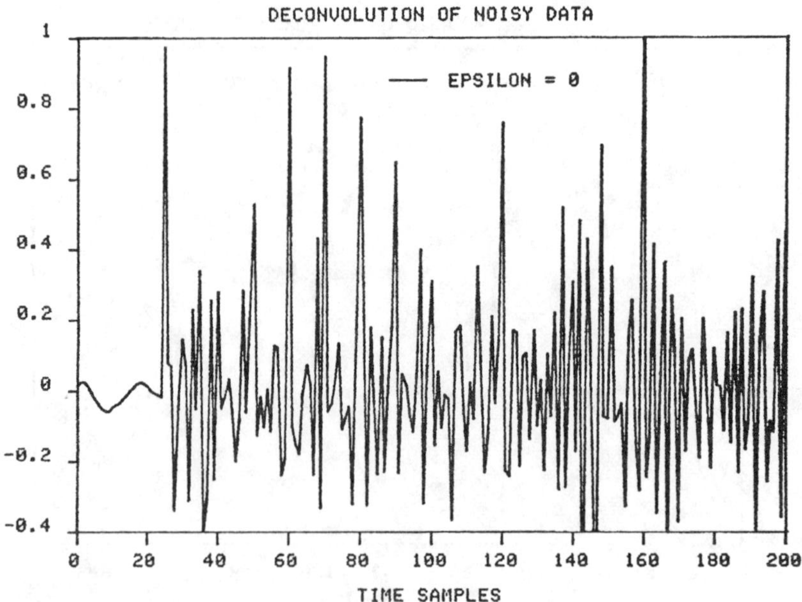



**Figure 5.47** Deconvolution of Noisy Data $y_n$, using $\epsilon = 0$

the basic nature of $x_n$ can still be seen. Inspecting Figure 5.39 we note the large spikes that are present in the impulse response $h_n$; these can cause the amplification of any additive noise component. Indeed, the noise reduction ratio of the filter $h_n$ is $\mathbf{h}^T\mathbf{h} = 612$, thus it will tend to amplify even small amounts of noise. To study the effect of noise, we have added a noise term $v_n$, as in Eq. (5.14.22), with variance equal to $10^{-4}$ (this corresponds to just 1% of the amplitude $a_0$); the composite signal $y_n$ is shown in Figure 5.46. One can barely see the noise. Yet, after filtering with the inverse filter $h_n$ of Figure (5.39), the noise component is amplified to a great extent. The result of deconvolving the noisy $y_n$ with $h_n$ is shown in Figure 5.47. To reduce the effects of noise, the prewhitening parameter $\epsilon$ must be chosen to be nonzero. Even a small nonzero value of $\epsilon$ can have a beneficial effect. Figures 5.48 and 5.49 show the deconvolved signal $x_n$ when the filter $h_n$ was designed with the choices $\epsilon = 0.0001$ and $\epsilon = 0.001$, respectively, of the parameter $\epsilon$. Note the trade-off between the noise reduction and the loss of resolution in the recovered spikes of $x_n$.

Based on the studies of Robinson and Treitel [19], Oldenburg [110], and others, the following summary of the use of the above deconvolution method may be made:

1. If the signal $f_n$ to be spiked is a minimal-phase signal, the optimal spiking delay must be chosen at the origin $i = 0$. The optimality of this choice is not actually seen until the filter order $M$ is sufficiently high. The reason for this choice has to do with the minimal-delay property of such signals which implies that most of their energy is concentrated at the beginning; therefore, they may be more easily compressed to spikes with zero delay.
2. If $f_n$ is a mixed-delay signal, as in the above example, then the optimal spiking delay will have some intermediate value.
3. Even if the shape of $f_n$ is not accurately known, the deconvolution procedure based on the approximate $f_n$ might have some partial success in deconvolving the replicas of $f_n$.
4. In the presence of noise in the data $y_n$ to deconvolved, some improvement may result by introducing a nonzero value for the prewhitening parameter $\epsilon$, where effectively the sample autocorrelation $\hat{R}_{ff}(0)$ is replaced by $(1 + \epsilon)\hat{R}_{ff}(0)$. The trade-off is a resulting loss of resolution.

The deconvolution problem of Eqs. (5.14.20) and (5.14.22) has been approached by a wide variety of other methods. Typically, a finite number of samples $y_n$; $n = 0,1, \ldots ,N$ is available. Collecting these into a vector $\mathbf{y} = [y_0,y_1, \ldots ,y_N]^T$, we write Eq. (5.14.22) in an obvious vectorial form

$$\mathbf{y} = F\mathbf{x} + \mathbf{v} \tag{5.14.23}$$

Instead of determining an approximate inverse filter for the blurring function $F$, an alternative method is to attempt to determine the best—in some sense—vector

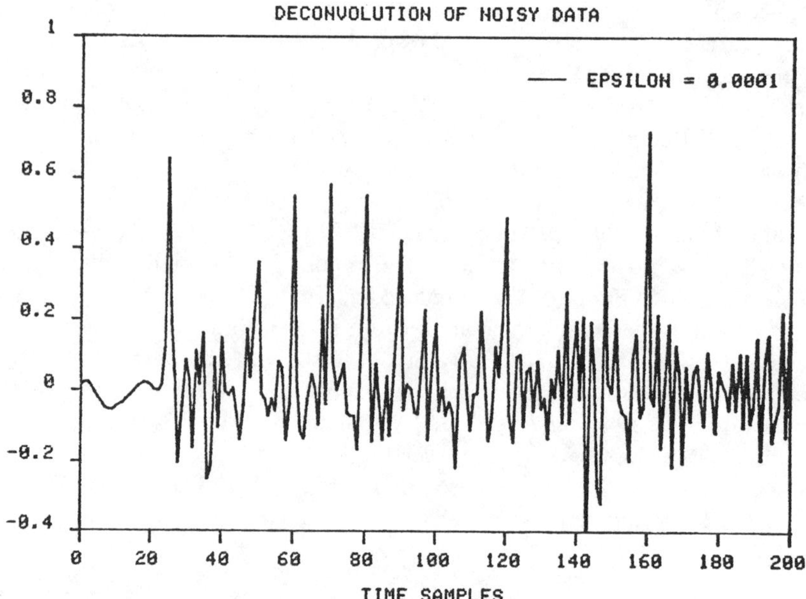



**Figure 5.48**  Deconvolution of Noisy Data $y_n$, using $\epsilon = 0.0001$

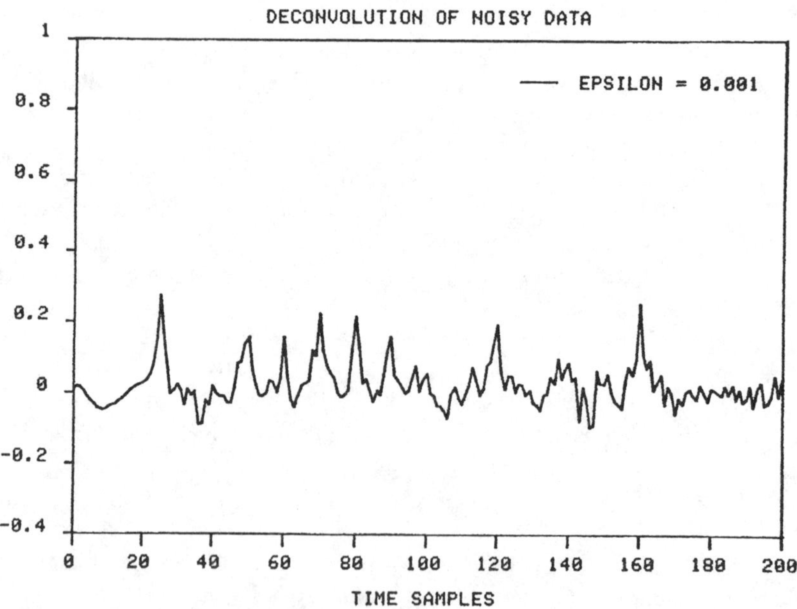



**Figure 5.49**  Deconvolution of Noisy Data $y_n$, using $\epsilon = 0.001$

**x** which is compatible with these equations. A popular method is based on the least-squares criterion [112,113]

$$\mathcal{E} = \sum_{n=0}^{N} v_n^2 = \mathbf{v}^T\mathbf{v} = (\mathbf{y} - F\mathbf{x})^T(\mathbf{y} - F\mathbf{x}) = \min \qquad (5.14.24)$$

That is, **x** is chosen so as to minimize $\mathcal{E}$. Setting the derivative with respect to **x** to zero gives the standard least-squares solution

$$\hat{\mathbf{x}} = (F^TF)^{-1}F^T\mathbf{y}$$

A prewhitening term can be added to the right of the performance index to stabilize the indicated inverse

$$\mathcal{E} = \mathbf{v}^T\mathbf{v} + \lambda\mathbf{x}^T\mathbf{x}$$

with solution $\hat{\mathbf{x}} = (F^TF + \lambda I)^{-1}F^T\mathbf{y}$.

Another approach that has been used recently with success is based on the $L_1$-norm criterion

$$\mathcal{E} = \sum_{n=0}^{N} |v_n| = \min \qquad (5.14.25)$$

This quantity is referred to as the $L_1$ norm of the vector **v**. The minimization of this norm with respect to **x** may be formulated as a *linear programming* problem [114–119]. It has been observed that this method performs very well in the presence of noise, and it tends to ignore a few "bad" data points—that is, those for which the noise value $v_n$ might be abnormally high—in favor of the good points, whereas the standard least-squares method based on the $L_2$-norm (5.14.24) will spend all its efforts trying to minimize the few large terms in the sum (5.14.24), and might not result in as good an estimate of **x** as it would if the few bad data points were to be ignored.

Another class of deconvolution methods are *iterative* methods. Such methods, like the linear programming method mentioned above, offer the additional option of enforcing *a priori constraints* that may be known to be satisfied by **x**, for example, *positivity, band-limiting,* or *time-limiting* constraints. The imposition of such constraints can improve the restoration process dramatically. The interested reader is referred to the review article [120].

## Problems

### Problem 5.1:

(a) Following the methods of Section 5.1, show that the optimal filter for predicting $D$ steps into the future—i.e., estimating $y(n + D)$ on the basis of $\{y(m); m \leq n\}$—is given by

$$H(z) = \frac{1}{B(z)}\,[z^D B(z)]_+$$

(b) Express $[z^D B(z)]_+$ in terms of $B(z)$ itself and the first $D - 1$ impulse response coefficients $b_m$; $m = 1, 2, \ldots, D - 1$ of $B(z)$.

(c) For the two random signals $y_n$ defined in Examples 5.1.1 and 5.1.2, find the optimal prediction filters for $D = 2$ and $D = 3$, and write the corresponding I/O filtering equations.

**Problem 5.2:**

Consider the order-$p$ autoregressive sequence $y_n$ defined by the difference equation (5.2.3). Show that a direct consequence of this difference equation is that the projection of $y_n$ onto the subspace spanned by the entire past $\{y_{n-i}; 1 \leqslant i < \infty\}$ is the same as the projection of $y_n$ onto the subspace spanned only by the past $p$ samples $\{y_{n-i}; 1 \leqslant i \leqslant p\}$.

**Problem 5.3:**

(a) Show that the performance index (5.3.2) may be written as

$$\mathcal{E} = E[e_n^2] = \mathbf{a}^T R \mathbf{a}$$

where $\mathbf{a} = [1, a_1, \ldots, a_p]^T$ is the order-$p$ prediction-error filter, and $R$ the autocorrelation matrix of $y_n$; that is, $R_{ij} = E[y_{n-i} y_{n-j}]$.

(b) Derive Eq. (5.3.7) by minimizing the index $\mathcal{E}$ with respect to the weights $\mathbf{a}$, subject to the linear constraint that $a_0 = 1$, and incorporating this constraint by means of a Lagrange multiplier.

**Problem 5.4:**

Take the inverse $z$-transform of Eq. (5.3.16) and compare the resulting equation with Eq. (5.3.14).

**Problem 5.5:**

Verify that Eqs. (5.3.20) and (5.3.21) are inverses of each other.

**Problem 5.6:**

A fourth order all-pole random signal process $y(n)$ is represented by the following set of signal model parameters (reflection coefficients and input variance):

$$\{\gamma_1, \gamma_2, \gamma_3, \gamma_4, \sigma_\epsilon^2\} = \{0.5, -0.5, 0.5, -0.5, 40.5\}$$

(a) Using the Levinson recursion, find the prediction error filter $A_4(z)$.

(b) Determine $\sigma_y^2 = R_{yy}(0)$. Using intermediate results from part (a), determine the autocorrelation lags $R_{yy}(k)$; $k = 1, 2, 3, 4$.

**Problem 5.7:**

The first five lags of the autocorrelation function of a fourth order autoregressive random sequence $y(n)$ are

$$\{R(0), R(1), R(2), R(3), R(4)\} = \{256, 128, -32, -16, 22\}$$

Determine the best prediction-error filters and the corresponding mean-squared errors of orders $p = 1,2,3,4$ by using Levinson's algorithm in matrix form.


**Problem 5.8:**

The fourth order prediction-error filter and mean-squared prediction error of a random signal have been determined to be

$$A_4(z) = 1 - 1.25z^{-1} + 1.3125z^{-2} - z^{-3} + 0.5z^{-4}, \qquad E_4 = 0.81$$

Using the subroutine RLEV, determine the autocorrelation lags $R(k)$; $0 \leqslant k \leqslant 4$, the four reflection coefficients, and all the lower order prediction-error filters.


**Problem 5.9:**

Verify the results of Example (5.3.1) using the subroutines LEV, FRWLEV, BKWLEV, and RLEV, as required.


**Problem 5.10:**

(a) Given the five signal samples

$$\{y_0, y_1, y_2, y_3, y_4\} = \{1, -1, 1, -1, 1\}$$

compute the corresponding sample autocorrelation lags $\hat{R}(k)$; $k = 0,1, \ldots ,4$, and send them through the routine LEV to determine the fourth order prediction error filter $A_4(z)$.

(b) Predict the sixth sample in this sequence.

(c) Repeat (a) and (b) for the sequence of samples $\{1,2,3,4,5\}$.


**Problem 5.11:**

Draw the lattice realization of the analysis and synthesis filters $A_4(z)$ and $1/A_4(z)$ obtained in Problems 5.6, 5.7, and 5.8.


**Problem 5.12:**

Test the minimal phase property of the two polynomials

$$A(z) = 1 - 1.08z^{-1} + 0.13z^{-2} + 0.24z^{-3} - 0.5z^{-4}$$
$$A(z) = 1 + 0.18z^{-1} - 0.122z^{-2} - 0.39z^{-3} - 0.5z^{-4}$$


**Problem 5.13:**

Consider a generalized version of the simulation example discussed in Section 5.8, defined by

$$x(n) = s(n) + v_1(n), \qquad y(n) = v_2(n)$$

where

$$s(n) = \sin(\omega_0 n + \phi)$$

$$v_1(n) = a_1 v_1(n - 1) + v(n)$$

$$v_2(n) = a_2 v_2(n - 1) + v(n)$$

where $v(n)$ is zero-mean, unit-variance, white noise, and $\phi$ a random phase independent of $v(n)$. This ensures that the $s(n)$ component is uncorrelated with $v_1(n)$ and $v_2(n)$.

(a) Show that

$$R_{xy}(k) = \frac{a_1^k}{1 - a_1 a_2} \qquad k \geq 0$$

$$R_{yy}(k) = \frac{a_2^k}{1 - a_2^2} \qquad k \geq 0$$

(b) Show that the infinite-order Wiener filter for estimating $x(n)$ on the basis of $y(n)$ has a (causal) impulse response

$$h_0 = 1, \; h_k = (a_1 - a_2)a_1^{k-1} \qquad \text{for } k \geq 1$$

(c) Next, consider the order-$M$ FIR Wiener filter. Send the theoretical correlations of part (a) for $k = 0, 1, \ldots, M$ through the subroutine FIRW to obtain the theoretical $M$th order Wiener filter realized both in the direct and the lattice forms. Draw these realizations. Compare the theoretical values of the weights **h**, **g**, and **γ** with the simulated values presented in Section 5.8 that correspond to the choice of parameters $M = 4$, $a_1 = -0.5$, and $a_2 = 0.8$. Also compare the answer for **h** with the first $M + 1$ samples of the infinite order Wiener filter impulse response of part (b).

(d) Repeat (c) with $M = 6$.


***Problem 5.14:***
A closed form solution of Problem 5.13 can be obtained as follows:

(a) Show that the inverse of the $(M + 1) \times (M + 1)$ autocorrelation matrix defined the autocorrelation lags $R_{yy}(k)$, $k = 0, 1, \ldots, M$, of Problem 5.13(a) is given by

$$R_{yy}^{-1} = \begin{bmatrix} 1 & -a_2 & 0 & \cdots & 0 & 0 \\ -a_2 & b & -a_2 & \cdots & 0 & 0 \\ 0 & -a_2 & b & \cdots & 0 & 0 \\ \vdots & \vdots & \vdots & \ddots & \vdots & \vdots \\ 0 & 0 & 0 & & b & -a_2 \\ 0 & 0 & 0 & & -a_2 & 1 \end{bmatrix}$$

where $b = 1 + a_2^2$.

(b) Using this inverse, show that the optimal $M$th order Wiener filter has impulse response

$$h_0 = 1, \quad h_k = (a_1 - a_2) \, a_1^{k-1} \text{ for } 1 \leqslant k \leqslant M - 1, \text{ and } h_M = \frac{(a_1 - a_2)}{1 - a_1 a_2} \, a_1^{M-1}$$

(c) Show that the lattice weights $\mathbf{g}$ can be obtained from $\mathbf{h}$ by the backward substitution

$$g_M = h_M, \text{ and } g_m = a_2 g_{m+1} + h_m, \ m = M - 1, M - 2, \ldots ,1,0$$

(d) For $M = 4$, $a_1 = -0.5$, $a_2 = 0.8$, compute the numerical values of $\mathbf{h}$ and $\mathbf{g}$ using the above expressions and compare them with those of Problem 5.13(c).

### Problem 5.15: *Computer Experiment*
Consider the noise canceling example of Section 5.8 and Problem 5.13, defined by the choice of parameters

$$\omega_0 = 0.075\pi \text{ [rads/sample]}, \quad \phi = 0, \ a_1 = -0.5, \ a_2 = 0.8, \ M = 4$$

(a) Generate 100 samples of the signals $x(n)$, $s(n)$, and $y(n)$. On the same graph, plot $x(n)$ and $s(n)$ versus $n$. Plot $y(n)$ versus $n$.

(b) Using these samples, compute the sample correlations $\hat{R}_{yy}(k)$, $\hat{R}_{xy}(x)$, for $k = 0,1, \ldots ,M$, and compare them with the theoretical values obtained in Problem 5.13(a).

(c) Send these lags through the routine FIRW to get the optimal Wiener filter weights $\mathbf{h}$ and $\mathbf{g}$, and the reflection coefficients $\boldsymbol{\gamma}$. Draw the lattice and direct form realizations of the Wiener filter.

(d) Filter $y_n$ through the Wiener filter realized in the lattice form, and plot the output $e(n) = x(n) - \hat{x}(n)$ versus $n$.

(e) Repeat (d) using the direct form realization of the Wiener filter.

(f) Repeat (d) when $M = 6$.

### Problem 5.16:
The following six samples

$$\{y_0,y_1,y_2,y_3,y_4,y_5\} = \{4.684, \ 7.247, \ 8.423, \ 8.650, \ 8.640, \ 8.392\}$$

have been generated by sending zero-mean unit-variance white-noise through the difference equation

$$y_n = a_1 y_{n-1} + a_2 y_{n-2} + \epsilon_n$$

where $a_1 = 1.70$, $a_2 = -0.72$. Iterating Burg's method by hand, obtain estimates of the model parameters $a_1$, $a_2$, and $\sigma_\epsilon^2$.

### Problem 5.17:
Derive Eq. (5.9.11).

***Problem 5.18:*** Computer Experiment

Ten samples from a fourth order autoregressive process $y(n)$ are given. It is desired to extract the model parameters $\{a_1,a_2,a_3,a_4,\sigma_\epsilon^2\}$ as well as the equivalent parameter set $\{\gamma_1,\gamma_2,\gamma_3,\gamma_4,\sigma_\epsilon^2\}$

    (a) Determine these parameters using Burg's method.

    (b) Repeat using the Yule-Walker method.

Note: The exact parameter values by which the above simulated samples were generated are

$$\{a_1,a_2,a_3,a_4,\sigma_\epsilon^2\}$$

$$= \{-2.2137,\ 2.9403,\ -2.1697,\ 0.9606,\ 1\}$$

| $n$ | $y(n)$ |
|---|---|
| 0 | 4.503 |
| 1 | $-10.841$ |
| 2 | $-24.183$ |
| 3 | $-25.662$ |
| 4 | $-14.390$ |
| 5 | 1.453 |
| 6 | 10.980 |
| 7 | 13.679 |
| 8 | 15.517 |
| 9 | 15.037 |

***Problem 5.19:*** Computer Experiment

A fourth order autoregressive process is defined by the difference equation

$$y_n + a_1 y_{n-1} + a_2 y_{n-2} + a_3 y_{n-3} + a_4 y_{n-4} = \epsilon_n$$

where $\epsilon_n$ is zero-mean, unit-variance, white gaussian noise. The filter parameters $\{a_1,a_2,a_3,a_4\}$ are chosen such that the prediction error filter

$$A(z) = 1 + a_1 z^{-1} + a_2 z^{-2} + a_3 z^{-3} + a_4 z^{-4}$$

has zeros at the locations

$$0.99\ \exp(\pm 0.2\pi j) \quad\text{and}\quad 0.99\ \exp(\pm 0.4\pi j)$$

    (a) Determine $\{a_1,a_2,a_3,a_4\}$.

    (b) Using a random number generator for $\epsilon_n$, generate a realization of $y_n$ consisting of 50 samples. To avoid transient effects, be sure to let the filter run for a while. For instance, discard the first 500 or 1000 outputs and keep the last 50.

    (c) Compute the sample autocorrelation of $y_n$ based on the above block of data.

    (d) Solve the normal equations by means of Levinson's algorithm to determine the Yule-Walker estimates of the model parameters $\{a_1,a_2,a_3,a_4,\sigma_\epsilon^2\}$ and compare them with the exact values.

    (e) Compute the corresponding Yule-Walker spectrum and plot it together with the exact autoregressive spectrum versus frequency. Be sure to allow for a sufficiently dense grid of frequencies to be able to resolve the narrow peaks of this example. Plot all spectra in decibels.

    (f) Using the same finite block of $y_n$ data, determine estimates of the model parameters $\{a_1,a_2,a_3,a_4,\sigma_\epsilon^2\}$ using Burg's method, and compare them with the Yule-Walker estimates and with the exact values.

    (g) Compute the corresponding Burg spectrum and plot it together with the exact spectrum versus frequency.

(h) Using the same block of $y_n$ data, compute the ordinary periodogram spectrum and plot it together with the exact spectrum.

(i) Window the $y_n$ data with a Hamming window and then compute the corresponding periodogram spectrum and plot it together with the exact spectrum.

(j) Repeat parts (b) through (i) using a longer realization of length 100.

(k) Repeat parts (b) through (i) using a length-200 realization of $y_n$.

(l) Evaluate the various results of this experiment.


***Problem 5.20:***
Show that the classical Bartlett spectrum of Eq. (5.11.6) can be written in the compact matrix form of Eq. (5.11.7).


***Problem 5.21:***
Show that in the limit of large $M$, the first sidelobe of the smearing function $W(\omega)$ of Eq. (5.11.10) is approximately 13 dB down from the main lobe.


***Problem 5.22:*** Computer Experiment

(a) Reproduce the spectra shown in Figures 5.19 through 5.22.

(b) For the AR case, let $M = 6$, and take the SNRs of both sinusoids to be 6 dB, but change the sinusoid frequencies to

$$\omega_1 = 0.5 + \Delta\omega, \qquad \omega_2 = 0.5 - \Delta\omega$$

where $\Delta\omega$ is variable. Study the dependence of bias of the spectral peaks on the frequency separation $\Delta\omega$ by computing and plotting the spectra for various values of $\Delta\omega$. (Normalize all spectra to 0 dB at the sinusoid frequency $\omega_1$).


***Problem 5.23:***
Derive Equation (5.11.30).


***Problem 5.24:***
Let

$$R = \sigma_v^2 I + \sum_{i=1}^{L} P_i \mathbf{s}_{\omega_i} \mathbf{s}_{\omega_i}^\dagger$$

be the autocorrelation matrix of Eq. (5.11.8). Show that the inverse $R^{-1}$ can be computed recursively as follows:

$$R_k^{-1} = R_{k-1}^{-1} - \frac{R_{k-1}^{-1} \mathbf{s}_{\omega_k} \mathbf{s}_{\omega_k}^\dagger R_{k-1}^{-1}}{\mathbf{s}_{\omega_k}^\dagger R_{k-1}^{-1} \mathbf{s}_{\omega_k} + P_k^{-1}}$$

for $k = 1, \ldots, L$, initialized by $R_0 = \sigma_v^2 I$.

**_Problem 5.25:_**

Consider the case of one sinusoid ($L = 1$) in noise and arbitrary filter order $M > 2$, so that the $(M + 1) \times (M + 1)$ autocorrelation matrix is

$$R = \sigma_v^2 I + P_1 s_{\omega_1} s_{\omega_1}^\dagger$$

(a) Show that the ($L = 1$) $-$ dimensional signal subspace is spanned by the eigenvector

$$e = s_{\omega_1}$$

and determine the corresponding eigenvalue.

(b) Show that the $M + 1 - L = M$ dimensional noise subspace is spanned by the $M$ linearly independent eigenvectors

$$
e_1 = \begin{bmatrix} 1 \\ -e^{j\omega_1} \\ 0 \\ 0 \\ \vdots \\ 0 \\ 0 \\ 0 \end{bmatrix} \updownarrow M-1, \quad
e_2 = \begin{bmatrix} 0 \\ 1 \\ -e^{j\omega_1} \\ 0 \\ \vdots \\ 0 \\ 0 \\ 0 \end{bmatrix},
$$

$$
e_3 = \begin{bmatrix} 0 \\ 0 \\ 1 \\ -e^{j\omega_1} \\ \vdots \\ 0 \\ 0 \\ 0 \end{bmatrix}, \dots, \quad
e_M = \begin{bmatrix} 0 \\ 0 \\ 0 \\ 0 \\ \vdots \\ 0 \\ 1 \\ -e^{j\omega_1} \end{bmatrix}
$$

all belonging to the minimum eigenvalue $\sigma_v^2$.

(c) Show that the eigenpolynomial $A(z)$ corresponding to an arbitrary linear combination of the $M$ noise eigenvectors

$$a = e_1 + c_1 e_2 + c_2 e_3 + \cdots + c_{M-1} e_M$$

can be factored in the form

$$A(z) = (1 - e^{j\omega_1} z^{-1})(1 + c_1 z^{-1} + c_2 z^{-2} + \cdots + c_{M-1} z^{-M+1})$$

exhibiting one zero at the desired sinusoid frequency $\omega_1$ on the unit circle, and $M - 1$ additional spurious zeros with arbitrary locations that depend on the particular choice of the coefficients $c_i$.

**Problem 5.26:**
The constraint (5.11.31) can be incorporated into the performance index (5.11.32) by means of a Lagrange multiplier

$$\mathcal{E} = \mathbf{a}^{\dagger} R \mathbf{a} + \lambda(1 - \mathbf{a}^{\dagger} \mathbf{a})$$

Show that the minimization of $\mathcal{E}$ is equivalent to the Pisarenko eigenvalue problem of Eq. (5.11.29), with the multiplier $\lambda$ playing the role of the eigenvalue. Show that the minimum of $\mathcal{E}$ is the minimum eigenvalue.


**Problem 5.27:**
Show Equation (5.12.11).


**Problem 5.28:** Computer Experiment
To simulate Eq. (5.12.7), the signal amplitudes $A_i(n)$ may be generated by

$$A_i(n) = A_i \exp(j\phi_{in})$$

where $\phi_{in}$ are independent random phases distributed uniformly over the interval $[0, 2\pi]$, and $A_i$ are deterministic amplitudes related to the assumed signal to noise ratios (SNR) in units of decibels by

$$(\text{SNR})_i = 10 \log_{10} \left[ \frac{|A_i|^2}{\sigma_v^2} \right], \qquad i = 1, 2, \ldots, L$$

(a) Consider one plane wave incident on an array of seven sensors from an angle $\theta_1 = 30°$. The sensors are equally spaced at half-wavelength spacings; i.e., $d = \lambda/2$. For each of the following values of the SNR of the wave

$$\text{SNR} = 0 \text{ dB}, \quad 10 \text{ dB}, \quad 20 \text{ dB}$$

generate $N = 1000$ snapshots of Eq. (5.12.7) and compute the empirical spatial correlation matrix across the array by

$$R = \frac{1}{N} \sum_{n=0}^{N-1} \mathbf{y}(n)^{*} \mathbf{y}(n)^{T}$$

Compute and plot on the same graph the three spatial spectra: Bartlett, autoregressive (AR), and maximum likelihood (ML), versus wavenumber $k$.
(b) Repeat for two plane waves incident from angles $\theta_1 = 25°$ and $\theta_2 = 35°$, and with equal powers of 30 dB.
(c) Repeat part (b) for angles $\theta_1 = 28°$ and $\theta_2 = 32°$.
(d) Repeat part (c) by gradually decreasing the (common) SNR of the two plane waves to the values of 20 dB, 10 dB, and 0 dB.
(e) For parts (a) through (d), also plot all the theoretical spectra.
Note: The empirical $R$ is Hermitian but not Toeplitz. If you find it difficult to compute $R^{-1}$ required in the AR and ML spectra, you may work with a Toeplitz

version of $R$ obtained by replacing each diagonal by the average of the entries along that diagonal, and then develop a complex-valued version of the routine LEV to obtain $R^{-1}$.

**Problem 5.29:**

Consider $L$ plane waves incident on a linear array of M + 1 sensors (L ≤ M) in the presence of spatially coherent noise. As discussed in section 5.12, the corresponding covariance matrix is given by

$$R = \sigma_v^2 Q + \sum_{i=1}^{L} P_i \mathbf{s}_{k_i} \mathbf{s}_{k_i}^\dagger$$

where the waves are assumed to be mutually uncorrelated.

(a) Show that the generalized eigenvalue problem

$$R\mathbf{a} = \lambda Q \mathbf{a}$$

has (1) an $(M + 1 - L)$-dimensional noise subspace spanned by $M + 1 - L$ linearly independent degenerate eigenvectors all belonging to the eigenvalue $\lambda = \sigma_v^2$ and (2) an $L$-dimensional signal subspace with $L$ eigenvalues greater than $\sigma_v^2$.

(b) Show that any two eigenvectors $\mathbf{a}_1$ and $\mathbf{a}_2$ belonging to distinct eigenvalues $\lambda_1$ and $\lambda_2$ are orthogonal to each other with respect to the inner product defined by the matrix $Q$; that is, show that

$$\mathbf{a}_1^\dagger Q \mathbf{a}_2 = 0$$

(c) Show that the $L$-dimensional signal subspace is spanned by the $L$ vectors

$$Q^{-1}\mathbf{s}_{k_i}; \qquad i = 1, 2, \ldots, L$$

(d) Show that any vector $\mathbf{a}$ in the noise subspace corresponds to a polynomial $A(z)$ that has $L$ of its $M$ zeros on the unit circle at locations

$$z_i = \exp(jk_i); \qquad i = 1, 2, \ldots, L$$

The remaining $M - L$ zeros can have arbitrary locations.

**Problem 5.30:**

The previous problem suggests the following approach to the problem of "selectively nulling" some of the sources and not nulling others. Suppose $L_1$ of the sources are not to be nulled and have known SNRs and directions of arrival, and $L_2$ of the sources are to be nulled. The total number of sources is then

$L = L_1 + L_2$, and assuming incoherent background noise, the incident field will have covariance matrix

$$R = \sigma_v^2 I + \sum_{i=1}^{L_1} P_i \mathbf{s}_{k_i} \mathbf{s}_{k_i}^\dagger + \sum_{i=L_1+1}^{L_1+L_2} P_i \mathbf{s}_{k_i} \mathbf{s}_{k_i}^\dagger$$

Define $Q$ by means of

$$\sigma_v^2 Q = \sigma_v^2 I + \sum_{i=1}^{L_1} P_i \mathbf{s}_{k_i} \mathbf{s}_{k_i}^\dagger$$

so that we may write $R$ as follows

$$R = \sigma_v^2 Q + \sum_{i=L_1+1}^{L_1+L_2} P_i \mathbf{s}_{k_i} \mathbf{s}_{k_i}^\dagger$$

Then, the nulling of the $L_2$ sources at wavenumbers $k_i$; $i = L_1 + 1, \ldots, L_1 + L_2$, can be effected by the $(M + 1 - L_2)$-dimensional "noise" subspace of the generalized eigenvalue problem

$$R\mathbf{a} = \lambda Q\mathbf{a}$$

having minimum eigenvalue equal to $\sigma_v^2$.

(a) As an example, consider the case $M = 2$, $L_1 = L_2 = 1$. Then,

$$R = \sigma_v^2 Q + P_2 \mathbf{s}_{k_2} \mathbf{s}_{k_2}^\dagger, \qquad \sigma_v^2 Q = \sigma_v^2 I + P_1 \mathbf{s}_{k_1} \mathbf{s}_{k_1}^\dagger$$

Show that the $(M + 1 - L_2 = 2)$-dimensional noise subspace is spanned by the two eigenvectors

$$\mathbf{e}_1 = \begin{bmatrix} 1 \\ -e^{jk_2} \\ 0 \end{bmatrix}, \qquad \mathbf{e}_2 = \begin{bmatrix} 0 \\ 1 \\ -e^{jk_2} \end{bmatrix}$$

(b) Show that an arbitrary linear combination

$$\mathbf{a} = \mathbf{e}_1 + \rho_1 \mathbf{e}_2$$

corresponds to a filter $A(z)$ having one zero at the desired location $z_2 = \exp(jk_2)$, and a spurious zero with arbitrary location.

(c) Show that the $(L_2 = 1)$-dimensional signal subspace is spanned by the vector

$$\mathbf{e}_3 = Q^{-1} \mathbf{s}_{k_2}$$

and that the corresponding generalized eigenvalue is

$$\lambda = \sigma_v^2 + P_2 \mathbf{s}_{k_2}^\dagger Q^{-1} \mathbf{s}_{k_2}$$

(d) Verify the orthogonality properties $e_i^\dagger Q e_3 = 0$, $i = 1,2$, for the three eigenvectors $e_1$, $e_2$, $e_3$ defined in parts (a) and (c).

(e) As another example, consider the case $M = 3$, and $L_1 = L_2 = 1$. Show that the $(M + 1 - L_2 = 3)$-dimensional noise subspace is spanned by the three eigenvectors

$$
e_1 = \begin{bmatrix} 1 \\ -e^{jk_2} \\ 0 \\ 0 \end{bmatrix}, \quad
e_2 = \begin{bmatrix} 0 \\ 1 \\ -e^{jk_2} \\ 0 \end{bmatrix}, \quad
e_3 = \begin{bmatrix} 0 \\ 0 \\ 1 \\ -e^{jk_2} \end{bmatrix}
$$

and the signal eigenvector is $e_4 = Q^{-1}s_{k_2}$. Generalize this part and part (a), to the case of arbitrary $M$ and $L_1 = L_2 = 1$.

(f) As a final example that corresponds to a unique noise eigenvector, consider the case $M = 2$, and $L_1 = 1$, $L_2 = 2$, so that

$$
R = \sigma_v^2 Q + P_2 s_{k_2} s_{k_2}^\dagger + P_3 s_{k_3} s_{k_3}^\dagger, \qquad \sigma_v^2 Q = \sigma_v^2 I + P_1 s_{k_1} s_{k_1}^\dagger
$$

with $k_2$ and $k_3$ to be nulled. Show that the $(M + 1 - L_2 = 1)$-dimensional noise subspace is spanned by

$$
a = e_1 = \begin{bmatrix} 1 \\ -(e^{jk_2} + e^{jk_3}) \\ e^{jk_2}e^{jk_3} \end{bmatrix}
$$

and that the corresponding polynomial $A(z)$ factors into the two desired zeros

$$
A(z) = (1 - e^{jk_2}z^{-1})(1 - e^{jk_3}z^{-1})
$$

**Problem 5.31:**
Using the continuity equations at an interface, derive the transmission matrix equation (5.13.2) and the energy conservation equation (5.13.4).

**Problem 5.32:**
Show Eq. (5.13.6).

**Problem 5.33:**
Figure 5.25 defines the scattering matrix $S$. Explain how the principle of linear superposition may be used to show the general relationship

$$
\begin{bmatrix} E'_+ \\ E_- \end{bmatrix} = S \begin{bmatrix} E_+ \\ E'_- \end{bmatrix}
$$

between incoming and outgoing fields.

***Problem 5.34:***
Show the two properties of the matrix $\psi_m(z)$ stated in Eqs. (5.13.13) and (5.13.14).

***Problem 5.35:***
Show Eqs. (5.13.25).

***Problem 5.36:***
The reflection response of a stack of four dielectrics has been found to be

$$ R(z) = \frac{-0.25 + 0.0313z^{-1} + 0.2344z^{-2} - 0.2656z^{-3} + 0.25z^{-4}}{1 - 0.125z^{-1} + 0.0664z^{-3} - 0.0625z^{-4}} $$

Determine the reflection coefficients $\{\rho_0, \rho_1, \rho_2, \rho_3, \rho_4\}$.

***Problem 5.37:***
Let $R_M(z)$ and $T_M(z)$ denote the reflection and transmission responses of $M$ slabs, defined by Eqs. (5.13.27). Using the lattice relationships of Eq. (5.13.23), derive a recursive relationship between $R_M(z)$, $T_M(z)$ and $R_{M-1}(z)$, $T_{M-1}(z)$. Using this recursive relationship, obtain the reflection coefficients $\{\rho_0, \rho_1, \rho_2, \rho_3, \rho_4\}$ of the previous problem.

***Problem 5.38:*** Computer Experiment
It is desired to probe the structure of a stack of dielectrics from its reflection response. To this end, a unit impulse is sent incident on the stack and the reflection response is measured as a function of time.

It is known in advance (although this is not necessary) that the stack consists of four equal travel-time slabs stacked in front of a semi-infinite medium.

Thirteen samples of the reflection response are collected as shown here. Determine the reflection coefficients $\{\rho_0, \rho_1, \rho_2, \rho_3, \rho_4\}$ by means of the dynamic predictive deconvolution procedure.

| $k$ | $R(k)$ |
|---|---|
| 0 | $-0.2500$ |
| 1 | $0.0000$ |
| 2 | $0.2344$ |
| 3 | $-0.2197$ |
| 4 | $0.2069$ |
| 5 | $0.0103$ |
| 6 | $0.0305$ |
| 7 | $-0.0237$ |
| 8 | $0.0093$ |
| 9 | $-0.0002$ |
| 10 | $0.0035$ |
| 11 | $-0.0017$ |
| 12 | $0.0004$ |

***Problem 5.39:*** Computer Experiment
Generate the results of Figures 5.27–5.28 and 5.36–5.37.

***Problem 5.40:*** Computer Experiment
This problem illustrates the use of the dynamic predictive deconvolution method in the design of broadband terminations of transmission lines. The termination is constructed by the cascade of $M$ equal travel-time segments of transmission

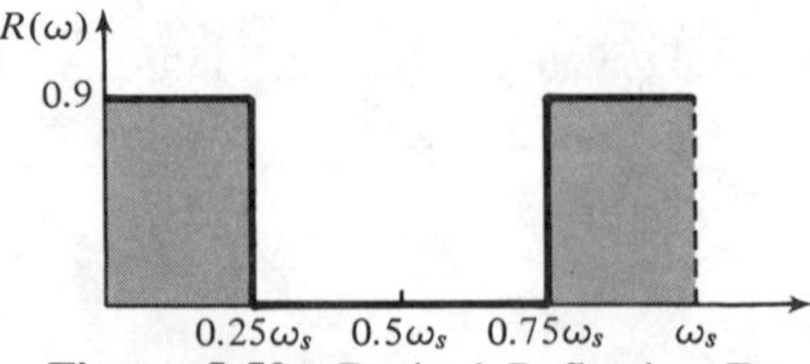

**Figure 5.50**  Desired Reflection Frequency Response

lines such that the overall reflection response of the structure approximates the desired reflection response. The characteristic impedances of the various segments are obtainable from the reflection coefficients $\{\rho_0, \rho_1, \ldots, \rho_M\}$. The reflection response $R(\omega)$ of the structure is a periodic function of $\omega$ with period $\omega_s = 2\pi/T_2$, where $T_2$ is the two-way travel time delay of each segment. The design procedure is illustrated by the following example: The desired frequency response $R(\omega)$ is defined over one Nyquist period by

$$R(\omega) = \begin{cases} 0 & \text{for} \quad 0.25\omega_s \leq \omega \leq 0.75\omega_s \\ 0.9 & \text{for} \quad 0 \leq \omega < 0.25\omega_s \text{ and } 0.75\omega_s < \omega \leq \omega_s \end{cases}$$

as shown in Fig. 5.50.

(a) Using the Fourier series method of designing digital filters [73], design a $N = 21$-tap filter with impulse response $R(k)$; $k = 0, 1, \ldots, N - 1$, whose frequency response approximates the desired response defined above. Window the designed reflection impulse response $R(k)$ by a length-$N$ Hamming window. Plot the magnitude frequency response of the windowed reflection series over one Nyquist interval $0 \leq \omega \leq \omega_s$.

(b) For $M = 6$, send the $N$ samples of the windowed reflection series through the dynamic predictive deconvolution routine DPD to obtain the polynomials $A_M(z)$ and $B_M(z)$ and the reflection coefficients $\{\rho_0, \rho_1, \rho_2, \ldots, \rho_M\}$. Plot the magnitude response of the structure; that is, plot

$$|R(\omega)| = \left| \frac{B_M(z)}{A_M(z)} \right| \qquad z = \exp(j\omega T_2) = \exp\left(2\pi j \frac{\omega}{\omega_s}\right)$$

and compare it with the windowed response of part (a). To facilitate the comparison, plot both responses of parts (a) and (b) on the same graph.

(c) Repeat part (b) for $M = 2$, $M = 3$, and $M = 10$.

(d) Repeat parts (a) through (c) for $N = 31$ reflection series samples.

(e) Repeat parts (a) through (c) for $N = 51$.


***Problem 5.41:***

Show that the performance matrix $P$ defined by Eq. (5.14.11) has trace equal to $M + 1$.

***Problem 5.42:*** Computer Experiment
Reproduce the results of Figures 5.38 through 5.49.

## *References*


1. G. P. Box and G. M. Jenkins, *Time Series Analysis, Forecasting, and Control*, San Francisco, Holden-Day, 1970.

2. P. Whittle, *Prediction and Regulation*, New York, Van Nostrand Reinhold, 1963.

3. J. Makhoul, Linear Prediction: A Tutorial Review, *Proc. IEEE*, **63**, 56 (1975).

4. N. Levinson, The Wiener RMS Error Criterion in Filter Design and Prediction, *J. Math. Physics*, **25**, 261 (1947).

5. J. Durbin, The Fitting of Time Series Models, *Rev. Inst. Int. Stat.*, **28**, 344–348 (1973).

6. J. D. Markel and A. H. Gray, Jr., *Linear Prediction of Speech*, New York, Springer-Verlag, 1976.

7. E. A. Robinson, *Multichannel Time-Series Analysis with Digital Computer Programs*, San Francisco, Holden-Day, 1967.

8. E. A. Robinson and S. Treitel, *Geophysical Signal Analysis*, Englewood Cliffs, Prentice-Hall, 1980.

9. E. A. Robinson and S. Treitel, Maximum Entropy and the Relationship of the Partial Autocorrelation to the Reflection Coefficients of a Layered System, *IEEE Trans. Acoust., Speech, Signal Process.*, **ASSP-28**, 22 (1980).

10. S. M. Kay and S. L. Marple, Spectrum Analysis—A Modern Perspective, *Proc. IEEE*, **69**, 1380–1419 (1981).

11. S. Haykin, Ed., *Nonlinear Methods of Spectral Analysis*, New York, Springer-Verlag, 1979.

12. F. L. Ramsey, Characterization of the Partial Autocorrelation Function," *Ann. Stat.*, **2**, 1296–1301 (1974)

13. M. Morf, A. Vieira, and T. Kailath, Covariance Characterization by Partial Autocorrelation Matrices, *Ann. Stat.*, **6**, 643–645 (1978).

14. A. J. Berkhout, Stability and Least-Squares Estimation, *Automatica*, **11**, 637–638 (1975).

15. A. Vieira and T. Kailath, Another Approach to the Schur-Cohn Criterion, *IEEE Trans. Circuits and Systems*, **CAS-24**, 218–220 (April 1977).

16. S. A. Tretter, *Introduction to Discrete-Time Signal Processing*, New York, Wiley, 1976).

17. N. I. Achiezer, *The Classical Moment Problem*, Edinburgh, Oliver and Boyd, 1965.

18. G. Szegö, *Orthogonal Polynomials*, Providence, RI, American Mathematical Society, 1958.

19. E. A. Robinson and S. Treitel, Digital Signal Processing in Geophysics, in A. Oppenheim, Ed., *Applications of Digital Signal Processing*, Englewood Cliffs, Prentice-Hall, 1978.

20. S. Treitel and E. A. Robinson, The Design of High-Resolution Digital Filters, *IEEE Trans. Geosc. Electron.*, **GE-4,** 25–38 (1966).

21. J. Claerbout, *Fundamentals of Geophysical Data Processing,* New York, McGraw-Hill, 1976.

22. I. C. Gohberg and I. A. Fel'dman, *Convolution Equations and Projection Methods for their Solution,* Providence, RI, American Mathematical Society, 1974.

23. W. F. Trench, An Algorithm for the Inversion of Finite Toeplitz Matrices, *J. Soc. Ind. Appl. Math.*, **12,** 515–522 (1964).

24. S. Zohar, Toeplitz Matrix Inversion: The Algorithm of W. F. Trench, *J. Assoc. Comput. Mach.*, **16,** 592–601 (1969).

25. S. Zohar, The Solution of a Toeplitz Set of Linear Equations, *J. Assoc. Comput. Mach.*, **21,** 272–276 (1974).

26. T. Kailath, A. Vieira, and M. Morf, Inverses of Toeplitz Operators, Innovations and Orthogonal Polynomials, *SIAM Rev.*, **20,** 106–119 (1978).

27. J. Le Roux and C. J. Gueguen, A Fixed Point Computation of Partial Correlation Coefficients, *IEEE Trans. Acoust., Speech, Signal Process.*, **ASSP-25,** 257–259 (1977).

28. H. M. Ahmed, J. M. Delosme, and M. Morf, Highly Concurrent Computing Structures for Matrix Arithmetic and Signal Processing, *Computer Magazine,* **15,** 65–82 (Jan. 1982).

29. H. T. Kung, Why Systolic Architectures?, *Computer Magazine,* **15,** 37–46 (Jan. 1982).

30. S. Y. Kung and Y. H. Hu, A Highly Concurrent Algorithm and Pipelined Architecture for Solving Toeplitz Systems, *IEEE Trans. Acoust., Speech, Signal Process.*, **ASSP-31,** 66–75 (1983).

31. S. Y. Kung and Y. H. Hu, Highly Concurrent Toeplitz Eigen-System Solver for High Resolution Spectral Estimation, *Proc. IEEE Int. Conf. Acoust., Speech, Signal Process.*, pp. 1422–1425 (1983).

32. J. P. Burg, Maximum Entropy Spectral Analysis, presented at *37th Annual Int. SEG Meeting,* Oklahoma City, (1967).

33. D. Childers, Ed., *Modern Spectrum Analysis,* New York, IEEE Press, 1978.

34. E. R. Kanasewich, *Time Sequence Analysis in Geophysics,* Edmonton, University of Alberta Press, 1975.

35. D. E. Smylie, G. K. C. Clarice, and T. J. Ulrich, Analysis of Irregularities in the Earth's Rotation, in *Methods of Computational Physics,* Vol. 13, New York, Academic, 1973, pp. 391–430.

36. T. J. Ulrich and R. W. Clayton, Time Series Modelling and Maximum Entropy, *Phys. Earth Planet. Inter.*, **12,** 188–200 (1976).

37. M. Morf, B. Dickinson, T. Kailath, and A. Vieira, Efficient Solution of Covariance Equations for Linear Prediction, *IEEE Trans. Acoust., Speech, Signal Process.*, **ASSP-25,** 429–435 (1977).

38. B. S. Atal and S. Hanauer, Speech Analysis and Synthesis by Linear Prediction of the Speech Wave, *J. Acoust. Soc. Amer.*, **50,** 637 (1971).

39. F. Itakura and S. Saito, A Statistical Method for Estimation of Speech Spectral Density and Formant Frequencies, *Electr. Commun.*, **53-A,** 36–43 (1970).

40. R. Schafer and L. Rabiner, Digital Representation of Speech Signals, *Proc. IEEE*, **63**, 66 (1975).

41. L. R. Rabiner and R. W. Schafer, *Digital Processing of Speech Signals*, Englewood Cliffs, Prentice-Hall, 1978.

42. J. D. Markel and A. H. Gray, Jr., Roundoff Noise Characteristics of a Class of Orthogonal Polynomial Structures, *IEEE Trans. Acoust., Speech, Signal Process.*, **ASSP-23**, 473–486 (1975).

43. R. Viswanathan and J. Makhoul, Quantization Properties of Transmission Parameters in Linear Predictive Systems, *IEEE Trans. Acoust., Speech, Signal Process.*, **ASSP-23**, 309–321 (1975).

44. A. Isaksson, A. Wennberg, and L. H. Zetterberg, Computer Analysis of EEG Signals with Parametric Models, *Proc. IEEE*, **69**, 451–461 (1981).

45. W. Gersch, Spectral Analysis of EEG's by Autoregressive Decomposition of Time Series, *Math. Biosci.*, **7**, 205–222 (1970).

46. C. D. McGillem, J. I. Aunon, and D. G. Childers, Signal Processing In Evoked Potential Research: Applications of Filtering and Pattern Recognition, *CRC Critical Reviews of Bioengineering*, **6**, 225–265 (October 1981).

47. A. Isaksson and A. Wennberg, Spectral Properties of Nonstationary EEG Signals, Evaluated by Means of Kalman Filtering: Application Examples from a Vigilance Test, in P. Kellaway and I. Petersen, Eds., *Quantitative Analysis Studies in Epilepsy*, New York, Raven Press, 1976.

48. G. Bodenstein and H. M. Praetorius, Feature Extraction from the Electroencephalogram by Adaptive Segmentation, *Proc. IEEE*, **65**, 642–652 (1977).

49. T. Bohlin, Analysis of EEG Signals with Changing Spectra using a Short-Word Kalman Estimator, *Math. Biosci.*, **35**, 221–259 (1977).

50. F. H. Lopes da Silva, Analysis of EEG Nonstationarities, in W. A. Cobb and H. Van Duijn, Eds., *Contemporary Clinical Neurophysiology* (EEG Suppl. No. 34), Amsterdam, Elsevier, 1978.

51. Z. Rogowski, I. Gath, and E. Bental, On the Prediction of Epileptic Seizures, *Biol. Cybernetics*, **42**, 9–15 (1981).

52. F. Itakura, Minimum Prediction Residual Principle Applied to Speech Recognition, *IEEE Trans. Acoust., Speech, Signal Process.*, **ASSP-23**, 67–72 (1975).

53. J. M. Tribolet, L. R. Robiner, and M. M. Sondhi, Statistical Properties of an LPC Distance Measure, *IEEE Trans. Acoust., Speech, Signal Process.*, **ASSP-27**, 550–558 (1979).

54. P. de Souza and P. J. Thompson, LPC Distance Measures and Statistical Tests with Particular Reference to the Likelihood Ratio, *IEEE Trans. Acoust., Speech, Signal Process.*, **ASSP-30**, 304–315 (1982).

55. R. M. Gray, et al., Distortion Measures for Speech Processing, *IEEE Trans. Acoust., Speech, Signal Process.* **ASSP-28**, 367–376 (1980).

56. J. L. Flanagan, Talking with Computers: Synthesis and Recognition of Speech by Machines, *IEEE Trans. Biomed. Eng.*, **BME-29**, 223–232 (1982).

57. L. Dusek, T. B. Schalk, and M. McMahan, Voice Recognition Joins Speech on Programmable Board, *Electronics*, **56**(8), 128–132 (April 1983).

58. W. Gersch and D. R. Sharpe, Estimation of Power Spectra with Finite Order Autoregressive Models, *IEEE Trans. Autom. Control.*, **AC-13**, 367–369 (1973).

59. O. L. Frost, Power Spectrum Estimation, in G. Tacconi, Ed., *Aspects of Signal Processing*, Boston, Reidel, 1977.

60. P. R. Gutowski, E. A. Robinson, and S. Treitel, Spectral Estimation: Fact or Fiction?, *IEEE Trans. Geosci. Electron.*, **GE-16,** 80–84 (1978).

61. *Proc. IEEE,* **70**(9) (September 1982), Special Issue on Spectral Estimation.

62. A. Papoulis, Maximum Entropy and Spectral Estimation: A Review, *IEEE Trans. Acoust., Speech, Signal Process.*, **ASSP-29,** 1176–1186 (1981).

63. E. A. Robinson, A Historical Perspective of Spectrum Estimation, *Proc. IEEE,* **70,** 885–907 (1982).

64. J. Capon, High Resolution Frequency Wavenumber Spectrum Analysis, *Proc. IEEE,* **57,** 1408–1418 (1969).

65. J. Capon, Maximum Likelihood Spectral Estimation, in S. Haykin, Ed., *Nonlinear Methods of Spectral Analysis,* New York, Springer-Verlag, 1979.

66. R. T. Lacoss, Data Adaptive Spectral Analysis Methods, *Geophysics,* **36,** 661–675 (1971).

67. V. F. Pisarenko, The Retrieval of Harmonics from a Covariance Function, *Geoph. J. R. Astron. Soc.,* **33,** 347–366 (1973).

68. E. H. Satorius and J. R. Zeidler, Maximum Entropy Spectral Analysis of Multiple Sinusoids in Noise, *Geophysics,* **43,** 1111–1118 (1978).

69. D. W. Tufts and R. Kumaresan, Singular Value Decomposition and Improved Frequency Estimation Using Linear Prediction, *IEEE Trans. Acoust., Speech, Signal Process.*, **ASSP-30,** 671–675 (1982).

70. D. W. Tufts and R. Kumaresan, Estimation of Frequencies of Multiple Sinusoids: Making Linear Prediction Perform like Maximum Likelihood, *Proc. IEEE,* **70,** 975–989 (1982).

71. S. L. Marple, Frequency Resolution of Fourier and Maximum Entropy Spectral Estimates, *Geophysics,* **47,** 1303–1307 (1982).

72. M. Quirk and B. Liu, On the Resolution of Autoregressive Spectral Estimation, *Proc. IEEE Int. Conf. Acoust., Speech, Signal Process.*, 1095–1097 (1983).

73. A. V. Oppenheim and R. W. Schafer, *Digital Signal Processing,* Englewood Cliffs, Prentice-Hall, 1975.

74. W. C. Knight, R. G. Pridham, and S. M. Kay, Digital Signal Processing for Sonar, *Proc. IEEE,* **69,** 1451–1506 (1981).

75. W. F. Gabriel, Spectral Analysis and Adaptive Array Superresolution Techniques, *Proc. IEEE,* **68,** 654–666 (1980).

76. R. N. McDonough, Application of the Maximum Likelihood Method and the Maximum Entropy Method to Array Processing, in S. Haykin, Ed., *Nonlinear Methods of Spectral Analysis,* New York, Springer-Verlag, 1979.

77. D. H. Johnson, The Application of Spectral Estimation Methods to Bearing Estimation Problems, *Proc. IEEE,* **70,** 1018–1028 (1982).

78. H. Cox, Resolving Power and Sensitivity to Mismatch of Optimum Array Processors, *J. Acoust. Soc. Am.,* **54,** 771–785 (1973).

79. A. J. Berni, Angle-of-Arrival Estimation Using an Adaptive Antenna Array, *IEEE Trans. Aerosp. Electron. Syst.,* **AES-11,** 278–284 (March 1975).

80. S. S. Reddi, Multiple Source Location—A Digital Approach, *IEEE Trans. Aerosp. Electron. Syst.*, **AES-15,** 95–105 (1979).

81. G. Bienvenu and L. Kopp, Adaptivity to Background Noise Spatial Coherence for High Resolution Passive Methods, *Proc. IEEE Int. Conf. Acoust., Speech, Signal Process.*, 307–310 (1980).

82. A Cantoni and L. Godara, Resolving the Directions of Sources in a Correlated Field Incident on an Array, *J. Acoust. Soc. Am.*, **67,** 1247–1255 (1980).

83. T. Thorvaldsen, Maximum Entropy Spectral Analysis in Antenna Spatial Filtering, *IEEE Trans. Antennas Propag.*, **AP-28,** 552–560 (1980).

84. D. Bordelon, Complementarity of the Reddi Method of Source Direction Estimation with those of Pisarenko and Cantoni and Godara, I., *J. Acoust., Soc. Am.*, **69,** 1355–1359 (May 1981).

85. R. Schmidt, Multiple Emitter Location and Signal Parameter Estimation, *Proc. RADC Spectral Estimation Workshop*, Rome, NY (1979), p. 243.

86. T. P. Bronez and J. A. Cadzow, An Algebraic Approach to Superresolution Adaptive Array Processing, *IEEE Trans. Aerosp. Electron. Syst.*, **AES-19,** 123–133 (1983).

87. R. Kumaresan and D. W. Tufts, Estimating the Angles of Arrival of Multiple Plane Waves, *IEEE Trans. Aerosp. Electron. Syst.*, **AES-19,** 134–139 (1983).

88. D. H. Johnson and S. R. DeGraaf, Improving the Resolution of Bearing in Passive Sonar Arrays by Eigenvalue Analysis, *IEEE Trans. Acoust., Speech, Signal Process.* **ASSP-30,** 638–647 (1982).

89. T. E. Barnard, Two Maximum Entropy Beamforming Algorithms for Equally Spaced Line Arrays, *IEEE Trans. Acoust., Speech, Signal Process.*, **ASSP-30,** 175–189 (1980).

90. T. S. Durrani and K. C. Sharman, Extraction of an Eigenvector-Oriented 'Spectrum' for the MESA Coefficients, *IEEE Trans. Acoust., Speech, Signal Process.*, **ASSP-30,** 649–651 (1982).

91. S. DeGraaf and D. H. Johnson, Capability of Array Processing Algorithms to Estimate Source Bearings, *Proc. IEEE Int. Conf. Acoust., Speech, Signal Process.*, 344–347 (1983).

92. H. Wakita, Direct Estimation of the Vocal Tract Shape by Inverse Filtering of Acoustic Speech Waveforms, *IEEE Trans. Audio Electroacoust.*, **AU-21,** 417 (1973).

93. J. A. Ware and K. Aki, Continuous and Discrete Inverse Scattering Problems in a Stratified Elastic Medium. I. Plane Waves at Normal Incidence, *J. Acoust. Soc. Am.*, **45,** 91 (1969).

94. L. C. Wood and S. Treitel, Seismic Signal Processing, *Proc. IEEE*, **63,** 649–661 (1975).

95. P. L. Goupillaud, An Approach to Inverse Filtering of Near-Surface Layer Effects from Seismic Records, *Geophysics*, **26,** 754–760 (1961).

96. J. F. Claerbout, Synthesis of a Layered Medium from its Acoustic Transmission Response, *Geophysics*, **33,** 264–269 (1968).

97. F. Koehler and M. T. Taner, Direct and Inverse Problems Relating Reflection Coefficients and Reflection Response for Horizontally Layered Media, *Geophysics*, **42,** 1199–1206 (1977).

98. E. A. Robinson and S. Treitel, The Fine Structure of the Normal Incidence Synthetic Seismogram, *Geophys. J. R. Astron. Soc.*, **53,** 289–309 (1978).

99. S. Treitel and E. A. Robinson, Maximum Entropy Spectral Decomposition of a Seismogram into its Minimum Entropy Component Plus Noise, *Geophysics*, **46,** 1108 (1981).

100. J. M. Mendel and F. Habibi-Ashrafi, A Survey of Approaches to Solving Inverse Problems for Lossless Layered Media Systems, *IEEE Trans. Geosci. Electron.*, **GE-18,** 320–330 (1980).

101. R. G. Newton, Inversion of Reflection Data for Layered Media: A Review of Exact Methods, *Geophys. J. R. Astron. Soc.*, **65,** 191–215 (1981).

102. E. A. Robinson, A Spectral Approach to Geophysical Inversion by Lorentz, Fourier, and Radon Transforms, *Proc. IEEE*, **70,** 1039–1054 (1982).

103. J. G. Berryman and R. R. Greene, Discrete Inverse Methods for Elastic Waves in Layered Media, *Geophysics*, **45,** 213–233 (1980).

104. F. J. Dyson, Old and New Approaches to the Inverse Scattering Problem, in E. H. Lieb, B. Simon, and A. S. Wightman, Eds., *Studies in Mathematical Physics*, Princeton, Princeton University Press, 1976.

105. K. M. Case, Inverse Scattering, Orthogonal Polynomials, and Linear Estimation, in I. C. Gohberg and M. Kac, Eds., *Topics in Functional Analysis, Advances in Mathematics Supplementary Studies*, Vol. 3, New York, Academic, 1978.

106. P. Dewilde, A. Vieira, and T. Kailath, On the Generalized Szegö-Levinson Realization Algorithm for Optimal Linear Predictors Based on a Network Synthesis Approach, *IEEE Trans. Circuits Syst.*, **CAS-25,** 663–675 (1978).

107. P. Dewilde and H. Dym, Schur Recursions, Error Formulas, and Convergence of Rational Estimators for Stationary Stochastic Sequences, *IEEE Trans. Inf. Theory*, **IT-27,** 446–461 (1981).

108. P. Dewilde, J. T. Fokkema, and I. Widya, Inverse Scattering and Linear Prediction: The Continuous Time Case, in M. Hazewinkel and J. C. Willems, Eds., *Stochastic Systems: The Mathematics of Filtering and Identification and Applications*, Boston, Reidel, 1981.

109. M. T. Silvia and E. A. Robinson, *Deconvolution of Geophysical Time Series in the Exploration for Oil and Natural Gas,* Amsterdam, Elsevier, 1979.

110. D. W. Oldenburg, A Comprehensive Solution to the Linear Deconvolution Problem,'' *Geophys. J. R. Astron. Soc.*, **65,** 331–357 (1981).

111. S. Treitel and L. R. Lines, Linear Inverse Theory and Deconvolution, *Geophysics*, **47,** 115 (1982).

112. S. Twomey, *Introduction to the Mathematics of Inversion in Remote Sensing and Indirect Measurements*, Amsterdam, Elsevier, 1977.

113. B. R. Frieden, Image Enhancement and Restoration, in T. S. Huang, Ed., *Picture Processing and Digital Filtering*, New York, Springer-Verlag, 1975.

114. J. F. Claerbout and F. Muir, Robust Modeling with Erratic Data, *Geophysics*, **38,** 826–844 (1973).

115. H. L. Taylor, S. C. Banks, and J. F. McCoy, Deconvolution with the $L_1$ Norm, *Geophysics*, **44,** 39–52 (1979).

116. R. Mammone and G. Eichmann, Superresolving Image Restoration Using Linear Programming, *Applied Optics*, **21**, 496–501 (1982).

117. R. Mammone and G. Eichmann, Restoration of Discrete Fourier Spectra Using Linear Programming, *J. Optical Soc. Am.*, **72**, 987–992 (1982).

118. I. Barrodale and F. D. K. Roberts, An Improved Algorithm for the Discrete $L_1$ Linear Approximation, *SIAM J. Numer. Anal.*, **10**, 839 (1973).

119. I. Barrodale and F. D. K. Roberts, Algorithm 478: Solution of an Overdetermined System of Equations in the $L_1$ Norm, *Commun. ACM*, **17**, 319 (1974).

120. R. W. Schafer, R. M. Mersereau, and M. A. Richards, Constrained Iterative Restoration Algorithms, *Proc. IEEE*, **69**, 432–450 (1981).

# 6

---

# Adaptive Filters

## 6.1 Adaptive Implementation of Wiener Filters

We review briefly the solution of the Wiener filtering problem.

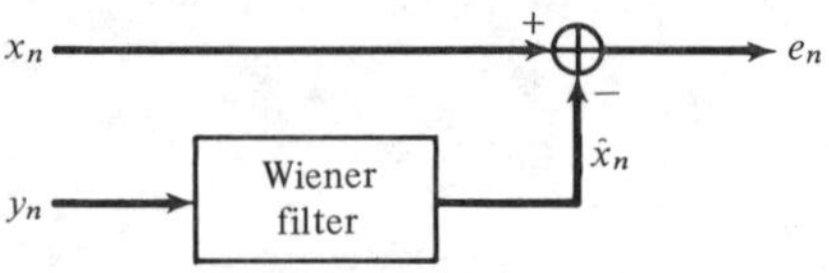



The general solution does not place any a priori restriction on the order of the Wiener filter. In general, an infinite number of weights is required to achieve the lowest estimation error. However, in adaptive implementations we must insist in advance that the number of filter weights be finite. This is so because the adaptation algorithm adapts each weight individually. Obviously, we cannot adapt an infinite number of weights.

We will assume then, that the optimal Wiener filter is an FIR filter, say with $M + 1$ weights

$$\mathbf{h} = [h_0, h_1, \ldots, h_M]^T \rightarrow H(z) = h_0 + h_1 z^{-1} + h_2 z^{-2} + \cdots + h_M z^{-M}$$

This filter processes the available observations $y_n$ to produce the estimate

$$\hat{x}_n = \sum_{m=0}^{M} h_m y_{n-m} = h_0 y_n + h_1 y_{n-1} + h_2 y_{n-2} + \cdots + h_M y_{n-M}$$

The weights $h_m$ are chosen optimally so that the mean-squared estimation error is minimized; that is,

$$\mathscr{E} = E[e_n^2] = \text{minimum}, \qquad e_n = x_n - \hat{x}_n$$

This minimization criterion leads to the orthogonality equations, which are the determining equations for the optimal weights. Writing the estimate in vector notation

$$\hat{x}_n = [h_0, h_1, \ldots, h_M] \begin{bmatrix} y_n \\ y_{n-1} \\ \vdots \\ y_{n-M} \end{bmatrix} = \mathbf{h}^T \mathbf{y}(n)$$

we may write the orthogonality equations as

$$E[e_n y_{n-m}] = 0, \qquad 0 \leq m \leq M$$

or equivalently,

$$E[e_n \mathbf{y}(n)] = 0$$

These give the normal equations

$$E[(x_n - \hat{x}_n)\mathbf{y}(n)] = E[(x_n - \mathbf{h}^T \mathbf{y}(n))\mathbf{y}(n)] = 0$$

$$E[\mathbf{y}(n)\mathbf{y}(n)^T]\mathbf{h} = E[x_n \mathbf{y}(n)]$$

or

$$R\mathbf{h} = \mathbf{r}, \qquad R = E[\mathbf{y}(n)\mathbf{y}(n)^T], \qquad \mathbf{r} = E[x_n \mathbf{y}(n)]$$

The optimal weights are obtained then by

$$\mathbf{h} = R^{-1}\mathbf{r} \tag{6.1.1}$$

The corresponding minimal value of the estimation error is computed by

$$\mathscr{E} = E[e_n^2] = E[e_n(x_n - \mathbf{h}^T \mathbf{y}(n))] = E[e_n x_n]$$

$$= E[x_n^2] - \mathbf{h}^T E[\mathbf{y}(n)x_n] = E[x_n^2] - \mathbf{r}^T R^{-1}\mathbf{r}$$

The normal equations, and especially the orthogonality equations, have their usual *correlation canceling* interpretations. The signal $x_n$ being estimated can be written as

$$x_n = e_n + \hat{x}_n = e_n + \mathbf{h}^T \mathbf{y}(n)$$

It is composed of two parts, the term $e_n$, which because of the orthogonality equations is entirely uncorrelated with $\mathbf{y}(n)$, and the second term, which is correlated with $\mathbf{y}(n)$. In effect, the filter removes from $x_n$ any part of it that is correlated with the secondary input $\mathbf{y}(n)$; what is left, $e_n$, is uncorrelated with $\mathbf{y}(n)$. The Wiener filter acts as a correlation canceler. If the primary signal $x_n$

and the secondary signal $\mathbf{y}(n)$ are in any way correlated, the filter will cancel from the output $e_n$ any such correlations.

One difficulty with the above solution is that the statistical quantities $R$ and $\mathbf{r}$ must be known, or at least estimated, in advance. This is what is done in speech processing: The input speech is divided into fairly short segments, each segment is assumed to arise from a stationary process, the statistical correlations are estimated by sample correlations, and, finally, the optimal weights corresponding to each segment are computed. In a sense, this procedure is data-adaptive, but more precisely, it is block-by-block adaptive.

The *Widrow-Hoff least-mean-square* (LMS) adaptation algorithm [1,2] is an alternative algorithm that adapts the weights on a *sample-by-sample* basis. It does not require a priori knowledge of the correlation matrices. As processing of the signal $y_n$ takes place, the filter gradually "learns" the required correlations, and adjusts its weights accordingly. The weights eventually converge to their optimal values given by the Wiener solution (6.1.1). The only requirement for this algorithm is that the statistical properties of the input signals not be changing fast, so that the filter has time to converge to the optimal weights. In other words, the filter will converge if there is something to converge to. If, after convergence, the statistical properties of the signal $y_n$ should change, then the filter will respond to these changes by readjusting its weights to their new optimal values and so on. Thus the adaptation algorithm can track the nonstationarities in the input signal provided the statistics change slowly enough for the filter to converge in between changes.

The adaptive implementation is shown in Fig. 6.1.

In the LMS algorithm, the output error signal $e_n$ is *fed back* and used to adjust the weights. If there are any correlations between the primary $x_n$ and the secondary $\mathbf{y}(n)$ signals, the filter will adjust its coefficients so as to cancel these correlations from the output error signal $e_n$. The filter tries to decorrelate the output error signal $e_n$ from the secondary signal $\mathbf{y}(n)$. This is the main idea behind the so-called adaptive noise canceling schemes. For example, if the primary signal consists of a desired signal plus undesired noise, and if the secondary signal $y_n$ consists of noise which, if not identical with the noise in the primary, is related to it, then $y_n$ is correlated only with the noise part of the primary signal. The adaptive filter adapts itself to cancel such correlations. Effectively, the filter makes a copy of the noise in the primary and then proceeds to cancel it. In addition to the noise canceling problems, adaptive filters have

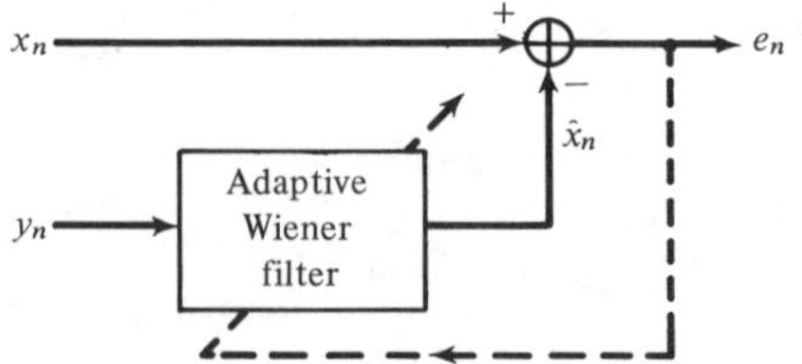


**Figure 6.1** Adaptive Wiener Filter

been applied widely to channel equalization; echo cancellation; radar, sonar, and seismic arrays; linear prediction and spectral analysis; and system identification.

## 6.2 *The Correlation Canceler Loop (CCL)*

To illustrate the basic principles behind adaptive filters, consider the simplest possible filter; that is, a filter with only one weight

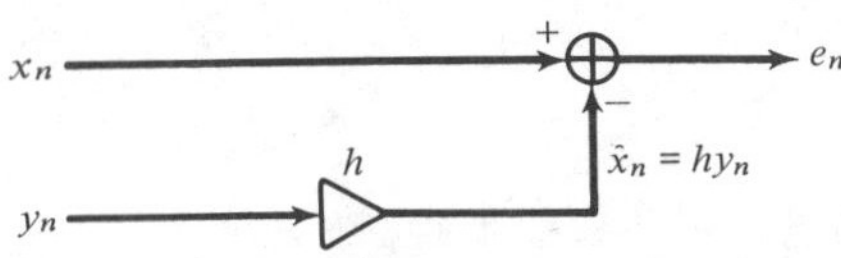


The weight $h$ must be selected optimally so as to produce the best possible estimate

$$\hat{x}_n = hy_n$$

of $x_n$. The estimation error is expressed as

$$\mathcal{E}(h) = E[e_n^2] = E[(x_n - hy_n)^2] = E[x_n^2] - 2hE[x_ny_n] + h^2E[y_n^2]$$
$$= E[x_n^2] - 2hr + h^2R \tag{6.2.1}$$

The minimization condition is

$$\frac{\partial\mathcal{E}(h)}{\partial h} = 2E\left[e_n\frac{\partial e_n}{\partial h}\right] = -2E[e_ny_n] = -2r + 2Rh = 0 \tag{6.2.2}$$

which gives the solution $h = R^{-1}r$, and also shows the correlation cancellation condition $E[e_ny_n] = 0$.

The adaptive implementation is based on solving the equation

$$\frac{\partial\mathcal{E}(h)}{\partial h} = 0 \tag{6.2.3}$$

iteratively, using a gradient-descent method. The dependence of the error $\mathcal{E}$ on the filter parameter is parabolic, with an absolute minimum occurring at the above optimal value $h = R^{-1}r$. This is shown below

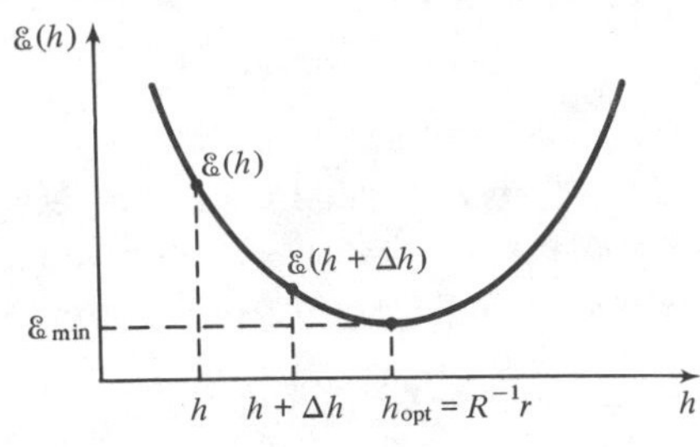

In the adaptive version, the filter parameter $h$ is made *time-dependent*, $h(n)$, and is updated from one time instant to the next as follows

$$h(n + 1) = h(n) + \Delta h(n) \qquad (6.2.4)$$

where $\Delta h(n)$ is a correction term that must be chosen properly in order to ensure that eventually the time-varying weight $h(n)$ will converge to the optimal value

$$h(n) \to h = R^{-1} r \qquad \text{as } n \to \infty$$

The filtering operation is now given by the still *linear* but *time noninvariant* form

$$\hat{x}_n = h(n) y_n \qquad (6.2.5)$$

The computation of the estimate at the next time instant should be made with the new weight; that is,

$$\hat{x}_{n+1} = h(n + 1) y_{n+1}$$

and so on. The simplest way to choose the correction term $\Delta h(n)$, is the *gradient-descent*, or steepest-descent, method. The essence of the method is this: It is required that the change $h \to h + \Delta h$ must move the performance index closer to its minimum than before; that is, $\Delta h$ must be such that

$$\mathcal{E}(h + \Delta h) \leq \mathcal{E}(h)$$

Therefore, if we always demand this, the repetition of the procedure will lead to smaller and smaller values of $\mathcal{E}$ until the smallest value has been attained. Assuming that $\Delta h$ is sufficiently small, we may expand to first order and obtain the condition

$$\mathcal{E}(h) + \Delta h \frac{\partial \mathcal{E}(h)}{\partial h} \leq \mathcal{E}(h)$$

If $\Delta h$ is selected as the *negative gradient* $-\mu(\partial \mathcal{E}/\partial h)$, then this inequality will be guaranteed; that is, if we choose

$$\Delta h = -\mu \frac{\partial \mathcal{E}(h)}{\partial h} \qquad (6.2.6)$$

then the inequality is indeed satisfied

$$\mathcal{E}(h) + \Delta h \frac{\partial \mathcal{E}(h)}{\partial h} = \mathcal{E}(h) - \mu \left| \frac{\partial \mathcal{E}(h)}{\partial h} \right|^2 \leq \mathcal{E}(h)$$

The adaptation parameter $\mu$ must be small enough to *justify* keeping only the *first order* terms in the above Taylor expansion. Applying this idea to our little adaptive filter, we choose the correction $\Delta h(n)$ according to Eq. (6.2.6), so that

$$h(n + 1) = h(n) + \Delta h(n) = h(n) - \mu \frac{\partial \mathcal{E}(h(n))}{\partial h} \qquad (6.2.7)$$

Inserting the expression for the gradient $\dfrac{\partial \mathcal{E}}{\partial h} = -2r + 2Rh$, we find

$$h(n + 1) = h(n) - \mu[-2r + 2Rh(n)]$$
$$= (1 - 2\mu R)h(n) + 2\mu r$$

This difference equation may be solved in closed form; for example, using $z$-transforms with any initial conditions $h(0)$ we find

$$h(n) = h + (1 - 2\mu R)^n(h(0) - h) \tag{6.2.8}$$

where $h$ is the optimal value $h = R^{-1}r$. The coefficient $h(n)$ will *converge* to its optimal value $h$, regardless of the starting value $h(0)$, provided $\mu$ is selected such that

$$|1 - 2\mu R| < 1$$

or $-1 < 1 - 2\mu R < 1$, or since $\mu$ must be positive (to be in the negative direction of the gradient), $\mu$ must satisfy

$$0 < \mu < \frac{1}{R} \tag{6.2.9}$$

To select $\mu$ one must have some a priori knowledge of the magnitude of the input variance $R = E[y_n^2]$. Such choice for $\mu$ will guarantee convergence, but the *speed of convergence* is controlled by how close the number $1 - 2\mu R$ is to one. The closer it is to unity the slower the speed of convergence. As $\mu$ is selected closer to zero, the closer $1 - 2\mu R$ moves to one, and thus the slower the convergence rate. The adaptation parameter must be selected small enough to guarantee convergence but not too small to cause a very *slow* convergence.


## 6.3   *The Widrow-Hoff LMS Adaptation Algorithm*

The purpose of the discussion in Section 6.2 was to show how the original Wiener filtering problem could be recast in an iterative form. From the practical point of view, this reformulation is still not computable since the adaptation of the weights requires a priori knowledge of the correlations $R$ and $r$. In the Widrow-Hoff algorithm the above adaptation algorithm is replaced with one that is computable [1,2]. The gradient that appears in Eq. (6.2.7)

$$h(n + 1) = h(n) - \mu\,\frac{\partial \mathcal{E}(h(n))}{\partial h}$$

is replaced by an "*instantaneous*" gradient by *ignoring* the expectation instructions; that is,

$$\frac{\partial \mathcal{E}(h(n))}{\partial h} = -2E[e_n y_n] = -2r + 2Rh(n) = -2E[x_n y_n] + 2E[y_n^2]h(n)$$

is *replaced* by

$$\frac{\partial \mathcal{E}}{\partial h} = -2e_n y_n = -2x_n y_n + 2y_n^2 h(n) \tag{6.3.1}$$

so that the weight-adjustment algorithm becomes

$$h(n + 1) = h(n) + 2\mu e_n y_n \tag{6.3.2}$$

In summary, the required computations are done in the following order

1. At time $n$, the filter weight $h(n)$ is available.
2. Compute the filter output $\hat{x}_n = h(n)y_n$.
3. Compute the estimation error $e_n = x_n - \hat{x}_n$.
4. Compute the next filter weight $h(n + 1) = h(n) + 2\mu e_n y_n$.
5. Go to next time instant $n \rightarrow n + 1$.

The following remarks are in order:

1. The output error $e_n$ is fed back and used to *control* the adaptation of the filter weight.

2. The filter tries to decorrelate the secondary signal from the output $e_n$. This, is easily seen as follows: If the weight $h(n)$ has more or less reached its optimum value, then $h(n + 1) \simeq h(n)$, and the adaptation equation implies also approximately $e_n y_n \simeq 0$.

3. Actually the weight $h(n)$ never really reaches the theoretical limiting value $h = R^{-1}r$. Instead, it stabilizes about this value, and continuously fluctuates about it.

4. The approximation of ignoring the expectation instruction in the gradient is known as the *stochastic approximation*. It complicates the mathematical aspects of the problem considerably. Indeed, the difference equation

$$h(n + 1) = h(n) + 2\mu e_n y_n = h(n) + 2\mu(x_n - h(n)y_n)y_n$$

makes $h(n)$ depend on the random variable $y_n$ in highly nonlinear fashion, and it is very difficult to discuss even the average behavior of $h(n)$.

5. In discussing the average behavior of the weight $h(n)$, the following approximation is typically (almost invariably) made in the literature

$$\begin{aligned}
E[h(n + 1)] &= E[h(n)] + 2\mu E[x_n y_n] - 2\mu E[h(n)y_n^2] \\
&= E[h(n)] + 2\mu E[x_n y_n] - 2\mu E[h(n)]E[y_n^2] \\
&= E[h(n)] + 2\mu r - 2\mu E[h(n)]R
\end{aligned}$$

where in the last term, the expectation $E[h(n)]$ was factored out, as though $h(n)$ were independent of $y_n^2$. With this approximation, the average $E[h(n)]$ satisfies the same difference equation as before. Typically, the weight $h(n)$ will be fluctuating about the theoretical convergence curve as it converges to the optimal value, as shown here.

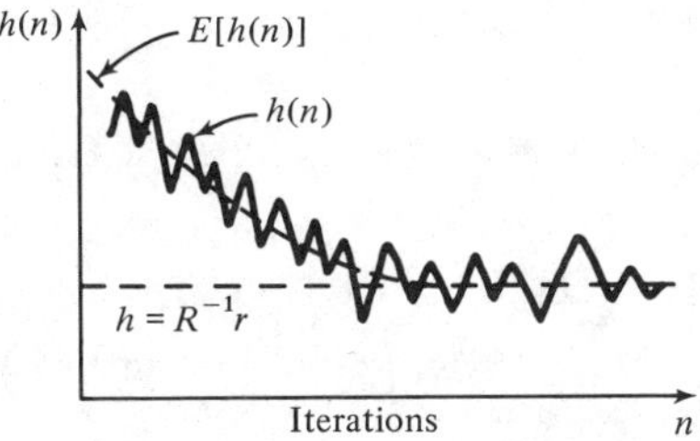


After convergence, the adaptive weight $h(n)$ continuously fluctuates about the Wiener solution $h$. A measure of these fluctuations is the quantity $E[(h(n) - h)^2]$. Under some restrictive conditions, this quantity has been calculated [3] to be

$$E[(h(n) - h)^2] = \mu\mathcal{E}_{min} \qquad \text{(for large } n\text{)}$$

Thus, the adaptation parameter $\mu$ controls the *size* of these fluctuations. This gives rise to the basic *trade-off* of the LMS algorithm: to obtain high accuracy in the converged weights (small fluctuations), a small value of $\mu$ is required, but this will slow down the convergence rate.

A realization of the CCL is shown in Fig. 6.2. The filtering part of the realization must be clearly distinguished from the feedback control loop that performs the adaptation of the filter weight.

Historically, the correlation canceler loop was introduced in adaptive antennas as a *sidelobe canceler* [4–7]. It is the simplest possible adaptive filter, and forms the elementary *building block* of more complicated, higher order adaptive filters.

We finish this section by presenting a simulation example of the CCL loop. The primary signal $x_n$ was defined by

$$x_n = -0.8y_n + u_n$$

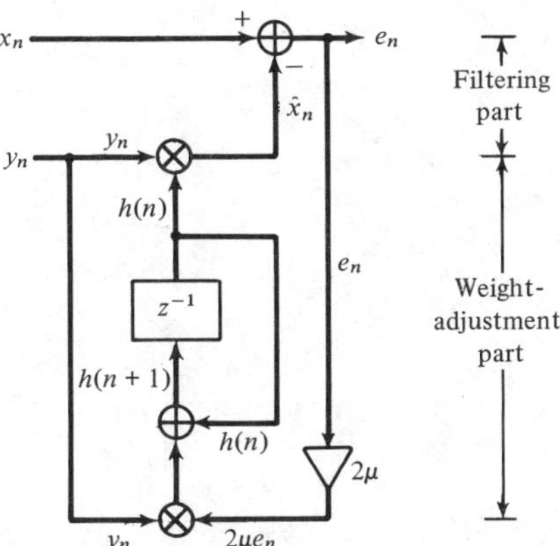


**Figure 6.2**   Correlation Canceler Loop

where the first term represents that part of $x_n$ which is correlated with $y_n$. The part $u_n$ is not correlated with $y_n$. The theoretical value of the CCL weight is found as follows:

$$r = E[x_n y_n] = -0.8E[y_n y_n] + E[u_n y_n] = -0.8R$$

where $R = E[y_n y_n]$. Thus, $h = R^{-1}r = -0.8$. The corresponding output of the CCL will be $\hat{x}_n = hy_n = -0.8y_n$, and therefore it will completely cancel the first term of $x_n$, leaving at the output $e_n = x_n - \hat{x}_n = u_n$. In the simulation, we generated 200 samples of a zero-mean, unit-variance, white-noise signal $y_n$, and another independent set of 200 samples of a zero-mean, unit-variance, white-noise signal $u_n$, and computed $x_n$. The adaptation algorithm was initialized, as is usually done, to zero initial weight $h(0) = 0$. Fig. 6.3 shows the transient behavior of the adaptive weight $h(n)$ as a function of the number of iterations $n$, for the two values of $\mu$, $\mu = 0.03$ and $\mu = 0.01$. Note that in both cases, the adaptive weight converges to the theoretical value $h = -0.8$, and that the smaller $\mu$ is slower but the fluctuations are also smaller. After the adaptive weight has reached its asymptotic value, the CCL begins to operate optimally, removing the correlated part of $x_n$ from the output $e_n$.

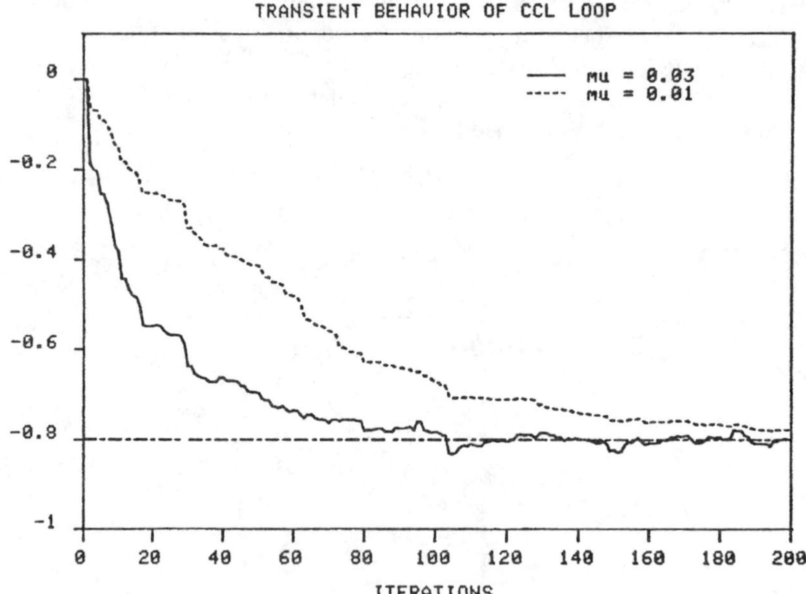


**Figure 6.3**   Transient Behavior of the Adaptive Weight $h(n)$

## 6.4   *The Adaptive Linear Combiner*

A straightforward generalization of the correlation canceler loop is the adaptive linear combiner, where one has available a main signal $x_n$, and a number of secondary signals $y_m(n)$, $m = 0,1,2, \ldots ,M$. These $(M + 1)$ secondary signals are to be linearly combined with appropriate weights $(h_0,h_1, \ldots ,h_M)$ to form an estimate of $x_n$:

$$\hat{x}_n = h_0 y_0(n) + h_1 y_1(n) + \cdots + h_M y_M(n) = [h_0,h_1, \cdots ,h_M] \begin{bmatrix} y_0(n) \\ y_1(n) \\ \vdots \\ y_M(n) \end{bmatrix}$$

$$= \mathbf{h}^T \mathbf{y}(n)$$

A realization of this is shown in Fig. 6.4.

The adaptive linear combiner is used in adaptive radar and sonar arrays [4–8]. It also encompasses the case of the ordinary FIR, or transversal, Wiener filter [2].

The optimal weights $h_m$ are those which minimize the estimation error squared

$$\mathcal{E} = E[e_n^2] = \min$$

The corresponding othogonality equations state that the estimation error be orthogonal (decorrelated) to each secondary signal $y_m(n)$

$$\frac{\partial \mathcal{E}}{\partial h_m} = 2E\left[ e_n \frac{\partial e_n}{\partial h_m} \right] = -2E[e_n y_m(n)] = 0, \qquad 0 \leq m \leq M$$

or in matrix form

$$E[e_n \mathbf{y}(n)] = 0 = E[x_n \mathbf{y}(n)] - E[\mathbf{y}(n)\mathbf{y}(n)^T]\mathbf{h} = \mathbf{r} - R\mathbf{h}$$

with optimal solution $\mathbf{h} = R^{-1}\mathbf{r}$.

The adaptive implementation is easily obtained by allowing the weights to

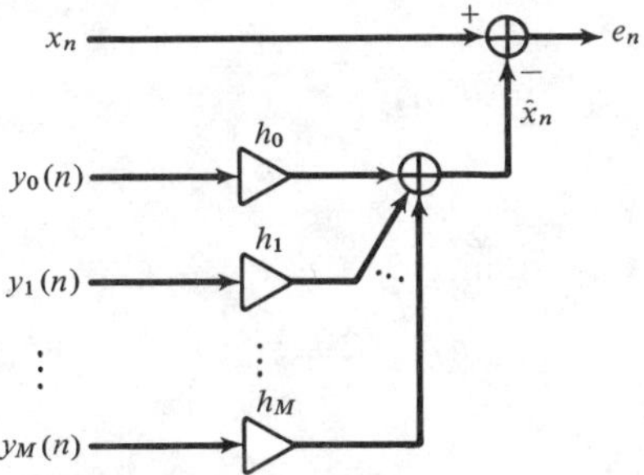


**Figure 6.4**   Linear Combiner

become time-dependent, $\mathbf{h}(n)$, and updating them in time according to the gradient descent algorithm

$$\mathbf{h}(n + 1) = \mathbf{h}(n) - \mu \frac{\partial \mathcal{E}(\mathbf{h}(n))}{\partial \mathbf{h}}$$

with instantaneous gradient

$$\frac{\partial \mathcal{E}}{\partial \mathbf{h}} = -2E[e_n \mathbf{y}(n)] \rightarrow -2e_n \mathbf{y}(n)$$

so that

$$\mathbf{h}(n + 1) = \mathbf{h}(n) + 2\mu e_n \mathbf{y}(n)$$

or component-wise

$$h_m(n + 1) = h_m(n) + 2\mu e_n y_m(n), \qquad \text{for } 0 \leq m \leq M \qquad (6.4.1)$$

The computational algorithm is outlined below

1. $\hat{x}_n = h_0(n)y_0(n) + h_1(n)y_1(n) + \cdots + h_M(n)y_M(n)$
2. $e_n = x_n - \hat{x}_n$
3. $h_m(n + 1) = h_m(n) + 2\mu e_n y_m(n), \; 0 \leq m \leq M$

It is evident that each weight $h_m(n)$ is being adapted by *its own* correlation canceler loop, while all weights use the *same* feedback error $e_n$ to control their loops. The case of two weights ($M = 1$) is shown in Fig. 6.5.

## 6.5  *Adaptive FIR Wiener Filter*

The adaptive transversal filter is a special case of the adaptive linear combiner. In this case, there is only *one* secondary signal $y_n$. The required $M + 1$ signals $y_m(n)$ are provided as *delayed replicas* of $y_n$; that is,

$$y_m(n) = y_{n-m}, \qquad \text{for } 0 \leq m \leq M \qquad (6.5.1)$$

A realization is shown in Fig. 6.6.
The estimate of $x_n$ is

$$\hat{x}_n = \sum_{m=0}^{M} h_m(n)y_{n-m} = h_0(n)y_n + h_1(n)y_{n-1} + \cdots + h_M(n)y_{n-M}$$

The time-varying filter weights $h_m(n)$ are continuously updated according to the gradient-descent LMS algorithm

$$h_m(n + 1) = h_m(n) + 2\mu e_n y_m(n), \qquad 0 \leq m \leq M$$

or

$$h_m(n + 1) = h_m(n) + 2\mu e_n y_{n-m}, \qquad \text{for } 0 \leq m \leq M \qquad (6.5.2)$$

Each weight is therefore updated by *its own CCL*.

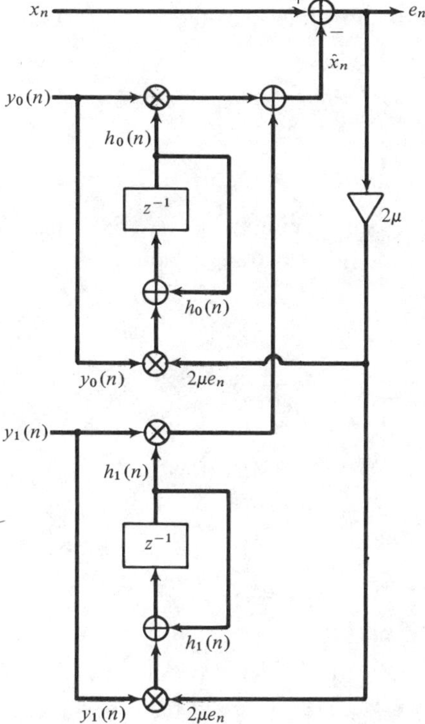

**Figure 6.5**  Adaptive Linear Combiner

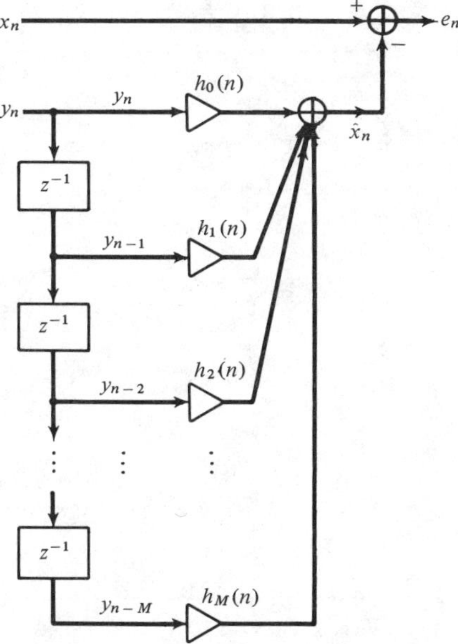

**Figure 6.6**  Adaptive FIR Wiener Filter

Again, we would like to summarize the computational steps in this case:

1. compute $\hat{x}_n = \sum_{m=0}^{M} h_m(n) y_{n-m}$

2. compute the error signal $e_n = x_n - \hat{x}_n$
3. adjust weights $h_m(n + 1) = h_m(n) + 2\mu e_n y_{n-m}, \ 0 \leq m \leq M$

Next, we present the same simulation example as that given in Section 6.3, but it is now approached with a two-tap adaptive filter ($M = 1$). The filtering equation is in this case

$$\hat{x}_n = h_0(n) y_n + h_1(n) y_{n-1}$$

The theoretical Wiener solution is found as follows: First note that

$$R_{xy}(k) = E[x_{n+k} y_n] = E[(-0.8 y_{n+k} + u_{n+k}) y_n] = -0.8 E[y_{n+k} y_n]$$
$$= -0.8 R_{yy}(k) = -0.8 R(k)$$

Thus, the cross-correlation vector $\mathbf{r}$ is $\mathbf{r} = -0.8 \begin{bmatrix} R(0) \\ R(1) \end{bmatrix}$ and the Wiener solution becomes

$$\mathbf{h} = R^{-1}\mathbf{r} = \begin{bmatrix} R(0) & R(1) \\ R(1) & R(0) \end{bmatrix}^{-1} \begin{bmatrix} -0.8 R(0) \\ -0.8 R(1) \end{bmatrix}$$
$$= \frac{-0.8}{R(0)^2 - R(1)^2} \begin{bmatrix} R(0) & -R(1) \\ -R(1) & R(0) \end{bmatrix} \begin{bmatrix} R(0) \\ R(1) \end{bmatrix} = \begin{bmatrix} -0.8 \\ 0 \end{bmatrix}$$

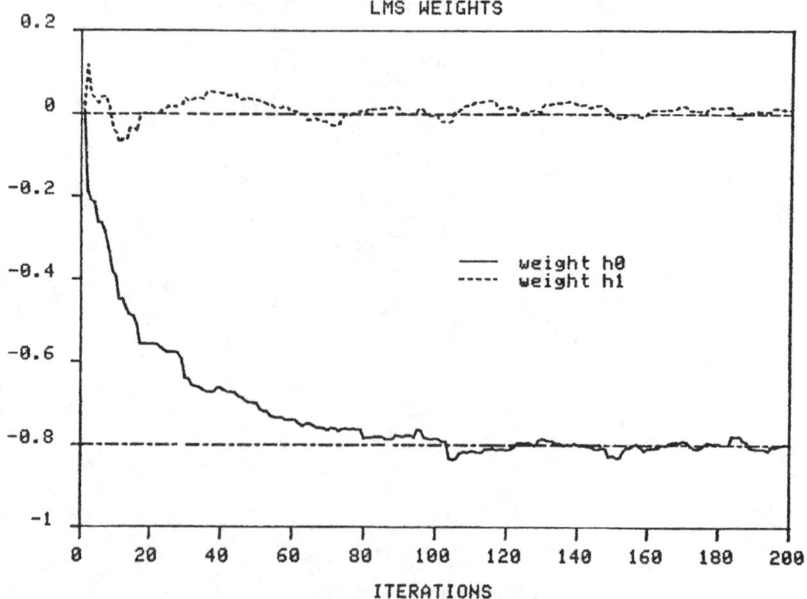


**Figure 6.7** Transient Behavior of FIR Adaptive Filter

We could have expected that $h_1$ is zero, since the signal $x_n$ does not depend on $y_{n-1}$ but only on $y_n$. The adaptive weights were both initialized to zero, and the value of $\mu$ was 0.03. Figure 6.7 shows the two adaptive weights $h_0(n)$ and $h_1(n)$ as a function of $n$, converging to their optimal values.

## 6.6 *Speed of Convergence*

The *convergence properties* of the LMS algorithm [2,3,9] may be discussed by restoring the expectation values where they should be

$$\frac{\partial \mathcal{E}}{\partial \mathbf{h}} = -2E[e_n \mathbf{y}(n)], \qquad \mathbf{y}(n) = \begin{bmatrix} y_0(n) \\ y_1(n) \\ \vdots \\ y_M(n) \end{bmatrix} = \begin{bmatrix} y_n \\ y_{n-1} \\ \vdots \\ y_{n-M} \end{bmatrix}$$

$$\begin{aligned}
\mathbf{h}(n + 1) &= \mathbf{h}(n) - \mu \frac{\partial \mathcal{E}}{\partial \mathbf{h}} \\
&= \mathbf{h}(n) + 2\mu E[e_n \mathbf{y}(n)] \\
&= \mathbf{h}(n) + 2\mu\{E[x_n \mathbf{y}(n)] - E[\mathbf{y}(n)\mathbf{y}(n)^T]\mathbf{h}(n)\} \\
&= \mathbf{h}(n) + 2\mu \mathbf{r} - 2\mu R \mathbf{h}(n)
\end{aligned}$$

or

$$\mathbf{h}(n + 1) = (I - 2\mu R)\mathbf{h}(n) + 2\mu \mathbf{r} \qquad (6.6.1)$$

where the covariance matrix $R$ is defined by

$$R = E[\mathbf{y}(n)\mathbf{y}(n)^T]$$
$$R_{ij} = E[y_i(n)y_j(n)] = E[y_{n-i}y_{n-j}] = R_{yy}(i - j), \qquad 0 \leq i,j \leq M$$

The difference equation (6.6.1) has solution

$$\mathbf{h}(n) = \mathbf{h} + (I - 2\mu R)^n(\mathbf{h}(0) - \mathbf{h})$$

Convergence requires that the quantity $(1 - 2\mu\lambda)$, for every eigenvalue $\lambda$ of $R$, have magnitude less than one

$$|1 - 2\mu\lambda| < 1, \qquad \text{or} -1 < 1 - 2\mu\lambda < 1, \qquad \text{or } 0 < \mu < \frac{1}{\lambda}$$

for every eigenvalue $\lambda$ of $R$. This condition is guaranteed, for example, if we require this inequality for $\lambda_{max}$, the maximum eigenvalue

$$0 < \mu < \frac{1}{\lambda_{max}} \qquad (6.6.2)$$

Note that $\lambda_{max}$ can be bounded from above by

$$\lambda_{max} < \text{Tr}[R] = \sum_{i=0}^{M} R(i,i) = \sum_{i=0}^{M} R(0) = (M + 1)R(0)$$

and one may require instead $\mu < [(M + 1)R(0)]^{-1}$. As for the *speed* of convergence, suppose that $\mu$ is selected half-way within its range (6.6.2), near $0.5/\lambda_{max}$, then the rate of convergence will depend on the *slowest* converging term of the form $(1 - 2\mu\lambda)^n$; that is, the term having $(1 - 2\mu\lambda)$ as close to one as possible. This occurs for the smallest eigenvalue $\lambda = \lambda_{min}$. Thus, the *slowest* converging term is effectively given by

$$(1 - 2\mu\lambda_{min})^n = \left(1 - \frac{\lambda_{min}}{\lambda_{max}}\right)^n$$

The ratio $\lambda_{min}/\lambda_{max}$ determines, therefore, the *speed of convergence*. The convergence is *fast* if $\lambda_{min}/\lambda_{max}$ is *large* (i.e., close to one). The convergence is *slow* if $\lambda_{min}/\lambda_{max}$ is *small* (i.e., there is a large spread in the eigenvalues). As we shall see shortly, a large spread in the eigenvalues of the covariance matrix $R$ corresponds to a *highly self-correlated* signal $y_n$. Thus, we obtain the general qualitative result that in situations where the secondary signal is *strongly self-correlated,* the convergence of the gradient-based LMS algorithm *will be slow.* In many applications, such as channel equalization, the convergence must be as quick as possible. Alternative adaptation schemes exist that combine the computational simplicity of the LMS algorithm with a fast speed of convergence. Examples are the adaptive gradient lattice, or the newer recursive least-squares lattice algorithms.

The possibility of accelerating the convergence rate may be seen by considering a more general version of the gradient descent algorithm in which the time update for the weight vector is chosen as

$$\Delta\mathbf{h} = -M\,\frac{\partial\mathcal{E}}{\partial\mathbf{h}} \tag{6.6.3}$$

where $M$ is a *positive definite* and *symmetric* matrix. The LMS steepest descent case is obtained as a special case of this when $M$ is proportional to the unit matrix $I$, $M = \mu I$. This choice guarantees convergence towards the minimum of the performance index $\mathcal{E}(\mathbf{h})$; indeed,

$$\mathcal{E}(\mathbf{h} + \Delta\mathbf{h}) = \mathcal{E}(\mathbf{h}) + \Delta\mathbf{h}^T\,\frac{\partial\mathcal{E}}{\partial\mathbf{h}} = \mathcal{E}(\mathbf{h}) - \left(\frac{\partial\mathcal{E}}{\partial\mathbf{h}}\right)^T M \left(\frac{\partial\mathcal{E}}{\partial\mathbf{h}}\right) \leq \mathcal{E}(\mathbf{h})$$

Since the performance index is

$$\mathcal{E} = E[e_n^2] = E[(x_n - \mathbf{h}^T\mathbf{y}(n))^2] = E[x_n^2] - 2\mathbf{h}^T\mathbf{r} + \mathbf{h}^T R\mathbf{h}$$

it follows that $\partial \mathcal{E}/\partial \mathbf{h} = -2(\mathbf{r} - R\mathbf{h})$, and the difference equation for the adaptive weights becomes

$$\mathbf{h}(n + 1) = \mathbf{h}(n) + \Delta\mathbf{h}(n)$$
$$= \mathbf{h}(n) + 2M(\mathbf{r} - R\mathbf{h}(n))$$

or

$$\mathbf{h}(n + 1) = (I - 2MR)\mathbf{h}(n) + 2M\mathbf{r} \tag{6.6.4}$$

with solution

$$\mathbf{h}(n) = \mathbf{h} + (I - 2MR)^n(\mathbf{h}(0) - \mathbf{h}) \tag{6.6.5}$$

where $\mathbf{h} = R^{-1}\mathbf{r}$ is the asymptotic value. It is evident from Eq. (6.6.4) or Eq. (6.6.5) that the choice of $M$ can *drastically* affect the speed of convergence. For example, if $M$ is chosen as

$$M = (2R)^{-1} \tag{6.6.6}$$

then $I - 2MR = 0$, and the convergence occurs in just one step! This choice of $M$ is equivalent to *Newton's method* of solving the system of equations

$$\mathbf{f}(\mathbf{h}) = \frac{\partial \mathcal{E}}{\partial \mathbf{h}} = 0$$

for the optimal weights. Indeed, Newton's method linearizes about each point $\mathbf{h}$ to get the next point; that is, $\Delta\mathbf{h}$ is selected such that

$$\mathbf{f}(\mathbf{h} + \Delta\mathbf{h}) \cong \mathbf{f}(\mathbf{h}) + \frac{\partial \mathbf{f}}{\partial \mathbf{h}} \Delta\mathbf{h} = 0$$

where we expanded to first order in $\Delta\mathbf{h}$. Solving for $\Delta\mathbf{h}$ we find

$$\Delta\mathbf{h} = -\left(\frac{\partial \mathbf{f}}{\partial \mathbf{h}}\right)^{-1} \mathbf{f}(\mathbf{h})$$

But since $\mathbf{f}(\mathbf{h}) = -2(\mathbf{r} - R\mathbf{h})$, we have $\partial \mathbf{f}/\partial \mathbf{h} = 2R$. Therefore, the choice $M = (2R)^{-1}$ corresponds precisely to Newton's update. Note that the property that Newton's method converges in one step is a well-known property valid for *quadratic* performance indices (in this case, the gradient $\mathbf{f}(\mathbf{h})$ is already linear in $\mathbf{h}$ and therefore Newton's local linearization is exact). The important feature of the choice (6.6.6) is that $M$ is proportional to the inverse of the covariance matrix; the 2 in Eq. (6.6.6) may be replaced with another number greater than one. The proportionality of $M$ to $R^{-1}$ makes the matrix $I - 2MR$ proportional to the unit matrix, with *equal* eigenvalues. Therefore, the disparity between the eigenvalues that could slow down the convergence rate is eliminated, and all eigenmodes converge at the *same* rate (which is faster the more $M$ resembles $(2R)^{-1}$). The implementation of such Newton-like methods requires knowledge of $R$, which we do not have. (If we did, we would simply compute the Wiener solution $\mathbf{h} = R^{-1}\mathbf{r}$.) However, as we shall see later, the so-called *least-squares algorithms* effectively provide an implementation of Newton-type methods, and that is the reason for their extremely fast convergence. *Adaptive lattice filters* also have very fast convergence properties. In that case, because of the ortho-gonalization of the successive lattice stages of the filter, the matrix $R$ is diagonal and the matrix $M$ can also be chosen to be diagonal so as to equalize and speed up the convergence rate of all the filter coefficients. Recursive least-squares and adaptive lattice filters are discussed in Sections 6.13 and 6.14, respectively.

Finally, we would like to demonstrate the previous statement that a strongly correlated signal $y_n$ has a large spread in the eigenvalue spectrum of its covariance matrix. For simplicity, consider the $2 \times 2$ case

$$R = E[\mathbf{y}(n)\mathbf{y}(n)^T] = E\left[\begin{pmatrix} y_n \\ y_{n-1} \end{pmatrix} \overbrace{y_n, y_{n-1}}\right] = \begin{bmatrix} R(0) & R(1) \\ R(1) & R(0) \end{bmatrix}$$

The two eigenvalues are easily found to be

$$\lambda_{min} = R(0) - |R(1)|$$

$$\lambda_{max} = R(0) + |R(1)|$$

and therefore, the ratio $\lambda_{min}/\lambda_{max}$ is given by

$$\frac{\lambda_{min}}{\lambda_{max}} = \frac{R(0) - |R(1)|}{R(0) + |R(1)|}$$

Since for an autocorrelation function we always have $|R(1)| \leq R(0)$, it follows that the largest value of $R(1)$ is $\pm R(0)$, implying that for highly correlated signals the ratio $\lambda_{min}/\lambda_{max}$ will be very close to zero.

## 6.7 *Adaptive Channel Equalizers*

Channels used in digital data transmissions can be modeled very often by linear time-invariant systems. The standard model for such a channel including channel noise is shown here.

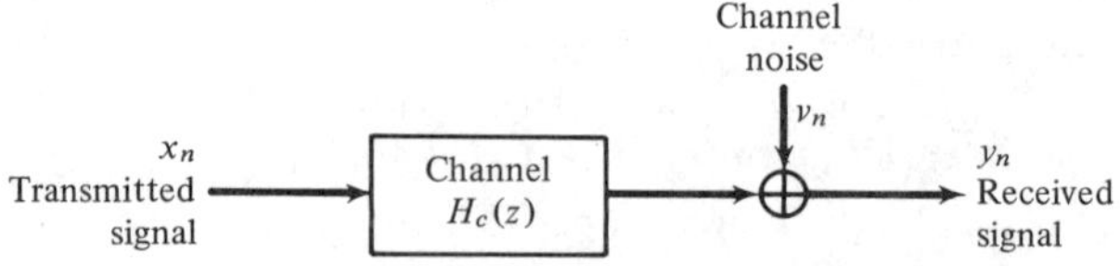


In the Figure, $H_c(z)$ is the transfer function for the channel and $v_n$ is the channel noise, assumed to be additive white gaussian noise. The transfer function $H_c(z)$ incorporates the effects of the modulator and demodulator filters, as well as the channel distortions. The purpose of a channel equalizer is to undo the distorting effects of the channel and recover, from the received waveform $y_n$, the signal $x_n$ that was transmitted. Typically, a channel equalizer will be a transversal filter with enough taps to approximate the inverse transfer function of the channel.

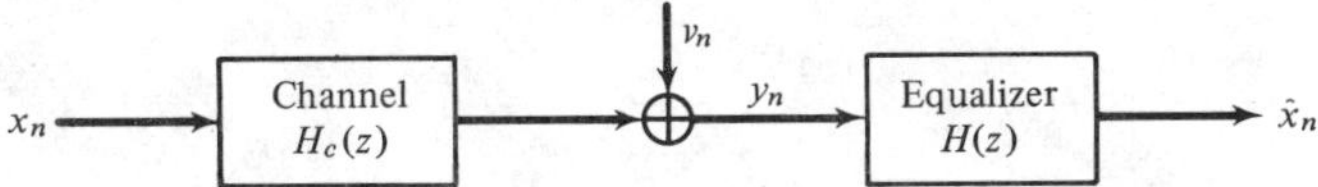


In the diagram shown here, $H(z)$ is the desired transfer function of the equalizer. In many situations, such in the telephone network, the channel is not known in advance, or it may be time-varying as in the case of multipath channels. Therefore it is desirable to design equalizers adaptively [10].

A channel equalizer, adaptive or not, is an optimal filter since it tries to produce as good an estimate $\hat{x}_n$ of the transmitted signal $x_n$ as possible. The Wiener filtering concepts that we developed thus far are ideally suited to this problem. This is shown below

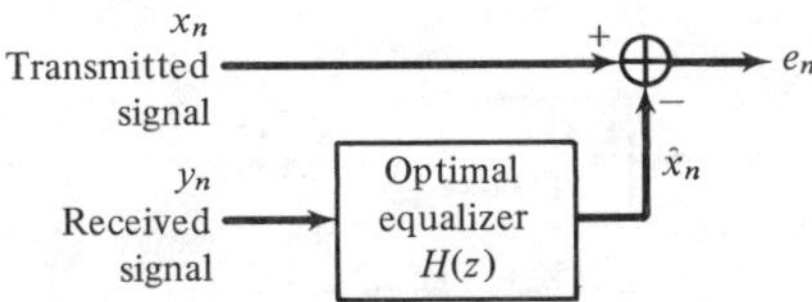


The design of the optimal filter requires two things: first, the autocorrelation of the received signal $y_n$, and second, the cross-correlation of the transmitted signal $x_n$ with the received signal. Since the transmitted signal is not available at the receiver, the following procedure is used: After the channel connection is established, a "training" sequence $x_n$, which is also known to the receiver, is transmitted over the channel. Then, the equalizer may be designed, and then the actual message transmitted. To appreciate the equalizer's action as an inverse filter, suppose that the training sequence $x_n$ is a white-noise sequence of variance $\sigma_x^2$. According to the theory developed in Chapter 4, the optimal filter estimating $x_n$ on the basis of $y_n$ is given by

$$H(z) = \frac{1}{\sigma_\epsilon^2 B(z)} \left[ \frac{S_{xy}(z)}{B(z^{-1})} \right]_+$$

where $B(z)$ is the spectral factor of $S_{yy}(z) = \sigma_\epsilon^2 B(z) B(z^{-1})$. To simplify the discussion, let us ignore the causal instruction:

$$H(z) = \frac{S_{xy}(z)}{\sigma_\epsilon^2 B(z) B(z^{-1})} = \frac{S_{xy}(z)}{S_{yy}(z)}$$

Since we have $Y(z) = H_c(z)X(z) + V(z)$, we find

$$S_{xy}(z) = S_{xx}(z)H_c(z^{-1}) + S_{xv}(z) = S_{xx}(z)H_c(z^{-1}) = \sigma_x^2 H_c(z^{-1})$$

$$S_{yy}(z) = H_c(z)H_c(z^{-1})S_{xx}(z) + S_{vv}(z) = H_c(z)H_c(z^{-1})\sigma_x^2 + \sigma_v^2$$

and the equalizer's transfer function is then

$$H(z) = \frac{S_{xy}(z)}{S_{yy}(z)} = \frac{\sigma_x^2 H_c(z^{-1})}{H_c(z)H_c(z^{-1})\sigma_x^2 + \sigma_v^2}.$$

It is seen that when the channel noise is weak (small $\sigma_v^2$), the equalizer essentially behaves as the inverse filter $1/H_c(z)$ of the channel.

In an adaptive implementation, we must use a filter with a finite number of weights. These weights are adjusted adaptively until they converge to their optimal values. Again, during this "training mode" a known pilot signal is sent over the channel and is received as $y_n$. At the receiving end, the pilot signal is locally generated and used in the adaptation algorithm. This implementation is shown below.

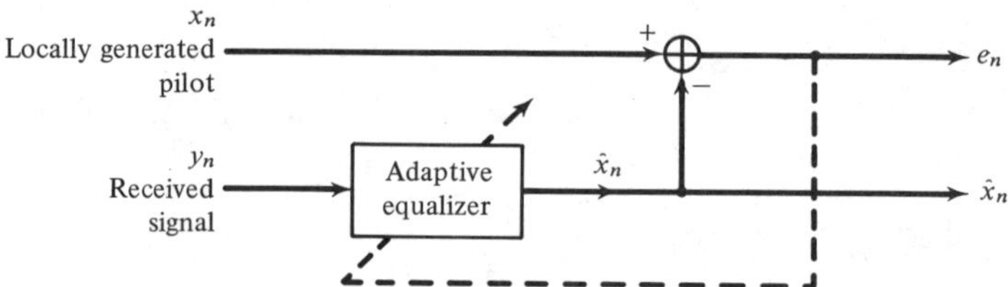


## 6.8 *Adaptive Echo Cancelers*

Consider two speakers A and B connected to each other by the telephone network. As a result of various impedance mismatches, when A's speech reaches B, it manages to "leak" through and echoes back to speaker A, as though it were B's speech.

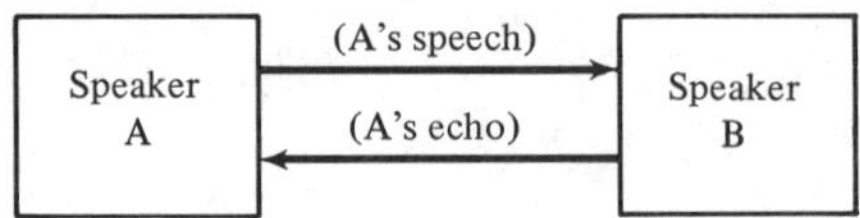


An echo canceler may be placed near B's end, as shown.

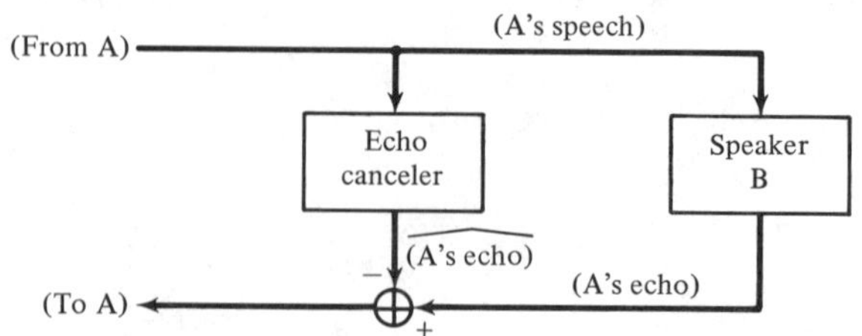

This should produce an *estimate* of A's echo through B's circuits, and then proceed to cancel it. Again, this is another case for which optimal filtering ideas are ideally suited. That is, it should produce the *best* possible estimate of A's echo. An adaptive echo canceler is an adaptive FIR filter placed as shown [11–14].

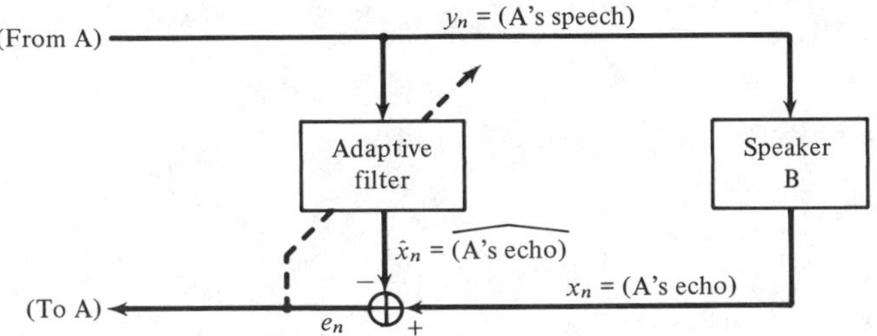


As always, the adaptive filter will adjust itself to cancel any correlations that might exist between its input (A's speech) and the main signal (A's echo).

## 6.9   *Adaptive Noise Canceling*

In many applications, two signals are available; one is composed of a desired signal plus undesired noise interference, and the other is composed only of noise interference which, if not identical with the noise part of the first signal, is *correlated* with it. This is shown in Fig. 6.8. An adaptive noise canceler [3] is an adaptive filter as shown in the Figure. Its operation as a correlation canceler should be clear. If the signals $x_n$ and $y_n$ are in any way correlated (i.e., the noise component of $x_n$ with $y_n$), then the filter will respond by adapting its weights until such correlations are canceled from the output $e_n$. It does so by producing the best possible replica of the noise component of $x_n$, and proceeding to cancel it. The output $e_n$ will now consist mainly of the desired signal.

An interesting property of the adaptive noise canceler is that when the secondary signal $y_n$ is purely sinusoidal at some frequency $\omega_0$, the adaptive filter behaves as a *notch filter* [3,15] at the sinusoid's frequency; that is, the transfer relationship between the primary input $x_n$ and the output $e_n$ becomes the *time-invariant* transfer function of a notch filter. This is a surprising property since

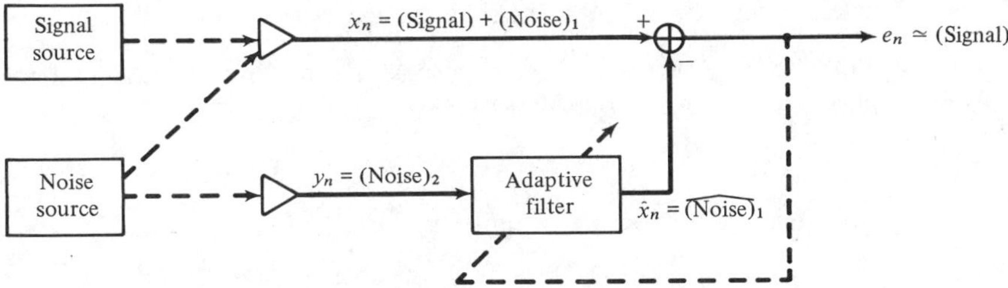


**Figure 6.8**   Adaptive Noise Canceler

the adaptation equations for the weights and the filtering I/O equation are in general time-noninvariant. To understand this effect, it proves convenient to work with complex-valued signals using a complex-valued reformulation of the LMS algorithm [16]. We make a short digression on this, first.

We assume that $x_n, y_n$, and the weights $\mathbf{h}(n)$ are complex valued. The performance index is replaced by

$$\mathcal{E} = E[e_n^* e_n] = \min$$

where the I/O filtering equation is still given by

$$\hat{x}_n = \sum_{m=0}^{M} h_m y_{n-m} = \mathbf{h}^T \mathbf{y}(n)$$

Since the weights $\mathbf{h}$ are complex, the index $\mathcal{E}$ depends on both the real and the imaginary parts of $\mathbf{h}$. Equivalently, we may think of $\mathcal{E}$ as a function of the two independent variables $\mathbf{h}$ and $\mathbf{h}^*$. A complex change in the weights $\Delta\mathbf{h}$ will change the index to

$$\mathcal{E}(\mathbf{h} + \Delta\mathbf{h}, \mathbf{h}^* + \Delta\mathbf{h}^*) = \mathcal{E}(\mathbf{h},\mathbf{h}^*) + \Delta\mathbf{h}^T \frac{\partial\mathcal{E}}{\partial\mathbf{h}} + \Delta\mathbf{h}^\dagger \frac{\partial\mathcal{E}}{\partial\mathbf{h}^*}$$

Choosing $\Delta\mathbf{h}$ to be proportional to the *complex conjugate* of the negative gradient; that is,

$$\Delta\mathbf{h} = -2\mu \frac{\partial\mathcal{E}}{\partial\mathbf{h}^*}$$

will move the index $\mathcal{E}$ towards its minimum value; indeed,

$$\mathcal{E}(\mathbf{h} + \Delta\mathbf{h}, \mathbf{h}^* + \Delta\mathbf{h}^*) = \mathcal{E}(\mathbf{h},\mathbf{h}^*) - 4\mu \left(\frac{\partial\mathcal{E}}{\partial\mathbf{h}}\right)^\dagger \left(\frac{\partial\mathcal{E}}{\partial\mathbf{h}}\right) \leq \mathcal{E}(\mathbf{h},\mathbf{h}^*)$$

Thus, the complex version of the LMS algorithm consists simply in replacing the *instantaneous gradient by its complex conjugate* [16]. We summarize the algorithm as follows:

1. Compute $\hat{x}_n = \mathbf{h}(n)^T \mathbf{y}(n)$
2. Compute $e_n = x_n - \hat{x}_n$
3. Update weights $\mathbf{h}(n + 1) = \mathbf{h}(n) + 2\mu e_n \mathbf{y}(n)^*$

Using this complex version, we now discuss the notching behavior of the adaptive filter. Suppose $y_n$ is sinusoidal

$$y_n = Ae^{j\omega_0 n} \tag{6.9.1}$$

at a given frequency $\omega_0$. Then, the weight-update equation becomes

$$h_m(n + 1) = h_m(n) + 2\mu e_n y_{n-m}^* = h_m(n) + 2\mu e_n A^* e^{-j\omega_0(n-m)}$$

for $m = 0,1, \ldots ,M$. The factor $e^{-j\omega_0(n-m)}$ suggests that we look for a solution of the form

$$h_m(n) = f_m(n)e^{-j\omega_0(n-m)}$$

Then, $f_m(n)$ must satisfy the difference equation

$$e^{-j\omega_0}f_m(n + 1) = f_m(n) + 2\mu A^*e_n \tag{6.9.2}$$

As a difference equation in $n$, this equation has *constant coefficients,* and therefore may be solved by $z$-transform techniques. Taking $z$-transforms of both sides we find

$$e^{-j\omega_0}zF_m(z) = F_m(z) + 2\mu A^*E(z)$$

which may be solved for $F_m(z)$ in terms of $E(z)$ to give

$$F_m(z) = E(z)\frac{2\mu A^*e^{j\omega_0}}{z - e^{j\omega_0}}$$

On the other hand, the I/O filtering equation from $y_n$ to the output $\hat{x}_n$ is

$$\hat{x}_n = \sum_{m=0}^{M} h_m(n)y_{n-m} = \sum_{m=0}^{M} f_m(n)e^{-j\omega_0(n-m)}Ae^{j\omega_0(n-m)} = \sum_{m=0}^{M} f_m(n)A$$

or, in the $z$-domain,

$$\hat{X}(z) = \sum_{m=0}^{M} F_m(z)A = E(z)\frac{2\mu(M + 1)|A|^2e^{j\omega_0}}{z - e^{j\omega_0}}$$

Finally, the I/O equation from $x_n$ to the output $e_n$ becomes

$$e_n = x_n - \hat{x}_n$$

or, in the $z$-domain,

$$E(z) = X(z) - \hat{X}(z) = X(z) - E(z)\frac{2\mu(M + 1)|A|^2e^{j\omega_0}}{z - e^{j\omega_0}}$$

which may be solved for the transfer function $E(z)/X(z)$

$$\frac{E(z)}{X(z)} = \frac{z - e^{j\omega_0}}{z - e^{j\omega_0}(1 - 2\mu(M + 1)|A|^2)} \tag{6.9.3}$$

This filter has a *zero* at $z = e^{j\omega_0}$ which corresponds to the *notch* at the frequency $\omega_0$. For sufficiently small values of $\mu$ and $A$, the filter is stable; its pole is at $z = e^{j\omega_0}(1 - 2\mu|A|^2(M + 1))$ and can be made to lie inside the unit circle. If the primary input $x_n$ happens to have a sinusoidal component at frequency $\omega_0$, this component will be completely notched away from the output. This will take place even when the sinusoidal reference signal is very weak (i.e., when $A$ is small). The implications of this property for *jamming by signal cancellation* in adaptive array processing have been discussed in [17]. The notching behavior of the adaptive noise canceler when the reference signal consists of a sinusoid plus noise has been discussed in [18].

## 6.10  *Adaptive Line Enhancer (ALE)*  [3,19–21]

A special case of adaptive noise canceling is when there is only *one* signal $x_n$ available which is contaminated by noise. In such a case, the signal $x_n$ provides *its own reference* signal $y_n$, which is taken to be a *delayed* replica of $x_n$: $y_n = x_{n-\Delta}$, as shown in Fig. 6.9. Will such arrangement be successful? The adaptive filter will respond by canceling any components of the main signal $x_n$ that are in any way correlated with the secondary signal $y_n = x_{n-\Delta}$. Suppose the signal $x_n$ consists of two components: a *narrowband* component that has *long-range* correlations such as a sinusoid, and a *broadband* component which will tend to have *short-range* correlations. One of these could represent the desired signal and the other an undesired interfering noise. Pictorially the autocorrelations of the two components could look as follows,

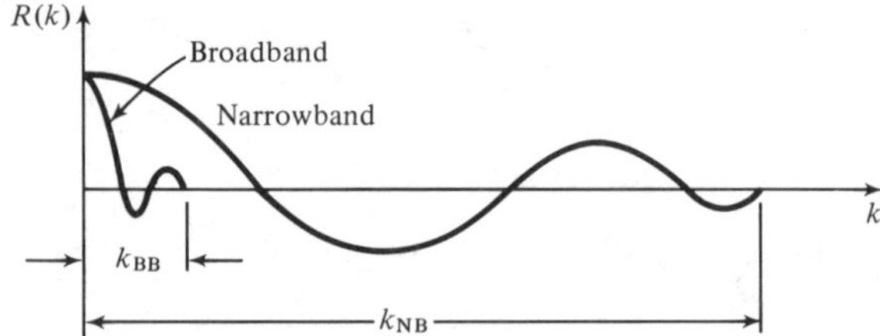


where $k_{NB}$ and $k_{BB}$ are effectively the self-correlation lengths of the narrowband and broadband components, respectively. Beyond these lags, the respective correlations die out quickly. Suppose the delay $\Delta$ is selected so that

$$k_{BB} \leqslant \Delta \leqslant k_{NB}$$

Since $\Delta$ is longer than the effective correlation length of the BB component, then the delayed replica $BB(n - \Delta)$ will be entirely uncorrelated with the BB part of the main signal. The adaptive filter will not be able to respond to this component. On the other hand, since $\Delta$ is shorter than the correlation length of the NB component, the delayed replica $NB(n - \Delta)$ that appears in the secondary

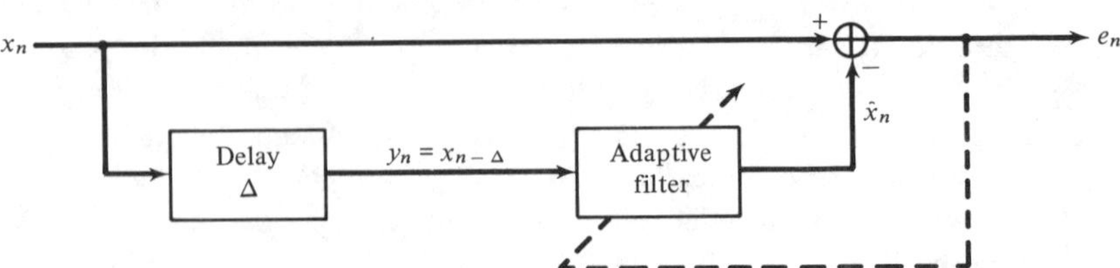


**Figure 6.9**  Adaptive Line Enhancer

input will still be correlated with the NB part of the main signal, and the filter will respond to cancel it. Thus, the filter outputs will be as shown:

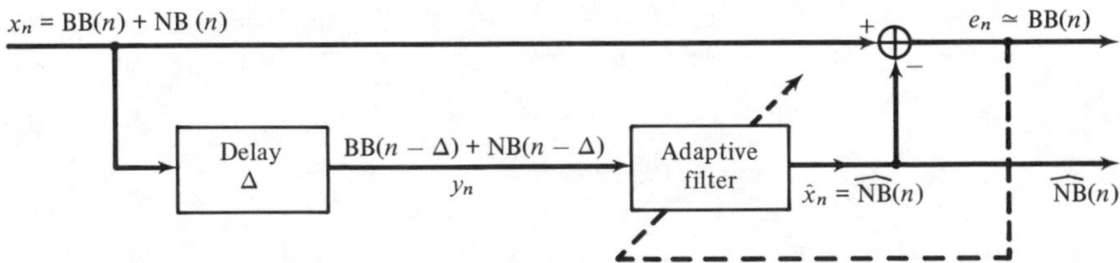


Note that if $\Delta$ is selected to be longer than both correlation lengths, the secondary input will become uncorrelated with the primary input, and the adaptive filter will turn itself off. In the opposite case, when the delay $\Delta$ is selected to be less than both correlation lengths, then both components of the secondary signal will be correlated with the primary signal, and therefore the adaptive filter will respond to cancel the primary $x_n$ completely. The computational algorithm for the ALE is

1. $\hat{x}_n = \Sigma_{m=0}^{M} h_m(n)y(n - m) = \Sigma_{m=0}^{M} h_m(n)x(n - m - \Delta)$
2. $e_n = x_n - \hat{x}_n$
3. $h_m(n + 1) = h_m(n) + 2\mu e_n x(n - m - \Delta), \qquad m = 0,1,2, \ldots ,M$

The Wiener solution for the steady-state weights is $\mathbf{h} = R^{-1}\mathbf{r}$, where the matrix $R$ and vector $\mathbf{r}$ are *both* expressible in terms of the autocorrelation of the signal $x_n$, as follows:

$$R_{ij} = E[y_{n-i}y_{n-j}] = E[x_{n-\Delta-i}x_{n-\Delta-j}] = R_{xx}(i - j)^{\cdot}$$

$$r_i = E[x_n y_{n-i}] = E[x_n x_{n-\Delta-i}] = R_{xx}(i + \Delta)$$

for $i,j = 0,1, \ldots ,M$. When the input signal consists of multiple sinusoids in additive white noise, the inverse $R^{-1}$ may be obtained using the methods of Section 5.11, thus resulting in a closed form expression for the steady-state optimal weights [21].

## 6.11   *Adaptive Linear Prediction*

A linear predictor is a special case of the ALE with the delay $\Delta = 1$. It is shown in Fig. 6.10, where to be consistent with our past notation on linear predictors we have denoted the main signal by $y_n$. The secondary signal, the input to the adaptive filter, is then $y_{n-1}$. Due to the special sign convention used for linear predictors, the adaptation algorithm now reads [22,23]

1. $\hat{y}_n = -\Sigma_{m=1}^{M} a_m(n)y_{n-m}$
2. $e_n = y_n - \hat{y}_n = y_n + a_1(n)y_{n-1} + a_2(n)y_{n-2} + \cdots + a_M(n)y_{n-M}$
3. $a_m(n + 1) = a_m(n) - 2\mu e_n y_{n-m}, \qquad 1 \leq m \leq M$

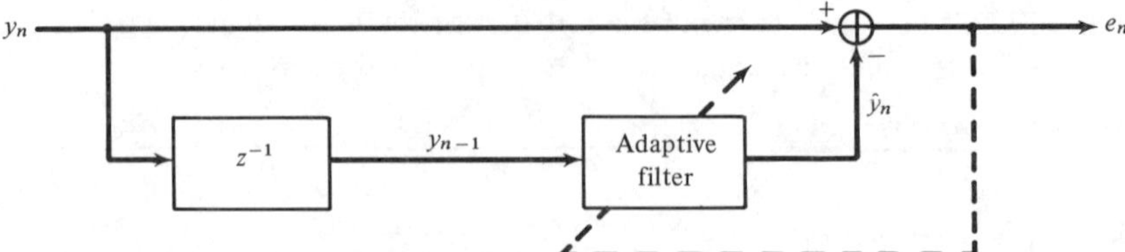

**Figure 6.10**  Adaptive Linear Predictor

The realization of Fig. 6.10 can be redrawn more explicitly as in Fig. 6.11.

Because of the importance of the adaptive predictor, we present a direct derivation of the LMS algorithm as it applies to this case. The weights $a_m$ are chosen optimally to minimize the mean output power of the filter; that is, the mean squared prediction error:

$$\mathcal{E} = E[e_n^2] = \mathbf{a}^T R \mathbf{a} = \min \qquad (6.11.1)$$

where $\mathbf{a} = [1,a_1,a_2, \ldots ,a_M]^T$ is the prediction error filter. The performance index (6.11.1) is minimized with respect to the $M$ weights $a_m$. The gradient with respect to $a_m$ is the $m$th component of the vector $2R\mathbf{a}$; namely,

$$\frac{\partial \mathcal{E}}{\partial a_m} = 2(R\mathbf{a})_m = 2(E[\mathbf{y}(n)\mathbf{y}(n)^T]\mathbf{a})_m = 2(E[\mathbf{y}(n)\mathbf{y}(n)^T\mathbf{a}])_m$$

$$= 2(E[\mathbf{y}(n)e_n])_m = 2E[y_{n-m}e_n]$$

The instantaneous gradient is obtained by *ignoring* the expectation instruction. This gives for the LMS time-update of the $m$th weight

$$\Delta a_m(n) = -\mu \frac{\partial \mathcal{E}}{\partial a_m} = -2\mu e_n y_{n-m} \qquad m = 1,2, \ldots ,M \quad (6.11.2)$$

The adaptive predictor may be thought of as an *adaptive whitening filter*, or an analysis filter which determines the LPC model parameters adaptively. As processing of the signal $y_n$ takes place, the autoregressive model parameters $a_m$ are extracted *on-line*. This is but one example of on-line system identification methods [24–26]. The extracted model parameters may be used in any desired

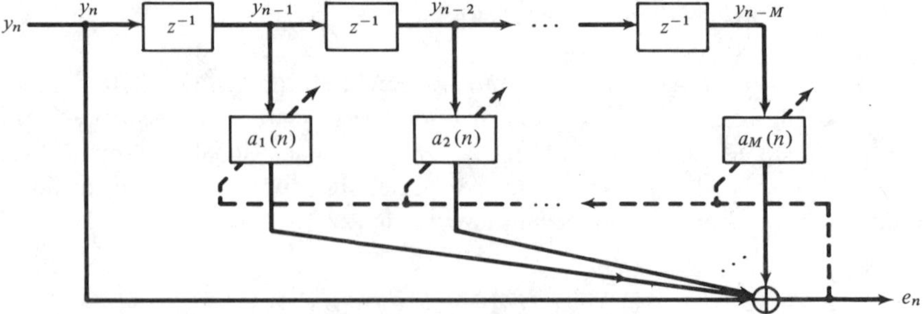

**Figure 6.11**  Direct Form Realization of Adaptive Predictor

way—for example, to provide the *autoregressive spectrum estimate* of the signal $y_n$. One of the advantages of the adaptive implementation is that it offers the possibility of *tracking* slow changes in the spectra of *nonstationary* signals. The only requirement for obtaining meaningful spectrum estimates is that the nonstationary changes of the spectrum be slow enough for the adaptive filter to have a chance to converge between changes. Typical applications are the *tracking of sinusoids in noise* whose frequencies may be slowly changing [22,23,27], or tracking the time development of the spectra of nonstationary EEG signals [28,29]. At *each* time instant $n$, the adaptive weights $a_m(n)$; $m = 1,2, \ldots ,M$ may be used to obtain an *instantaneous* autoregressive estimate of the spectrum of $y_n$, in the form

$$S_n(\omega) = \frac{1}{|1 + a_1(n)e^{-j\omega} + a_2(n)e^{-2j\omega} + \cdots + a_M(n)e^{-Mj\omega}|^2}$$

This is the adaptive implementation of the LP spectrum estimate discussed in Section 5.11. The same adaptive approach to LP spectrum estimation may also be used in the problem of *multiple source location,* discussed in Section 5.12. The only difference in the algorithm is to replace $y_{n-m}$ by $y_m(n)$—that is, by the signal recorded at the $m$th sensor at time $n$—and to use the complex-valued version of the LMS algorithm. For completeness, we summarize the computational steps in this case, following the notation of Section 5.12:

1. $e_n = y_0(n) + a_1(n)y_1(n) + \cdots + a_M(n)y_M(n)$
2. $a_m(n + 1) = a_m(n) - 2\mu e_n y_m(n)^*; \qquad m = 1,2, \ldots ,M$

At *each* time instant $n$, the corresponding *spatial spectrum* estimate may be computed by

$$S_n(k) = \frac{1}{|1 + a_1(n)e^{-jk} + a_2(n)e^{-2jk} + \cdots + a_M(n)e^{-Mjk}|^2}$$

where the wavenumber $k$ and its relationship to the angle of bearing was defined in Section 5.12. Figure 6.12 shows the corresponding adaptive array processing configuration.

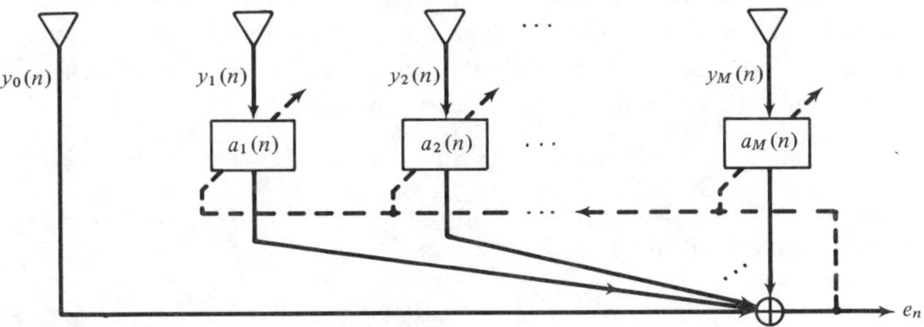


**Figure 6.12**  Adaptive Array Processor

The time-adaptive as well as the block-data adaptive methods of superresolution array processing have been reviewed in [30]. The above LMS algorithm for the array weights is effectively equivalent to the Howells-Applebaum algorithm [4,5,30].

## 6.12  *Adaptive Implementation of Pisarenko's Method*

In Section 5.11, we noted that the Pisarenko eigenvalue problem was equivalent to the minimization of the performance index

$$\mathcal{E} = E[e_n^* e_n] = \mathbf{a}^\dagger R \mathbf{a} = \min \tag{6.12.1}$$

subject to the quadratic constraint

$$\mathbf{a}^\dagger \mathbf{a} = 1 \tag{6.12.2}$$

where

$$e_n = \sum_{m=0}^{M} a_m y_{n-m} = [a_0, a_1, a_2, \ldots, a_M] \begin{bmatrix} y_n \\ y_{n-1} \\ y_{n-2} \\ \vdots \\ y_{n-M} \end{bmatrix} = \mathbf{a}^T \mathbf{y}(n)$$

The solution of the minimization problem shown in Eqs. (6.12.1) and (6.12.2) is the eigenvector $\mathbf{a}$ belonging to the *minimum* eigenvalue of the covariance matrix $R$. If there are $L$ sinusoids of frequencies $\omega_i$; $i = 1, 2, \ldots, L$, and we use a filter with $M$ weights, such that $M \geq L$, then the eigenpolynomial $A(z)$ corresponding to the minimum eigenvector $\mathbf{a}$ will have $L$ zeros on the unit circle at precisely the desired set of frequencies; that is,

$$A(z_i) = 0 \qquad \text{where } z_i = \exp(j\omega_i), \qquad i = 1, 2, \ldots, L$$

The adaptive implementation [31] of the Pisarenko eigenvalue problem is based on the above minimization criterion. The LMS gradient descent algorithm can be used to update the weights, but some care must be taken to satisfy the essential quadratic constraint (6.12.2) at *each* iteration of the algorithm. Any infinitesimal change $d\mathbf{a}$ of the weights must respect the constraint. This means the $d\mathbf{a}$ cannot be arbitrary but must satisfy the condition

$$d(\mathbf{a}^\dagger \mathbf{a}) = \mathbf{a}^\dagger d\mathbf{a} + d\mathbf{a}^\dagger \mathbf{a} = 0 \tag{6.12.3}$$

so that the new weight $\mathbf{a} + d\mathbf{a}$ still lies on the quadratic surface $\mathbf{a}^\dagger \mathbf{a} = 1$. The ordinary gradient of the performance index $\mathcal{E}$ is

$$\frac{\partial \mathcal{E}}{\partial \mathbf{a}^*} = R \mathbf{a}$$

Projecting this onto the surface $\mathbf{a}^\dagger \mathbf{a} = 1$ by the projection matrix $(I - \mathbf{a}\mathbf{a}^\dagger)$, where $I$ is the $(M + 1)$-dimensional unit matrix, we obtain the "constrained" gradient

$$\left(\frac{\partial \mathcal{E}}{\partial \mathbf{a}^*}\right)_c = (I - \mathbf{a}\mathbf{a}^\dagger)\frac{\partial \mathcal{E}}{\partial \mathbf{a}^*} = (I - \mathbf{a}\mathbf{a}^\dagger)\,R\mathbf{a} = R\mathbf{a} - \mathcal{E}\mathbf{a} \quad (6.12.4)$$

which is *tangent* to the constraint surface at the point $\mathbf{a}$. The vanishing of the constrained gradient is *equivalent* to the Pisarenko eigenvalue problem. The weight update can now be chosen to be proportional to the constrained gradient

$$\Delta\mathbf{a} = -\mu\left(\frac{\partial \mathcal{E}}{\partial \mathbf{a}^*}\right)_c = -\mu(R\mathbf{a} - \mathcal{E}\mathbf{a})$$

This choice guarantees that $\Delta\mathbf{a}$ satisfies Eq. (6.12.3); indeed, because of the projection matrix in front of the gradient, it follows that $\mathbf{a}^\dagger\Delta\mathbf{a} = 0$. Actually, since $\Delta\mathbf{a}$ is not infinitesimal, it will correspond to a *finite* motion along the tangent to the surface at the point $\mathbf{a}$. Thus, the new point $\mathbf{a} + \Delta\mathbf{a}$ will be slightly off the surface and must be *renormalized* to have unit norm. Using

$$R\mathbf{a} = E[\mathbf{y}(n)^*\mathbf{y}(n)^T]\mathbf{a} = E[\mathbf{y}(n)^*e_n], \quad \text{and} \quad \mathcal{E} = E[e_n^* e_n]$$

we write the update as

$$\Delta\mathbf{a} = -\mu(E[e_n\mathbf{y}(n)^*] - E[e_n^* e_n]\mathbf{a})$$

The LMS algorithm is obtained by *ignoring* the indicated ensemble expectation values. The weight adjustment procedure consists of two steps: first, shift the old weight $\mathbf{a}(n)$ by $\Delta\mathbf{a}(n)$, and then renormalize it to unit norm:

$$\mathbf{b}(n + 1) = \mathbf{a}(n) + \Delta\mathbf{a}(n) \qquad (6.12.5)$$
$$\mathbf{a}(n + 1) = \mathbf{b}(n + 1)/\|\mathbf{b}(n + 1)\|$$

where the weight update is computed by

$$\Delta\mathbf{a}(n) = -\mu[e_n\mathbf{y}(n)^* - e_n^* e_n\mathbf{a}(n)] \qquad (6.12.6)$$

In summary, the computational steps are as follows:

1. At time $n$, $\mathbf{a}(n)$ is available and normalized to unit norm
2. Compute the output $e_n = \Sigma_{m=0}^M a_m(n)y_{n-m} = \mathbf{a}(n)^T\mathbf{y}(n)$
3. Update the filter weights using Eqs. (6.12.5) and (6.12.6)
4. Go to the next time instant, $n \to n + 1$.

A realization of the adaptive filter is shown in Fig. 6.13. After a number of iterations, the algorithm may be stopped and the Pisarenko spectrum estimate computed

$$S_n(\omega) = \frac{1}{|a_0(n) + a_1(n)e^{-j\omega} + a_2(n)e^{-2j\omega} + \cdots + a_M(n)e^{-Mj\omega}|^2}$$

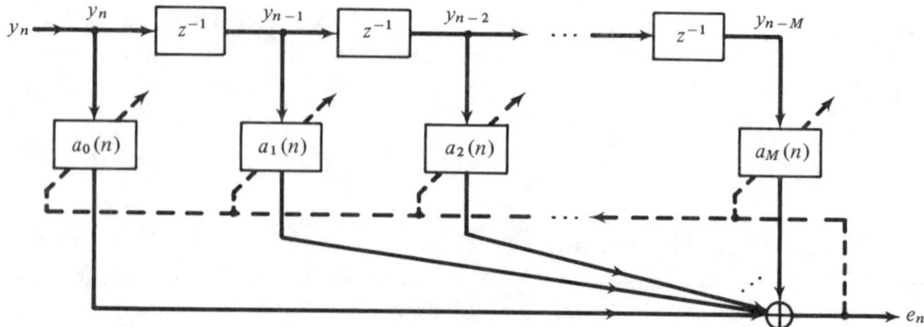

**Figure 6.13**  Adaptive Implementation of Pisarenko's Method

After convergence, $S_n(\omega)$ should exhibit very sharp peaks at the sought frequency angles $\omega_i$; $i = 1,2, \ldots ,L$. The convergence properties of this algorithm have been studied in [32]. Alternative adaptation schemes for the weights have been proposed in [33,34]. The algorithm may also be applied to the array problem of *multiple source location* [35]. Again, the only change is to replace $y_{n-m}$ by $y_m(n)$.

Both the adaptive prediction and the Pisarenko approaches to the two problems of extracting sinusoids in noise and multiple emitter location have a common aim; namely, to produce an adaptive filter $A(z)$ with zeros very near or on the unit circle at the desired frequency angles. Taking the inverse magnitude response as an estimate of the spectrum of the signal

$$S(\omega) = \frac{1}{|A(\omega)|^2}$$

is a simple device to get a curve that exhibits sharp spectral peaks at the desired frequencies. A satisfactory alternative approach would be simply to find the roots of the polynomial $A(z)$ and pick those that are closest to the unit circle. The phase angles of these roots are precisely the desired frequencies. In other words, the frequency information we are attempting to extract by means of the adaptive filter is more directly represented by the zeros of the filter than by its weights. It would be desirable then to develop methods by which these zeros can be estimated directly without having to submit the filter $A(z)$ to root-finding algorithms. In implementing this idea adaptively, we would like to adapt and track the zeros of the adaptive filter as they move about in the complex $z$-plane, converging to their final destinations which are the desired zeros. In this way, the frequency information can be extracted directly. Such "zero-tracking" adaptation algorithms have been proposed recently [36–38]. Even though the representations of the filter in terms of its zeros and in terms of its weights are mathematically equivalent, the zero representation may be more appropriate in some applications, in the sense that a better insight into the nature of the underlying processes may be gained from it than from the weight representation. As an example we mention the problem of predicting epileptic seizures by LPC modeling of the EEG signal where it was found [39] that the trajectories of the zeros of the prediction-error filter in the $z$-plane exhibited an unexpected behavior; namely, prior to the onset of a seizure, one of the zeros became the "most mobile" and moved towards the unit circle, whereas the other zeros of the filter did not move much. The trajectory of the most mobile zero could be used as a signature for the onset of the oncoming seizure. Such behavior could not be easily discerned by the frequency response or by the final zero location.

Next, we describe briefly the *zero-tracking algorithm* as it applies to the Pisarenko problem and present a simulation example. Its application to *adaptive prediction* and to *emitter location* has been discussed in [36,37]. For simplicity, we assume that the number of sinusoids that are present is the same as the order of the filter **a**; that is, $L = M$. The case $L < M$ will be discussed later on. The eigenpolynomial of the minimum eigenvector **a** may be factored into its zeros as follows:

$$A(z) = a_0 + a_1 z^{-1} + a_2 z^{-2} + \cdots + a_M z^{-M} \tag{6.12.7}$$
$$= a_0(1 - z_1 z^{-1})(1 - z_2 z^{-1}) \cdots (1 - z_M z^{-1})$$

where $a_0$ may be thought of as a normalization factor which guarantees the unit norm constraint (6.12.2), and $z_i = \exp(j\omega_i)$, $i = 1,2, \ldots ,M$ are the desired sinusoid zeros on the unit circle. In the adaptive implementation, the weights $a_m$ become time dependent $a_m(n)$ and are adapted from each time instant to the next until they converge to the asymptotic values defined by Eq. (6.12.7). At each $n$, the corresponding polynomial can be factored into its zeros as follows:

$$a_0(n) + a_1(n)z^{-1} + a_2(n)z^{-2} + \cdots + a_M(n)z^{-M}$$
$$= a_0(n)(1 - z_1(n)z^{-1})(1 - z_2(n)z^{-1}) \cdots (1 - z_M(n)z^{-1}) \tag{6.12.8}$$

where again the factor $a_0(n)$ ensures the unit norm constraint. In the zero-tracking algorithm, the weight update equation (6.12.5) is replaced by a *zero-update equation* of the form

$$z_i(n + 1) = z_i(n) + \Delta z_i(n), \qquad i = 1,2, \ldots ,M \tag{6.12.9}$$

where the zero updates $\Delta z_i(n)$ must be such that to ensure the convergence of the zeros to their asymptotic values $z_i$. One way to do this is to make the algorithm equivalent to the LMS algorithm. The functional dependence of $z_i(n)$ on $a_m(n)$ defined by Eq. (6.12.8) implies that if the weights $a_m(n)$ are changed by a small amount $\Delta a_m(n)$ given by Eq. (6.12.6), then a small change $\Delta z_i(n)$ will be induced on the corresponding zeros. This is given as follows:

$$\Delta z_i(n) = \sum_{m=0}^{M} \frac{\partial z_i(n)}{\partial a_m} \Delta a_m(n) \tag{6.12.10}$$

where the partial derivatives are given by [40]

$$\frac{\partial z_i(n)}{\partial a_m} = -\frac{1}{a_0(n)} \frac{z_i(n)^{M-m}}{\prod_{j \neq i} (z_i(n) - z_j(n))} \tag{6.12.11}$$

for $0 \leqslant m \leqslant M$. Equation (6.12.10) is strictly valid for infinitesimal changes, but for small $\mu$, it can be taken to be an adequate approximation for the purpose of computing $\Delta z_i(n)$. The advantage of this expression is that only the *current* zeros $z_i(n)$ are needed to compute $\Delta z_i(n)$. The complete algorithm is summarized as follows:

1. At time $n$, the zeros $z_i(n)$; $i = 1,2, \ldots ,M$ are available
2. Using convolution, compute the corresponding filter weights and normalize them to unit norm. That is, first convolve the factors of Eq. (6.12.8) to obtain the vector

$$\mathbf{b}(n)^T = [1, b_1(n),\ b_2(n),\ \ldots ,b_M(n)]$$

$$= [1, -z_1(n)] * [1, -z_2(n)] * \cdots * [1, -z_M(n)]$$

   and then normalize the vector $\mathbf{b}(n)$ to unit norm

$$\mathbf{a}(n) = \frac{\mathbf{b}(n)}{\|\mathbf{b}(n)\|}$$

3. Compute the filter output $e_n = \sum_{m=0}^{M} a_m(n) y_{n-m}$
4. Compute the LMS coefficient update $\Delta a_m(n)$ using Eq. (6.12.6). Compute the zero update $\Delta z_i(n)$ using Eqs. (6.12.10) and (6.12.11), and update the zeros using Eq. (6.12.9)

The algorithm may be initialized by a random selection of the initial zeros inside the unit circle in the $z$-plane. Next, we present a simulation example consisting of a fourth order filter and four sinusoids

$$y_n = v_n + e^{j\omega_1 n} + e^{j\omega_2 n} + e^{j\omega_3 n} + e^{j\omega_4 n}$$

with frequencies

$$\omega_1 = 0.25\pi, \qquad \omega_2 = -0.25\pi, \qquad \omega_3 = 0.75\pi, \qquad \omega_4 = -0.75\pi$$

and a zero-mean, unit-variance, white noise sequence $v_n$ (this corresponds to all sinusoids having 0 dB signal to noise ratio). The value of $\mu$ was 0.001. Figure 6.14 shows the adaptive trajectories of the four filter zeros as they converge onto the unit circle at the above frequency values. After convergence, the adaptive zeros remain within small neighborhoods about the asymptotic zeros. The diameter of these neighborhoods decreases with smaller $\mu$, but so does the speed of convergence [36]. The transient behavior of the zeros can be seen by plotting $z_i(n)$ versus iteration number $n$. Figure 6.15 shows the real and imaginary parts of the adaptive trajectory $z_3(n)$ converging to the real and imaginary parts of the asymptotic zero $z_3 = \exp(j\omega_3)$.

When the number $L$ of sinusoids is less than the order $M$ of the filter, only $L$ of the $M$ zeros $z_i(n)$ of the filter will be driven to the unit circle at the right frequency angles. The remaining $(M - L)$ zeros correspond to spurious degrees of freedom (the degeneracy of the minimum eigenvalue $\sigma_v^2$), and are affected

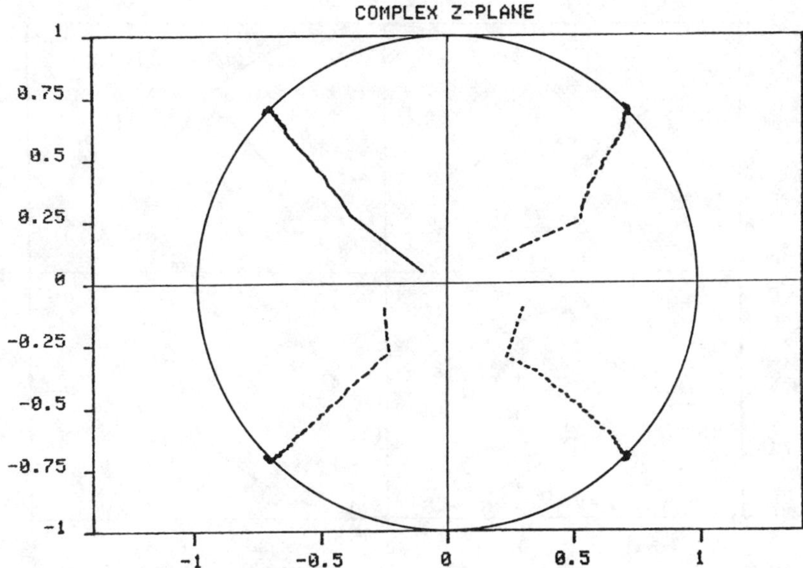

**Figure 6.14** $z$-Plane Trajectories of the Four Adaptive Zeros $z_i(n)$, $i = 1, 2, 3,$

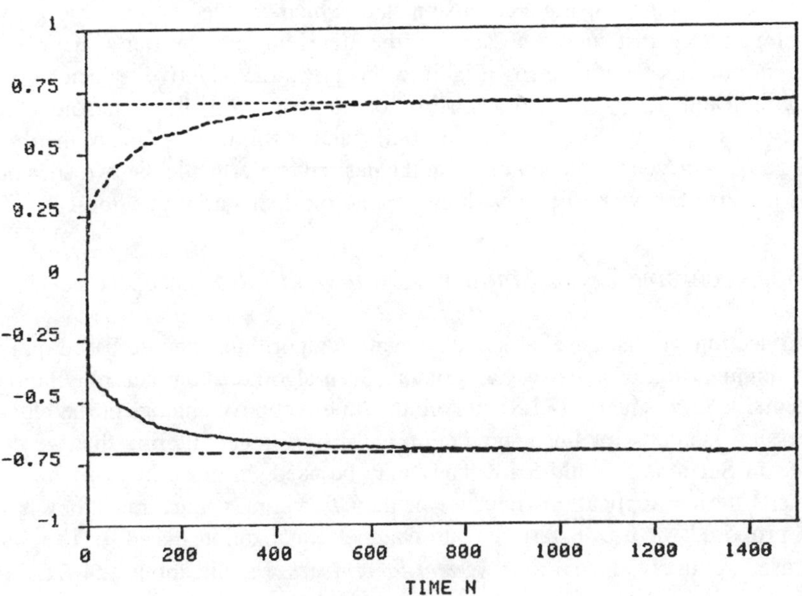

**Figure 6.15** $\mathrm{Re}(z_3(n))$ and $\mathrm{Im}(z_3(n))$ versus $n$

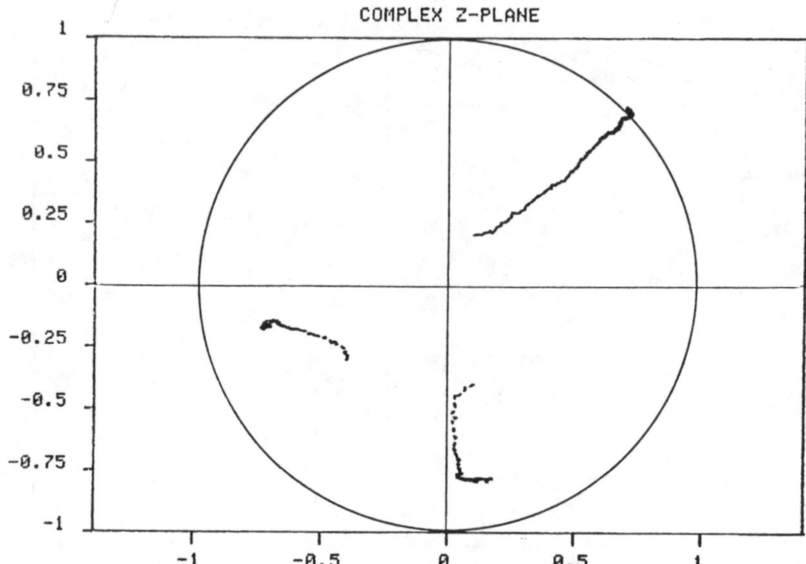


**Figure 6.16**  Single Sinusoid with Order-3 Adaptive Filter

by the adaptation process only insofar as the $M$ zero trajectories are not entirely independent of each other but are mutually coupled through Eq. (6.12.10). Where these spurious zeros converge depends on the particular initialization. For some special initial conditions it is possible for the spurious zeros to move close to the unit circle, thus causing a confusion as to which are the true sinusoid zeros. To safeguard against such a possibility, the algorithm may be run again with a different choice of initial zeros. Figure 6.16 shows the adaptive trajectory of a single sinusoid, $L = 1$, using a third order filter, $M = 3$. The sinusoid's frequency was $\omega_1 = 0.25\pi$, its SNR was 0 dB, and $\mu$ was 0.001. One of the three filter zeros is driven to the unit circle at the desired angle, while the two spurious zeros traverse fairly short paths which depend on their initial positions.

## 6.13  *Recursive Least-Squares Adaptive Filters*

In this section we discuss alternative adaptation algorithms that are based on the *exact* minimization of *least-squares* criteria. Such algorithms are generally known as recursive least-squares (RLS) and are the time-recursive analogs of the block-processing methods for linear prediction and FIR Wiener filtering that we discussed in Sections 5.9 and 5.14. They may be used, in place of LSM, in any Wiener filtering application. Because of their *fast convergence* rate, they have been proposed for use in fast start-up channel equalization [41–43]. They are also used routinely in real-time *system identification* applications [24–26,44].

Their main disadvantage is that they require a fair amount of computation (of order of $M^2$, for $M$-tap filters) per time update. In biomedical applications, they can be easily implemented on minicomputers [28,29]. In other applications, such as equalization of rapidly varying channels, they may be too costly for implementation. Recently however, the RLS algorithms have been reformulated in "*fast Kalman*" as well as "*recursive least-squares lattice*" (RLSL) forms [45–54] which admit very efficient implementations (of order $10M$ computations per time update). The RLSL algorithms combine the best of both the LMS and the RLS; namely, the computational efficiency of the former and the fast convergence of the latter. In this section, we only present the conventional RLS version and refer the reader to Friedlander's comprehensive review article [54] for more information on the RLSL.

We start with the RLS formulation of the FIR Wiener filtering problem. The estimation criterion $\mathcal{E} = E[e_n^2] = \min$ is replaced with a least-squares time average that includes all the estimation errors, from the initial time instant to the current time; that is, $e_k$; $k = 0, 1, \ldots, n$

$$\mathcal{E}_n = \sum_{k=0}^{n} e_k^2 = \min \tag{6.13.1}$$

where

$$e_k = x_k - \hat{x}_k$$

and $\hat{x}_k$ is the estimate of $x_k$ produced by the (order $M$) Wiener filter

$$\hat{x}_k = \sum_{m=0}^{M} h_m y_{k-m} = [h_0, h_1, \ldots, h_M] \begin{bmatrix} y_k \\ y_{k-1} \\ \vdots \\ y_{k-M} \end{bmatrix} = \mathbf{h}^T \mathbf{y}(k)$$

To better track possible nonstationarities in the signals, the performance index (6.13.1) is sometimes modified by introducing *exponential weighting*

$$\mathcal{E}_n = \sum_{k=0}^{n} \lambda^{n-k} e_k^2 = e_n^2 + \lambda e_{n-1}^2 + \cdots + \lambda^n e_0^2 \tag{6.13.2}$$

where the "forgetting factor" $\lambda$ is positive and less than one. The index (6.13.2) emphasizes the *most recent* observations and exponentially ignores the older ones. We will base our discussion on this criterion. The optimal weights $\mathbf{h}$ are chosen to minimize Eq. (6.13.2). Setting the derivative with respect to $\mathbf{h}$ to zero we find the least-squares analogs of the *orthogonality equations*

$$\frac{\partial \mathcal{E}_n}{\partial \mathbf{h}} = -2 \sum_{k=0}^{n} \lambda^{n-k} e_k \mathbf{y}(k) = 0$$

which may be recast in their *normal equation* form

$$\sum_{k=0}^{n} \lambda^{n-k}(x_k - \mathbf{h}^T\mathbf{y}(k))\mathbf{y}(k) = 0$$

$$\left[\sum_{k=0}^{n} \lambda^{n-k}\mathbf{y}(k)\mathbf{y}(k)^T\right]\mathbf{h} = \sum_{k=0}^{n} \lambda^{n-k}x_k\mathbf{y}(k)$$

Defining the quantities

$$R(n) = \sum_{k=0}^{n} \lambda^{n-k}\mathbf{y}(k)\mathbf{y}(k)^T$$

$$\mathbf{r}(n) = \sum_{k=0}^{n} \lambda^{n-k}x_k\mathbf{y}(k)$$

(6.13.3)

we write the normal equations as $R(n)\mathbf{h} = \mathbf{r}(n)$, with solution $\mathbf{h} = R(n)^{-1}\mathbf{r}(n)$. Note that the $n$-dependence of $R(n)$ and $\mathbf{r}(n)$ makes $\mathbf{h}$ depend on the current time $n$; we shall write, therefore,

$$\mathbf{h}(n) = R(n)^{-1}\mathbf{r}(n) = P(n)\mathbf{r}(n) \tag{6.13.4}$$

where $P(n) = R(n)^{-1}$. These are the least-squares analogs of the ordinary Wiener solution, with $R(n)$ and $\mathbf{r}(n)$ playing the role of the covariance matrix $E[\mathbf{y}(n)\mathbf{y}(n)^T]$ and cross-correlation vector $E[x_n\mathbf{y}(n)]$, respectively. The RLS algorithm is obtained by writing Eq. (6.13.4) *recursively* in $n$. The quantities (6.13.3) satisfy the recursions

$$R(n) = \lambda R(n-1) + \mathbf{y}(n)\mathbf{y}(n)^T \tag{6.13.5}$$

$$\mathbf{r}(n) = \lambda\mathbf{r}(n-1) + x_n\mathbf{y}(n) \tag{6.13.6}$$

Writing Eq. (6.13.5) in the equivalent inverted form

$$P(n) = [\lambda P(n-1)^{-1} + \mathbf{y}(n)\mathbf{y}(n)^T]^{-1}$$

and using the *matrix inversion lemma*, we find the update equation

$$P(n) = \frac{1}{\lambda}\left[P(n-1) - \frac{P(n-1)\mathbf{y}(n)\mathbf{y}(n)^TP(n-1)}{\lambda + \mathbf{y}(n)^TP(n-1)\mathbf{y}(n)}\right] \tag{6.13.7}$$

Inserting Eqs. (6.13.7) and (6.13.6) into Eq. (6.13.4), and using the *previous* optimal solution $\mathbf{h}(n-1) = P(n-1)\mathbf{r}(n-1)$, we can relate the current $\mathbf{h}(n)$ to the previous $\mathbf{h}(n-1)$. After some algebra we find

$$\mathbf{h}(n) = \mathbf{h}(n-1) + \mathbf{k}(n)e_{n/n-1} \tag{6.13.8}$$

where the *gain* vector $\mathbf{k}(n)$ and $e_{n/n-1}$ are defined as

$$\mathbf{k}(n) = \frac{P(n-1)\mathbf{y}(n)}{\lambda + \mathbf{y}(n)^TP(n-1)\mathbf{y}(n)}, \quad e_{n/n-1} = x_n - \mathbf{h}(n-1)^T\mathbf{y}(n) \tag{6.13.9}$$

Note that $e_{n/n-1}$ is not quite the same as the estimation error $e_n = x_n - \mathbf{h}(n)^T\mathbf{y}(n)$. Equation (6.13.8) is written in a Kalman filter form; namely, the quantity $\mathbf{h}(n - 1)$, which is the best filter based on all the data up to time $n - 1$, may be thought of as a reasonable prediction of the next best filter $\mathbf{h}(n)$; the quantity $e_{n/n-1}$ is the tentative estimation error made on the basis of the prediction $\mathbf{h}(n - 1)$; and finally, Eq. (6.13.8) represents the correction of the prediction $\mathbf{h}(n - 1)$ to obtain the new best filter $\mathbf{h}(n)$. Using Eq. (6.13.7) and some algebra, we can rewrite the gain vector $\mathbf{k}(n)$ in the alternative form

$$\mathbf{k}(n) = P(n)\mathbf{y}(n) = R(n)^{-1}\mathbf{y}(n) \qquad (6.13.10)$$

Then, Eq. (6.13.8) is written as

$$\mathbf{h}(n) = \mathbf{h}(n - 1) + R(n)^{-1}\mathbf{y}(n)e_{n/n-1}$$

which differs from the LMS algorithm by the presence of $R(n)^{-1}$ in front of the weight correction term. Since $R(n)$ is a measure of the covariance matrix $E[\mathbf{y}(n)\mathbf{y}(n)^T]$, the presence of $R(n)^{-1}$ makes the RLS algorithm behave like Newton's method, hence its fast convergence properties. The matrix update Eq. (6.13.7) can be done with $O(M^2)$ operations and represents the main computational burden of the RLS algorithm. The algorithm is summarized as follows:

1. At time instant $n$, the quantities $\mathbf{h}(n - 1)$ and $P(n - 1)$ are available, and so are $x_n$ and the data vector $\mathbf{y}(n)$ stored in the tapped delay line of the filter
2. Using Eqs. (6.13.7) through (6.13.9), compute $\mathbf{k}(n)$, $\mathbf{h}(n)$, $P(n)$ and the estimate $\hat{x}_n = \mathbf{h}(n)^T\mathbf{y}(n)$ and estimation error $e_n = x_n - \hat{x}_n$
3. Go to $n \to n + 1$

The algorithm may be initialized by taking $R(-1) = 0$, which would imply $P(-1) = \infty$. Instead, we may use a very *large* number for $P(-1)$. The algorithm is quite insensitive to the paricular choice of $P(-1)$. The initial weights $\mathbf{h}(-1)$ may be set to zero. The algorithm has a stronger dependence on the forgetting factor $\lambda$. As a *measure* of the effective memory of the performance index (6.13.2), we may take the quantity

$$\bar{n} = \frac{\displaystyle\sum_{n=0}^{\infty} n\lambda^n}{\displaystyle\sum_{n=0}^{\infty} \lambda^n} = \frac{\lambda}{1 - \lambda}$$

Smaller $\lambda$s correspond to shorter memory $\bar{n}$, and can track better the nonstationary changes of the underlying signals. On the other hand if $\lambda$ is too small, that is, if $\bar{n}$ is shorter than the effective duration of each stationary segment, the performance index will not be taking full advantage of all available samples (which could extend over the entire stationary segment). As a consequence, the computed weights $\mathbf{h}(n)$ will exhibit more noisy behavior. This is demonstrated in the simulations below. In particular, if the signals are stationary, the best value of $\lambda$ is unity.

*Adaptive prediction* may also be formulated in a similar manner. The performance index is still given by Eq. (6.13.2) but now $e_k$ represents the prediction error

$$e_k = y_k - \hat{y}_k$$

$$\hat{y}_k = - \sum_{m=1}^{M} a_m(n)y_{k-m} = -\mathbf{a}(n)^T \mathbf{y}(k-1)$$

where $\mathbf{y}(k-1) = [y(k-1), y(k-2), \ldots , y(k-M)]^T$, so that

$$e_k = y_k + \mathbf{a}(n)^T \mathbf{y}(k-1), \qquad \text{for } k = 0,1, \ldots ,n \qquad (6.13.11)$$

Minimizing the performance index with respect to $\mathbf{a}(n)$ yields the solution

$$\mathbf{a}(n) = -R(n)^{-1}\mathbf{r}(n) = -P(n)\mathbf{r}(n)$$

where $P(n) = R(n)^{-1}$ and

$$R(n) = \sum_{k=0}^{n} \lambda^{n-k}\mathbf{y}(k-1)\mathbf{y}(k-1)^T, \qquad \mathbf{r}(n) = \sum_{k=0}^{n} \lambda^{n-k}y_k\mathbf{y}(k-1)$$

Following the same procedure as above, the equation for the prediction coefficients $\mathbf{a}(n)$ may be written recursively, as follows

$$\mathbf{a}(n) = \mathbf{a}(n-1) - \mathbf{k}(n)e_{n/n-1} \qquad (6.13.12)$$

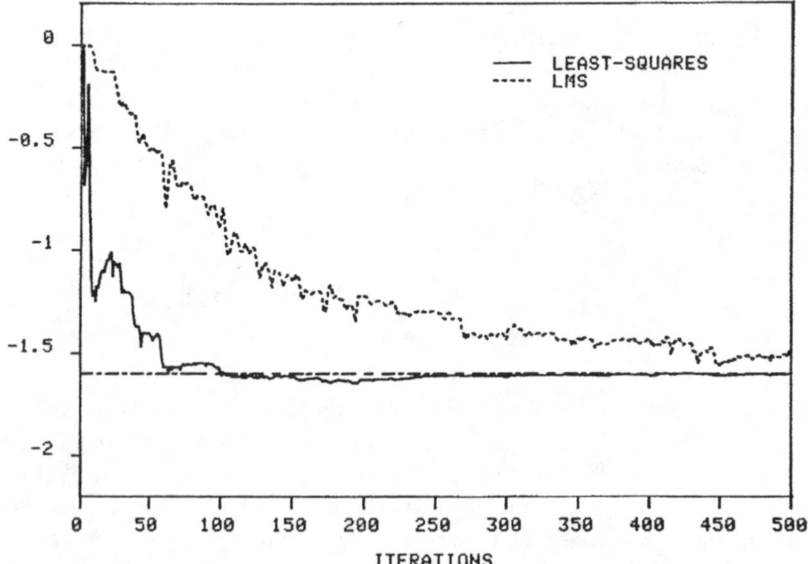


**Figure 6.17**  Transient Behavior of the Adaptive Weight $a_1(n)$ versus $n$, Adapted According to the RLS and LMS Algorithms

where

$$e_{n/n-1} = y_n + \mathbf{a}(n - 1)^T \mathbf{y}(n - 1) \qquad (6.13.13)$$

$$\mathbf{k}(n) = \frac{P(n - 1)\mathbf{y}(n - 1)}{\lambda + \mathbf{y}(n - 1)^T P(n - 1)\mathbf{y}(n - 1)} \qquad (6.13.14)$$

and $P(n)$ is updated by

$$P(n) = \frac{1}{\lambda}\left[ P(n - 1) - \frac{P(n - 1)\mathbf{y}(n - 1)\mathbf{y}(n - 1)^T P(n - 1)}{\lambda + \mathbf{y}(n - 1)^T P(n - 1)\mathbf{y}(n - 1)} \right] \qquad (6.13.15)$$

We have carried out a simulation comparing the convergence of the RLS adaptive predictor to the LMS predictor. The sequence $y_n$ was generated by sending zero-mean, unit-variance, white noise $\epsilon_n$ through the second order autoregressive model

$$y_n + a_1 y_{n-1} + a_2 y_{n-2} = \epsilon_n$$

where the LPC model parameters were $a_1 = -1.6$, $a_2 = 0.8$. Figure 6.17 shows the transient behavior of the weight $a_1(n)$ as it converges to its asymptotic value $a_1 = -1.6$, computed according to RLS and according to LMS. The value of $\lambda$ was unity, and $P(-1) = 10$. The LMS adaptation parameter $\mu$ was 0.002. The LMS could have been accelerated a little more, but at the expense of increasing the noisiness of the weights due to the gradient noise. Figure 6.18

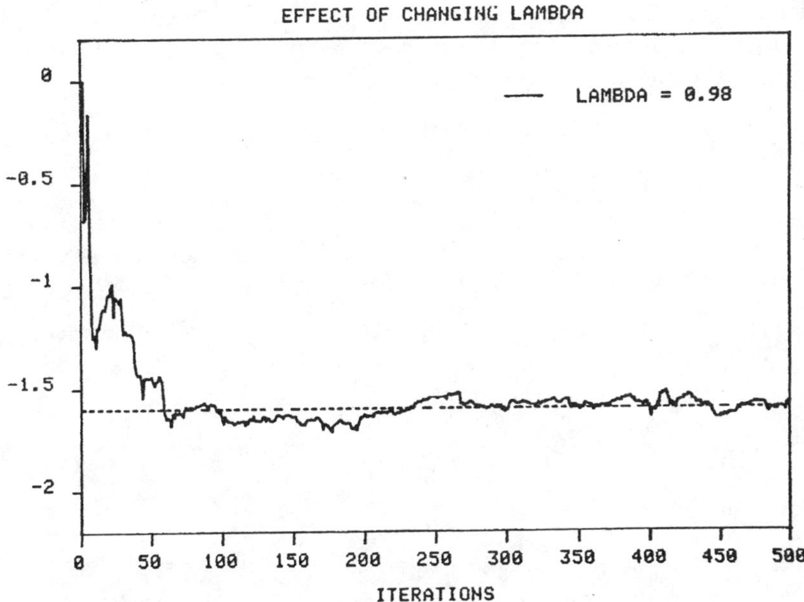


**Figure 6.18** Effect of Reducing the Forgetting Factor on the Adaptive Weight $a_1(n)$

shows the effect of decreasing the value of $\lambda$ to $\lambda = 0.98$. Note the more noisy nature of the converged weight. As mentioned above, this is because the performance index has an effective memory of $\bar{n} = \lambda/(1 - \lambda) = 49$, and does not take full advantage, in this stationary case, of all the available samples in computing the optimal weights.

## 6.14  *Adaptive Lattice Filters*

In this section we discuss the "*gradient adaptive lattice*" implementations of linear prediction and lattice Wiener filters [55–60]. They are based on a gradient-descent, LMS-like approach applied to the weights of the *lattice representations* rather than to the weights of the direct form representation. Taking advantage of the decoupling of the successive stages of the lattice, and properly choosing the adaptation constants $\mu$, all lattice weights can be made to converge fast and, in contrast to the LMS weights, with a convergence rate that is essentially *independent* of the *eigenvalue spread* of the *input* covariance matrix. The gradient lattice algorithms are very similar but not identical to the recursive least-squares lattice algorithms [45–54], and they share the same properties of fast convergence and computational efficiency with the latter. Typically, the gradient lattice converges somewhat more slowly than RLSL. Some comparisons between the two types of algorithms are given in [54,60].

We start by casting the ordinary lattice filter of linear prediction in a gradient-adaptive form, and then discuss the gradient-adaptive form of the lattice Wiener filter, the stationary version of which was presented in Section 5.8.

The lattice recursion for an $M$th order prediction-error filter of a stationary signal $y_n$ was found in Section 5.5 to be

$$e_{p+1}^{+}(n) = e_p^{+}(n) - \gamma_{p+1}e_p^{-}(n - 1)$$
$$e_{p+1}^{-}(n) = e_p^{-}(n - 1) - \gamma_{p+1}e_p^{+}(n) \tag{6.14.1}$$

for $p = 0,1, \ldots ,M - 1$, where $e_0^{+}(n) = e_0^{-}(n) = y_n$. The optimal value of the reflection coefficient $\gamma_{p+1}$ can be obtained by minimizing the performance index

$$\mathcal{E}_{p+1} = E[e_{p+1}^{+}(n)^2 + e_{p+1}^{-}(n)^2] \tag{6.14.2}$$

Differentiating with respect to $\gamma_{p+1}$ we find

$$\frac{\partial \mathcal{E}_{p+1}}{\partial \gamma_{p+1}} = 2E\left[ e_{p+1}^{+}(n) \frac{\partial e_{p+1}^{+}(n)}{\partial \gamma_{p+1}} + e_{p+1}^{-}(n) \frac{\partial e_{p+1}^{-}(n)}{\partial \gamma_{p+1}} \right]$$

and, using Eq. (6.14.1),

$$\frac{\partial \mathcal{E}_{p+1}}{\partial \gamma_{p+1}} = - 2E\left[ e_{p+1}^{+}(n)e_p^{-}(n - 1) + e_{p+1}^{-}(n)e_p^{+}(n) \right] \tag{6.14.3}$$

Inserting Eq. (6.14.1) into Eq. (6.14.3), we rewrite the latter as

$$\frac{\partial \mathcal{E}_{p+1}}{\partial \gamma_{p+1}} = -2(C_{p+1} - \gamma_{p+1}D_{p+1}) \qquad (6.14.4)$$

where

$$C_{p+1} = 2E[e_p^+(n)e_p^-(n-1)] \qquad (6.14.5)$$

$$D_{p+1} = E[e_p^+(n)^2 + e_p^-(n-1)^2] \qquad (6.14.6)$$

Setting the gradient (6.14.4) to zero, we find the *optimal value* of $\gamma_{p+1}$

$$\gamma_{p+1} = \frac{C_{p+1}}{D_{p+1}} = \frac{2E[e_p^+(n)e_p^-(n-1)]}{E[e_p^+(n)^2 + e_p^-(n-1)^2]} \qquad (6.14.7)$$

which, due to the assumed *stationarity*, agrees with Eq. (5.6.4). Replacing the numerator and denominator of Eq. (6.14.7) by time averages leads to Burg's method.

The *gradient lattice* is obtained by solving $(\partial \mathcal{E}_{p+1}/\partial \gamma_{p+1}) = 0$ iteratively by the *gradient descent* method

$$\gamma_{p+1}(n+1) = \gamma_{p+1}(n) - \mu_{p+1}\frac{\partial \mathcal{E}_{p+1}}{\partial \gamma_{p+1}(n)} \qquad (6.14.8)$$

where $\mu_{p+1}$ is a small positive adaptation constant. Before we drop the expectation instructions in Eq. (6.14.3), we use the result of Eq. (6.14.4) to discuss *qualitatively* the convergence rate of the algorithm. Inserting Eq. (6.14.4) into Eq. (6.14.8), we find

$$\gamma_{p+1}(n+1) = \gamma_{p+1}(n) + 2\mu_{p+1}(C_{p+1} - \gamma_{p+1}(n)D_{p+1})$$

or

$$\gamma_{p+1}(n+1) = (1 - 2\mu_{p+1}D_{p+1})\,\gamma_{p+1}(n) + 2\mu_{p+1}C_{p+1} \qquad (6.14.9)$$

Actually, if we replace $\gamma_{p+1}$ by $\gamma_{p+1}(n)$ in Eq. (6.14.1), the stationarity of the lattice is lost, and it is *not correct* to assume that $C_{p+1}$ and $D_{p+1}$ are independent of $n$. The implicit dependence of $D_{p+1}$ and $C_{p+1}$ on the (time-varying) reflection coefficients of the previous lattice stages makes Eq. (6.14.9) a nonlinear difference equation in the reflection coefficients. In the analogous discussion of the LMS case in Section 6.6, the corresponding difference equation for the weights was linear with *constant* coefficients—because of the tapped delay line structure, the stationarity of the input signal $\mathbf{y}(n)$ was not affected by the time-varying weights. Nevertheless, we will use Eq. (6.14.9) in a qualitative manner, replacing $D_{p+1}$ and $C_{p+1}$ by their constant asymptotic values, but only for the purpose of *motivating* the final choice of the adaptation parameter $\mu_{p+1}$. The solution of Eq. (6.14.9), then, is

$$\gamma_{p+1}(n) = \gamma_{p+1} + (1 - 2\mu_{p+1}D_{p+1})^n(\gamma_{p+1}(0) - \gamma_{p+1}) \qquad (6.14.10)$$

where $\gamma_{p+1}$ is the asymptotic value of the weight, given in Eq. (6.14.7). The stability of Eqs. (6.14.9) and (6.14.10) requires that

$$\left|1 - 2\mu_{p+1}D_{p+1}\right| < 1 \qquad (6.14.11)$$

If we choose $\mu_{p+1}$ as

$$2\mu_{p+1} = \frac{\alpha}{D_{p+1}} \qquad 0 < \alpha < 1 \qquad (6.14.12)$$

then $1 - 2\mu_{p+1}D_{p+1} = 1 - \alpha$ will satisfy Eq. (6.14.11). Note that $\alpha$ was chosen to be independent of the order $p$. This implies that all reflection coefficients $\gamma_{p+1}$ will essentially converge at the same rate. Using Eqs. (6.14.3) and (6.14.12), we write Eq. (6.14.8) as follows:

$$\gamma_{p+1}(n + 1) = \gamma_{p+1}(n) + \frac{\alpha}{D_{p+1}} E[e_{p+1}^+(n)e_p^-(n - 1)$$

$$+ e_{p+1}^-(n)e_p^+(n)] \qquad (6.14.13)$$

The *practical implementation* of this method consists of *ignoring* the expectation instruction, and using a *least-squares approximation* for $D_{p+1}$ of the form [55–57]

$$D_{p+1}(n) = (1 - \lambda) \sum_{k=0}^{n} \lambda^{n-k} [e_p^+(k)^2 + e_p^-(k - 1)^2] \qquad (6.14.14)$$

where $0 < \lambda < 1$. It may also be computed *recursively* by

$$D_{p+1}(n) = \lambda D_{p+1}(n - 1) + (1 - \lambda) [e_p^+(n)^2 + e_p^-(n - 1)^2] \qquad (6.14.15)$$

This quantity is a measure of $D_{p+1}$ of Eq. (6.14.6); indeed, taking expectations of both sides and assuming stationarity, we find

$$E[D_{p+1}(n)] = (1 - \lambda) \sum_{k=0}^{n} \lambda^{n-k} E[e_p^+(k)^2 + e_p^-(k - 1)^2]$$

$$= D_{p+1} (1 - \lambda) \sum_{k=0}^{n} \lambda^{n-k}$$

$$= D_{p+1} (1 - \lambda^{n+1})$$

which reduces to $D_{p+1}$ for large $n$. With the above changes, we obtain the adaptive version of Eq. (6.14.13),

$$\gamma_{p+1}(n + 1) = \gamma_{p+1}(n) + \frac{\alpha}{D_{p+1}(n)} [e_{p+1}^+(n)e_p^-(n - 1)$$

$$+ e_{p+1}^-(n)e_p^+(n)] \qquad (6.14.16)$$

It can be written in a slightly different form by defining the quantity

$$d_{p+1}(n) = \sum_{k=0}^{n} \lambda^{n-k} [e_p^+(k)^2 + e_p^-(k-1)^2]$$

$$= \lambda d_{p+1}(n-1) + [e_p^+(n)^2 + e_p^-(n-1)^2] \qquad (6.14.17)$$

and noting that $D_{p+1}(n) = (1 - \lambda)d_{p+1}(n)$. Defining the new parameter $\beta = \alpha/(1 - \lambda)$, we rewrite Eq. (6.14.16) in the form

$$\gamma_{p+1}(n+1) = \gamma_{p+1}(n) + \frac{\beta}{d_{p+1}(n)} [e_{p+1}^+(n)e_p^-(n-1)$$

$$+ e_{p+1}^-(n)e_p^+(n)] \qquad (6.14.18)$$

This is usually operated with $\beta = 1$ or, equivalently, $\alpha = 1 - \lambda$. This choice makes Eq. (6.14.18) equivalent to a recursive reformulation of Burg's method [54–57]. This may be seen as follows: Define the quantity $c_{p+1}(n)$ by

$$c_{p+1}(n) = \beta \sum_{k=0}^{n} \lambda^{n-k}[2e_p^+(k)e_p^-(k-1)] + (1 - \beta)d_{p+1}(n)$$

Then, inserting Eq. (6.14.1), with $\gamma_{p+1}$ replaced by $\gamma_{p+1}(n)$, into Eq. (6.14.18), we find after some algebra

$$\gamma_{p+1}(n+1) = \frac{c_{p+1}(n)}{d_{p+1}(n)}$$

Setting $\beta = 1$, we obtain

$$\gamma_{p+1}(n+1) = \frac{2 \sum\limits_{k=0}^{n} \lambda^{n-k} e_p^+(k)e_p^-(k-1)}{\sum\limits_{k=0}^{n} \lambda^{n-k}[e_p^+(k)^2 + e_p^-(k-1)^2]} \qquad (6.14.19)$$

which corresponds to Burg's method, and also guarantees that the magnitude of $\gamma_{p+1}(n+1)$ will remain less than one at each iteration. The adaptive lattice is depicted in Fig. 6.19.

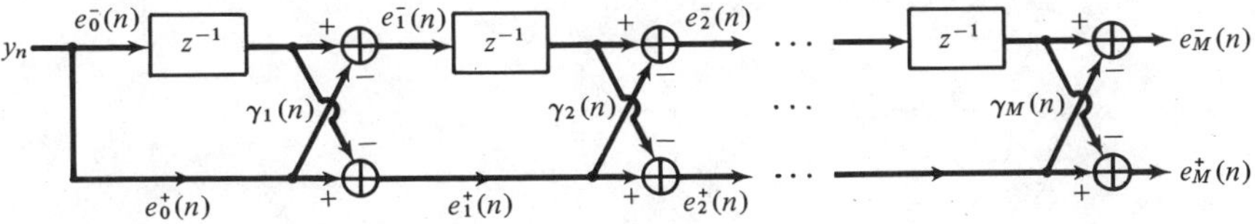


**Figure 6.19**   Adaptive Lattice Predictor

At each time instant $n$, the order recursions are

$$e_{p+1}^+(n) = e_p^+(n) - \gamma_{p+1}(n)e_p^-(n-1)$$

$$e_{p+1}^-(n) = e_p^-(n-1) - \gamma_{p+1}(n)e_p^+(n)$$

(6.14.20)

for $p = 0,1,\ldots,M-1$, with $\gamma_{p+1}(n)$ updated in time using Eq. (6.14.18) or Eq. (6.14.19). Initialize Eq. (6.14.20) by $e_0^+(n) = e_0^-(n) = y_n$. We summarize the computational steps as follows:

1. At time $n$, the coefficients $\gamma_{p+1}(n)$ and $d_{p+1}(n-1)$ are available
2. Iterate Eq. (6.14.20) for $p = 0,1,\ldots,M-1$
3. Using Eq. (6.14.17), compute $d_{p+1}(n)$, for $p = 0,1,\ldots,M-1$
4. Using Eq. (6.14.18), compute $\gamma_{p+1}(n+1)$, for $p = 0,1,\ldots,M-1$
5. Go to $n \to n+1$

Figure 6.20 shows the very fast transient behavior of the reflection coefficients for the same second order simulation example of Section 6.13. The asymptotic reflection coefficients, corresponding to the prediction error filter $A(z) = 1 - 1.6z^{-1} + 0.8z^{-2}$, are $\gamma_1 = 0.8889$ and $\gamma_2 = -0.8$. The values $\beta = 1$ and $\lambda = 0.99$ were used.

Next, we discuss the *adaptive lattice realization* of the *FIR Wiener filter* of Section 5.8. We use the same notation as in that section. The time-invariant

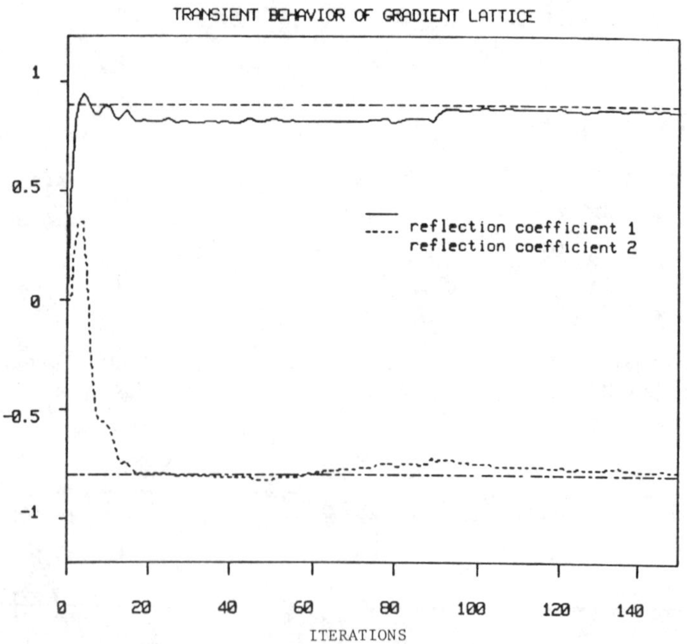


**Figure 6.20**  Transient Behavior of the Reflection Coefficients $\gamma_1(n)$ and $\gamma_2(n)$ versus $n$, Adapted According to the Gradient Lattice Algorithm

lattice weights $g_p$ are chosen optimally to minimize the mean-squared estimation error

$$\mathcal{E} = E[e_n^2] = \min \tag{6.14.21}$$

where $e_n = x_n - \hat{x}_n$, and

$$\hat{x}_n = \sum_{p=0}^{M} g_p e_p^-(n) = [g_0, g_1, \ldots, g_M] \begin{bmatrix} e_0^-(n) \\ e_1^-(n) \\ \vdots \\ e_M^-(n) \end{bmatrix} = \mathbf{g}^T \mathbf{e}^-(n) \tag{6.14.22}$$

The gradient with respect to $\mathbf{g}$ is

$$\frac{\partial \mathcal{E}}{\partial \mathbf{g}} = -2E[e_n \mathbf{e}^-(n)] \tag{6.14.23}$$

Inserting Eq. (6.14.22) into Eq. (6.14.23), we rewrite the latter as

$$\frac{\partial \mathcal{E}}{\partial \mathbf{g}} = -2(\mathbf{r} - R\mathbf{g}) \tag{6.14.24}$$

where $\mathbf{r}$ and $R$ are defined in terms of the *backward* lattice signals $e_p^-(n)$ as

$$\mathbf{r} = E[x_n \mathbf{e}^-(n)], \qquad R = E[\mathbf{e}^-(n)\mathbf{e}^-(n)^T] \tag{6.14.25}$$

The gradient-descent method applied to the weights $\mathbf{g}$ is

$$\mathbf{g}(n + 1) = \mathbf{g}(n) - M \frac{\partial \mathcal{E}}{\partial \mathbf{g}(n)} \tag{6.14.26}$$

where, following the discussion of Section 6.6, we have used a positive definite symmetric adaptation matrix $M$, to be chosen below. Then, Eq. (6.14.26) becomes

$$\mathbf{g}(n + 1) = (I - 2MR)\mathbf{g}(n) + 2M\mathbf{r} \tag{6.14.27}$$

The orthogonality of the backward prediction errors $\mathbf{e}^-(n)$ causes their covariance matrix $R$ to be *diagonal*

$$R = \text{diag}[E_0, E_1, \ldots, E_M] \tag{6.14.28}$$

where $E_p$ are the variances of $e_p^-(n)$

$$E_p = E[e_p^-(n)^2] \tag{6.14.29}$$

for $p = 0, 1, \ldots, M$. If we chose $M$ to be diagonal, $M = \text{diag}[\mu_0, \mu_1, \ldots, \mu_M]$, the state matrix $(I - 2MR)$ of Eq. (6.14.27) will also be diagonal and therefore Eq. (6.14.27) decouples into its individual components

$$g_p(n + 1) = (1 - 2\mu_p E_p)\, g_p(n) + 2\mu_p r_p \tag{6.14.30}$$

for $p = 0, 1, \ldots, M$, where $r_p = E[x_n e_p^-(n)]$. Its solution is

$$g_p(n) = g_p + (1 - 2\mu_p E_p)^n\, (g_p(0) - g_p) \tag{6.14.31}$$

where $g_p = r_p/E_p$ are the *optimal weights*. The convergence *rate* depends on the quantity $(1 - 2\mu_p E_p)$. Choosing $\mu_p$ by

$$2\mu_p = \frac{\alpha}{E_p} \qquad 0 < \alpha < 1 \qquad (6.14.32)$$

implies that all lattice weights $g_p(n)$ will have the *same* rate of convergence. Using Eqs. (6.14.32) and (6.14.23) we can rewrite Eq. (6.14.26) component-wise as follows

$$g_p(n + 1) = g_p(n) + \frac{\alpha}{E_p} E[e_n e_p^-(n)]$$

*Ignoring* the expectation instruction, and replacing $E_p$ by its *time average*

$$\begin{aligned} E_p(n) &= (1 - \lambda) \sum_{k=0}^{n} \lambda^{n-k} e_p^-(k)^2 \\ &= \lambda\, E_p(n - 1) + (1 - \lambda) e_p^-(n)^2 \end{aligned} \qquad (6.14.33)$$

we obtain the adaptation equation for the $p$th weight

$$g_p(n + 1) = g_p(n) + \frac{\alpha}{E_p(n)} e_n e_p^-(n) \qquad (6.14.34)$$

for $p = 0, 1, \ldots, M$. Defining

$$d_p^-(n) = \sum_{k=0}^{n} \lambda^{n-k} e_p^-(k)^2 = \lambda d_p^-(n - 1) + e_p^-(n)^2 \qquad (6.14.35)$$

and noting $E_p(n) = (1 - \lambda)\, d_p^-(n)$, we rewrite Eq. (6.14.34) as

$$g_p(n + 1) = g_p(n) + \frac{\beta}{d_p^-(n)} e_n e_p^-(n) \qquad (6.14.36)$$

where $\beta = \alpha/(1 - \lambda)$. Typically, Eq. (6.14.36) is operated with $\beta = 1$, or $\alpha = 1 - \lambda$ [55,57]. The realization of the adaptive lattice Wiener filter is shown in Fig. 6.21.

The combination of Eq. (6.14.18) and Eq. (6.14.36) *defines* the operation of the filter. The adaptation of the reflection coefficients $\gamma_{p+1}(n)$, provides a gradual *orthogonalization* of the backward prediction error signals $e_p^-(n)$, which in turn drive the adaptation equations for the lattice weights $g_p(n)$.

Such adaptive lattice filters may be used in *any* Wiener filtering application. Their attractive features are: (1) computational *efficiency*; (2) very *fast* rate of convergence, which is essentially *independent* of the eigenvalue spread of the input covariance matrix; (3) *modularity* of structure, which allows the addition of more lattice sections without affecting the previous ones; and (4) numerical stability and insensitivity to *coefficient quantization* [54–60].

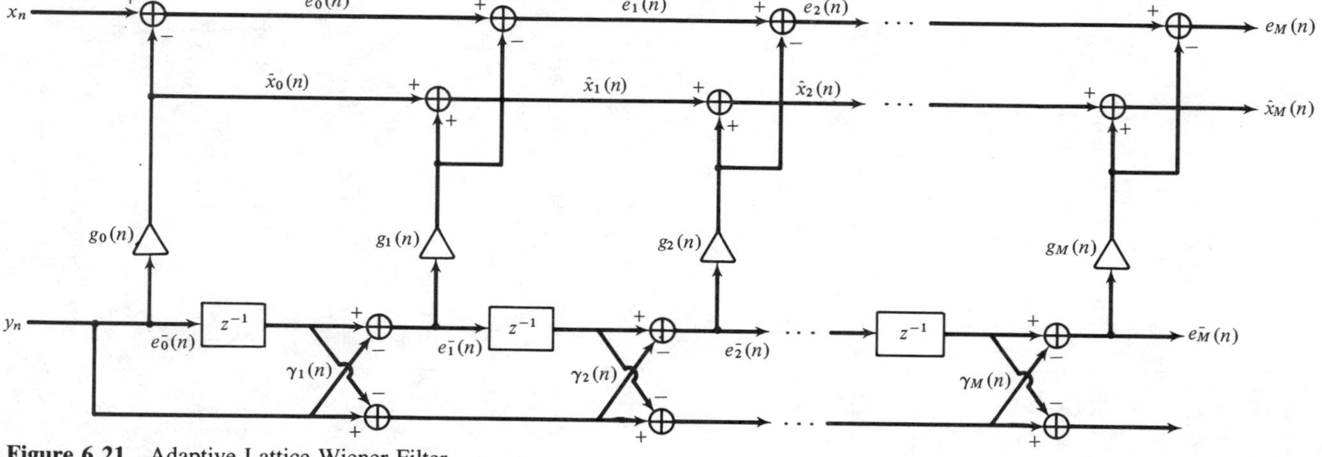


**Figure 6.21**   Adaptive Lattice Wiener Filter

## Problems

***Problem 6.1:*** Computer Experiment

(a) Reproduce the results of Fig. 6.3.

(b) On the same graph of part (a), plot the theoretical convergence curve of the weight $h(n)$ obtained by using Eq. (6.2.8).

(c) Using 10 different realizations of $x_n$ and $y_n$, compute 10 different realizations of the adaptive weight of Eq. (6.3.2). Compute the average weight over the 10 realizations and plot it versus $n$, together with the theoretical weight of Eq. (6.2.8). Use $\mu = 0.03$.

(d) Reproduce the results of Fig. 6.7.

***Problem 6.2:***

Verify that Eq. (6.6.5) is the solution of Eq. (6.6.4).

***Problem 6.3:***

Consider an adaptive filter with two taps:

$$\hat{x}_n = h_0(n)y_n + h_1(n)y_{n-1} = [h_0(n), h_1(n)]\begin{bmatrix} y_n \\ y_{n-1} \end{bmatrix} = \mathbf{h}(n)^T\mathbf{y}(n)$$

The optimal filter weights are found adaptively by the gradient descent algorithm

$$\mathbf{h}(n+1) = \mathbf{h}(n) - \mu\frac{\partial\mathcal{E}}{\partial\mathbf{h}(n)}$$

where $\mathcal{E} = E[e_n^2]$, and $e_n$ is the estimation error.

(a) Show that the above difference equation may be written as

$$\mathbf{h}(n+1) = \mathbf{h}(n) + 2\mu(\mathbf{r} - R\mathbf{h}(n))$$

where

$$\mathbf{r} = \begin{bmatrix} R_{xy}(0) \\ R_{xy}(1) \end{bmatrix}, \qquad R = \begin{bmatrix} R_{yy}(0) & R_{yy}(1) \\ R_{yy}(1) & R_{yy}(0) \end{bmatrix}$$

(b) Suppose $R_{xy}(0) = 10$, $R_{xy}(1) = 5$, $R_{yy}(0) = 3$, $R_{yy}(1) = 2$. Find the optimal weights $\mathbf{h} = \lim \mathbf{h}(n)$ as $n \rightarrow \infty$.

(c) Select $\mu = 1/6$. Explain why such a value is sufficiently small to guarantee convergence of the difference equation of part (a). What other values of $\mu$ also guarantee convergence?

(d) With $\mu = 1/6$, solve the difference equation of part (a) in closed form for $n = 0, 1, \ldots$. Discuss the rate of convergence.

***Problem 6.4:***
Consider a single CCL as shown in Fig. 6.2.

(a) Suppose the reference signal is set equal to a unit step signal; that is, $y(n) = u(n)$. Show that the CCL will behave as a time-invariant linear filter with input $x_n$ and output $e_n$. Determine the transfer function $H(z)$ from $x_n$ to $e_n$.

(b) Find and interpret the poles and zeros of $H(z)$.

(c) Determine the range of $\mu$-values for which $H(z)$ is stable.

***Problem 6.5:***
Repeat Problem 6.4 when the reference signal is the alternating unit step; that is, $y(n) = (-1)^n u(n)$.

***Problem 6.6:***
Using the transfer function of Eq. (6.9.3), derive an approximate expression for the 3-dB width of the notch. You may work to lowest order in $\mu$.

***Problem 6.7:*** Computer Experiment
Consider the noise canceling example discussed in Section 5.8 and in Problems 5.13–5.15 and defined by the following choice of parameters:

$$\omega_0 = 0.075\pi \text{ [rads/sample]}, \quad \phi = 0, \quad a_1 = -0.5, \quad a_2 = 0.8, \quad M = 4$$

(a) Generate a realization of the signals $x(n)$ and $y(n)$ and process them through the adaptive noise canceler of Section 6.9, using an $M$th order adaptive filter and adaptation parameter $\mu$. By trial and error select a value for $\mu$ that makes the LMS algorithm convergent, but not too small as to make the convergence too slow. Plot one of the filter weights $h_m(n)$ versus iteration number $n$, and compare the asymptotic value with the theoretical value obtained in Problem 5.14.

(b) After the weights have converged, plot 100 output samples of the error signal $e_n$, and observe the noise cancellation property.

(c) Repeat (a) and (b) using an adaptive filter of order $M = 6$.

***Problem 6.8:*** Computer Experiment
(a) Plot the magnitude of the frequency response of the adaptive noise canceler notch filter of Eq. (6.9.3) versus frequency $\omega$ ($z = e^{j\omega}$). Generate several such plots for various values of $\mu$ and observe the effect of $\mu$ on the width of the notch.

(b) Let $x(n) = e^{j\omega_0 n}$, and $y(n) = Ae^{j\omega_0 n}$ and select the parameters as

$$\omega_0 = 0.075\pi, \quad M = 2, \quad A = 0.01, \quad \mu = 0.1$$

Process $x(n)$ and $y(n)$ through the adaptive noise canceler of Section 6.9, and plot the output $e(n)$ versus $n$ and observe the cancellation of the signal $x(n)$ due to the notch filter created by the presence of the weak sinusoidal reference signal $y(n)$.

***Problem 6.9:*** Computer Experiment

Let $x(n) = x_1(n) + x_2(n)$, where $x_1(n)$ is a narrowband component defined by

$$x_1(n) = \sin(\omega_0 n + \phi), \qquad \omega_0 = 0.075\pi \ [\text{rads/sample}]$$

where $\phi$ is a random phase uniformly distributed over $[0, 2\pi]$, and $x_2(n)$ is a fairly broadband component generated by sending zero-mean, unit-variance, white noise $\epsilon(n)$ through the filter

$$x_2(n) = \epsilon(n) + 2\epsilon(n-1) + \epsilon(n-2)$$

(a) Compute the autorcorrelation functions of $x_1(n)$ and $x_2(n)$, and sketch them versus lag $k$. Based on this computation, select a value for the delay $\Delta$ to be used in the adaptive line enhancer discussed in Section 6.10.

(b) Generate a realization of $x(n)$ and process it through the ALE with an appropriately chosen adaptation parameter $\mu$. Plot the output signals $\hat{x}(n)$ and $e(n)$ and compare them with the components $x_1(n)$ and $x_2(n)$, respectively.


***Problem 6.10:***

The response of the ALE to an input sinusoid in noise can be studied as follows: Let the input be

$$x_n = A_1 \exp(j\omega_1 n + \phi) + v_n$$

where $\phi$ is a random phase independent of the zero-mean white noise $v_n$. The optimum Wiener filter weights of the ALE are given by

$$\mathbf{h} = R^{-1}\mathbf{r}$$

where $R_{ij} = R_{xx}(i - j)$ and $r_i = R_{xx}(i + \Delta)$, as discussed in Section 6.10.

(a) Using the methods of Section 5.11, show that the optimum filter $\mathbf{h}$ is given by

$$\mathbf{h} = \frac{e^{j\omega_1\Delta}}{\dfrac{\sigma_v^2}{P_1} + M + 1}\, \mathbf{s}_{\omega_1}$$

where the phasing vector $\mathbf{s}_{\omega_1}$ was defined in Section 5.11, and $P_1 = |A_1|^2$ is the power of the sinusoid.

(b) Show that the mean output power of the ALE is given by

$$E[|\hat{x}_n|^2] = \mathbf{h}^\dagger R \mathbf{h} = \sigma_v^2 \mathbf{h}^\dagger \mathbf{h} + P_1 |\mathbf{h}^\dagger \mathbf{s}_{\omega_1}|^2$$

(c) Show that the SNR at the output is enhanced by a factor $M + 1$ over the SNR at the input; that is, show that

$$(\text{SNR})_{\text{out}} = \frac{P_1 |\mathbf{h}^\dagger \mathbf{s}_{\omega_1}|^2}{\sigma_v^2 \mathbf{h}^\dagger \mathbf{h}} = \frac{P_1}{\sigma_v^2}(M + 1) = (M + 1)\,(\text{SNR})_{\text{in}}$$

(d) Derive an expression for the eigenvalue spread $\lambda_{\max}/\lambda_{\min}$ in terms of the parameters $\sigma_v^2$, $P_1$, and $M$.

(e) Show that if the delay $\Delta$ is removed; that is, $\Delta = 0$, then the optimal weight vector becomes equal to the unit vector

$$\mathbf{h} = [1,0,0, \ldots ,0]^T$$

and that this choice corresponds to complete cancellation of the input signal $x(n)$ from the output $e(n)$.

**Problem 6.11:** Computer Experiment
Consider the autoregressive process $y_n$ generated by the difference equation

$$y_n = -a_1 y_{n-1} - a_2 y_{n-2} + \epsilon_n$$

where $a_1 = -1.6$, $a_2 = 0.8$, and $\epsilon_n$ is zero-mean, unit-variance, white noise. Generate a realization of $y_n$ and process it through the LMS adaptive predictor of order 2, as discussed in Section 6.11. Use a value for the adaptation parameter $\mu$ of your own choice. Plot the adaptive prediction coefficients $a_1(n)$ and $a_2(n)$ versus $n$, and compare their converged values with the theoretical values given above.

**Problem 6.12:** Computer Experiment
A complex-valued version of the LMS adaptive predictor of Section 6.11 is defined by

$$e_n = y_n + a_1(n)y_{n-1} + a_2(n)y_{n-2} + \cdots + a_M(n)y_{n-M}$$

$$a_m(n + 1) = a_m(n) - 2\mu e_n y^*_{n-m}, \quad m = 1,2, \ldots ,M$$

Let $y_n$ consist of two complex sinusoids in zero-mean white noise

$$y_n = A_1 e^{j\omega_1 n} + A_2 e^{j\omega_2 n} + v_n$$

where the frequencies and the SNRs are

$$\omega_1 = 0.3\pi, \quad \omega_2 = 0.7\pi \text{ [rads/sample]}$$

$$10 \log_{10}[|A_1|^2/\sigma_v^2] = 10 \log_{10}[|A_2|^2/\sigma_v^2] = 20 \text{ dB}$$

(a) Generate a realization of $y_n$ and process it through an $M$th order LMS adaptive predictor using an adaptation constant $\mu$. Experiment with several choices of $M$ and $\mu$. In each case, stop the algorithm after convergence has taken place and plot the AR spectrum $S(\omega) = 1/|A(\omega)|^2$ versus frequency $\omega$. Discuss your results.

(b) Using the same realization of $y_n$, iterate the adaptive Pisarenko algorithm defined by Eqs. (6.12.5) and (6.12.6). After convergence of the Pisarenko weights, plot the Pisarenko spectrum estimate $S(\omega) = 1/|A(\omega)|^2$ versus frequency $\omega$.

(c) Repeat (a) and (b) when the SNR of the sinewaves is lowered to 0 dB. Compare the adaptive AR and Pisarenko methods.

***Problem 6.13:***
  (a) Verify the recursion equations (6.13.5) and (6.13.6).
  (b) Prove Eq. (6.13.7).
  (c) Show Eq. (6.13.8).
  (d) Show Eq. (6.13.10).
  (e) Show Eq. (6.13.12).

***Problem 6.14:*** Computer Experiment
Reproduce the results of Figs. 6.17 and 6.18. Repeat these graphs for various choices of the LMS adaptation parameter $\mu$ and RLS parameter $\lambda$.

***Problem 6.15:*** Computer Experiment
Reproduce the results of Fig. 6.20.

***Problem 6.16:*** Computer Experiment
Consider the noise canceling example defined in Problem 6.7.

  (a) Implement the adaptive noise canceler using an adaptive lattice FIR filter as discussed in Section 6.14. After convergence, plot the output signal $e_M(n)$ versus $n$.

  (b) Plot one of the adaptive lattice weights $g_M(n)$ versus $n$, and compare its asymptotic value with its theoretical value derived in Problem 5.14.

## *References*


1. B. Widrow and M. Hoff, Adaptive Switching Circuits, *IRE Wescon Conv. Rec.*, pt. 4, 96–104 (1960).

2. B. Widrow, Adaptive Filters, in R. Kalman and N. DeClaris, Eds., *Aspects of Network and System Theory*, New York, Holt, Rinehart and Winston, 1971.

3. B. Widrow, et al., Adaptive Noise Cancelling—Principles and Applications, *Proc. IEEE*, **63**, 1692–1716 (1975).

4. S. P. Applebaum, Adaptive Arrays, *IEEE Trans. Antennas Prop.*, **AP-24**, 585–598 (1976).

5. R. A. Monzingo and T. W. Miller, *Introduction to Adaptive Arrays*, New York, Wiley, 1980.

6. F. Gabriel, Adaptive Arrays—An Introduction, *Proc. IEEE*, **64**, 239–272 (1976).

7. B. Widrow, et al., Adaptive Antenna Systems, *Proc. IEEE*, **55**, 2143–2159 (1967).

8. A. M. Vural and M. T. Stark, A Summary and the Present Status of Adaptive Array Processing Techniques, 19th IEEE Conference on Decision and Control, 931–938 (1980).

9. B. Widrow, et al., Stationary and Nonstationary Learning Characteristics of the LMS Adaptive Filter, *Proc. IEEE*, **64**, 1151–1162 (1976).

10. R. W. Lucky, J. Salz, and E. J. Weldon, Jr., *Principles of Data Communication*, New York, McGraw-Hill, 1968.

11. N. A. M. Vierhoeckx, H. Elzen, F. Snijders, and P. Gerwen, Digital Echo Cancellation for Baseband Data Transmission, *IEEE Trans. Acoust., Speech, Signal Process.*, **ASSP-27,** 768–781 (1979).

12. M. M. Sondhi and D. A. Berkley, Silencing Echoes on the Telephone Network, *Proc. IEEE*, **66,** 948–963 (1980).

13. D. L. Duttweiler and Y. S. Chen, A Single Chip VLSI Echo Canceler, *Bell Syst. Tech. J.*, **59,** 149 (1980).

14. D. L. Duttweiler, Bell's Echo-Killer Chip, *IEEE Spectrum*, **17,** 34–37 (1980).

15. J. Glover, Adaptive Noise Cancelling Applied to Sinusoidal Interferences, *IEEE Trans. Acoust., Speech, Signal Process.*, **ASSP-25,** 484–491 (1977).

16. B. Widrow, J. McCool, and M. Ball, The complex LMS algorithm, *Proc. IEEE*, **63,** pp. 719–720 (1975).

17. B. Widrow, K. Duvall, R. Gooch, and W. Newman, Signal Cancellation Phenomena in Adaptive Antennas: Causes and Cures, *IEEE Trans. Antennas Prop.*, **AP-30,** 469–478 (1982).

18. M. J. Shensa, Non-Wiener Solutions of Adaptive Noise Canceller with a Noisy Reference, *IEEE Trans. Acoust., Speech, Signal Process.*, **ASSP-28,** 468–473 (1980).

19. J. R. Treichler, Transient and Convergent Behavior of the Adaptive Line Enhancer, *IEEE Trans. Acoust., Speech, Signal Process.*, **ASSP-27,** 53–62 (1979).

20. D. W. Tufts, L. J. Griffiths, B. Widrow, J. Glover, J. McCool, and J. Treichler, Adaptive Line Enhancement and Spectrum Analysis, *Proc. IEEE*, **65,** 169 (1977).

21. J. R. Zeidler, et al., Adaptive Enhancement of Multiple Sinusoids in Uncorrelated Noise, *IEEE Trans. Acoust., Speech, Signal Process.* **ASSP-26,** 240–254 (1978).

22. L. J. Griffiths, Rapid Measurement of Digital Instantaneous Frequency, *IEEE Trans. Acoust., Speech, Signal Process.*, **ASSP-23,** 207–222 (1975).

23. D. Morgan and S. Craig, Real-Time Linear Prediction Using the Least Mean Square Gradient Algorithm, *IEEE Trans. Acoust., Speech, Signal Process.*, **ASSP-24,** 494–507 (1976).

24. P. Eykhoff, *System Identification: Parameter and State Estimation*, New York, Wiley, 1974.

25. K. J. Astrom and P. Eykhoff, System Identification—A Survey, *Automatica*, **7,** 123–162 (1971).

26. G. C. Goodwin and R. L. Payne, *Dynamic System Identification, Experimental Design and Data Analysis*, New York, Academic, 1977.

27. W. S. Hodgkiss and J. A. Presley, Jr., Adaptive Tracking of Multiple Sinusoids whose Power Levels are Widely Separated, *IEEE Trans. Acoust., Speech, Signal Process.*, **ASSP-29,** 710–721 (1981).

28. A. Isaksson, A. Wennberg, and L. H. Zetterberg, Computer Analysis of EEG Signals with Parametric Models, *Proc. IEEE*, **69,** 451–461 (1981).

29. T. Bohlin, Analysis of EEG Signals with Changing Spectra using a Short-Word Kalman Estimator, *Math. Biosci.*, **35,** 221–259 (1977).

30. W. F. Gabriel, Spectral Analysis and Adaptive Array Superresolution Techniques, *Proc. IEEE*, **68,** 654–666 (June 1980).

31. P. A. Thompson, An Adaptive Spectral Analysis Technique for Unbiased Frequency

Estimation in the Presence of White Noise, *Proc. 13th Asilomar Conf. Circuits, Systems, and Computers*, 529 (Nov. 1979).

32. M. G. Larimore and R. J. Calvert, Convergence Studies of Thompson's Unbiased Adaptive Spectral Estimator, *Proc. 14th Asilomar Conf. Circuits, Systems, and Computers*, 258 (Nov. 1980).

33. V. U. Reddy, B. Egard, and T. Kailath, Least Squares Type Algorithm for Adaptive Implementation of Pisarenko's Harmonic Retrieval Method, *IEEE Trans. Acoust., Speech, Signal Process.*, **ASSP-30,** 399–405 (June 1982).

34. F. K. Soong and A. M. Petersen, On the High Resolution and Unbiased Frequency Estimates of Sinusoids in White Noise—A New Adaptive Approach, *Proc. IEEE Int. Conf. Acoust., Speech, Signal Process.*, 1362 (April 1982).

35. A. Cantoni and L. Godara, Resolving the Directions of Sources in a Correlated Field Incident on an Array, *J. Acoust., Soc. Am.*, **67,** 1247–1255 (1980).

36. S. J. Orfanidis and L. M. Vail, Zero-Tracking Adaptive Filters, Rutgers University Preprint, Dept. of Electrical Engineering (February 1983).

37. S. J. Orfanidis and L. M. Vail, Point-Source Location by Null-Tracking Adaptive Arrays, Rutgers University Preprint, Dept. of Electrical Engineering (March 1983).

38. S. J. Orfanidis and L. M. Vail, Zero-Tracking Adaptation Algorithms, *Proc. ASSP Spectrum Estimation Workshop, II*, Tampa, FL (November 1983).

39. Z. Rogowski, I. Gath, and E. Bental, On the Prediction of Epileptic Seizures, *Biol. Cybernetics*, **42,** 9–15 (1981).

40. A. V. Oppenheim and R. W. Schafer, *Digital Signal Processing*, Englewood Cliffs, Prentice-Hall, 1975.

41. D. Godard, Channel Equalization Using a Kalman Filter for Fast Data Transmission, *IBM J. Res. Dev.*, **18,** 267–273, 1974.

42. R. D. Gitlin and F. R. Magee, Self-Orthogonalizing Adaptive Equalization Algorithms, *IEEE Trans. Commun.*, **COM-25,** 666 (1977).

43. R. W. Chang, A New Equalizer Structure for Fast Start-up Digital Communication, *Bell Syst. Tech. J.*, **50,** 1969–2014 (1971).

44. J. Mendel, *Discrete Techniques of Parameter Estimation*, New York, Marcel Dekker, 1973.

45. D. D. Falconer and L. Ljung, Application of Fast Kalman Estimation to Adaptive Equalization, *IEEE Trans. Commun.*, **COM-26,** 1439–1446 (1976).

46. L. Ljung, M. Morf, and D. Falconer, Fast Calculations of Gain Matrices for Recursive Estimation Schemes, *Int. J. Control*, **27,** 1–19 (1978).

47. M. Morf and D. T. L. Lee, Recursive Least-Squares Ladder Forms for Fast Parameter Tracking, *Proc. 17th IEEE Conf. Decision Contr.*, 1326 (1979).

48. E. H. Satorius and M. J. Shensa, Recursive Lattice Filters—A Brief Overview, *Proc. 19th IEEE Conf. Decision Contr.*, 955–959 (1980).

49. D. Lee, M. Morf, and B. Friedlander, Recursive Square-Root Ladder Estimation Algorithms, *IEEE Trans. Acoust., Speech, Signal Process.*, **ASSP-29,** 627–641 (1981).

50. M. J. Shensa, Recursive Least-Squares Lattice Algorithms: A Geometrical Approach, *IEEE Trans. Autom. Control*, **AC-26,** 695–702 (1981).

51. E. H. Satorius and J. D. Pack, Application of Least-Squares Lattice Algorithms to Channel Equalization, *IEEE Trans. Commun.*, **COM-29**, 136–142 (1981).

52. E. Schichor, Fast Recursive Estimation Using the Lattice Structure, *Bell Syst. Tech. J.*, **61**, 97–115 (1981).

53. M. S. Mueller, Least-Squares Algorithms for Adaptive Equalizers, *Bell Syst. Tech. J.*, **60**, 1905–1925 (1981).

54. B. Friedlander, Lattice Filters for Adaptive Processing, *Proc. IEEE*, **70**, 829–867 (1982).

55. L. J. Griffiths, A Continuously-Adaptive Filter Implemented as a Lattice Structure, *Int. Conf. Acoust., Speech, Signal Processing*, Hartford CT 87–90 (1977).

56. J. Makhoul, A Class of All-Zero Lattice Digital Filters: Properties and Applications, *IEEE Trans. Acoust., Speech, Signal Process.*, **ASSP-26**, 304–314 (1978).

57. E. H. Satorius and S. T. Alexander, Channel Equalization Using Adaptive Lattice Algorithms, *IEEE Trans. Commun.*, **COM-27**, 899–905 (1979).

58. C. J. Gibson and S. Haykin, Learning Characteristics of Adaptive Lattice Filtering Algorithms, *IEEE Trans. Acoust., Speech, Signal Process.*, **ASSP-28**, 681–691 (1980).

59. M. L. Honig and D. G. Messerschmidt, Convergence Properties of the Adaptive Digital Lattice Filter, *IEEE Trans. Acoust., Speech, Signal Process.*, **ASSP-29**, 642–653 (1981).

60. R. S. Medaugh and L. J. Griffiths, A Comparison of Two Fast Linear Predictors, *Proc. IEEE Int. Conf. Acoust., Speech, Signal Processing*, Atlanta, GA (March 1981), 293–296.

# Appendix

This appendix contains a number of FORTRAN 77 subroutines designed to be used together as a pool of subroutines. These routines are listed below. The comment statements within each subroutine explain its usage.

LEV
: Levinson's Algorithm. It generates all the prediction error filters and prediction errors up to a given order, from the autocorrelation lags.

BKWLEV
: Backward Levinson Recursion. It generates all the lower order lattice polynomials from a given minimal-phase polynomial. It can also be used to test stability, by the Schur recursion.

FRWLEV
: Forward Levinson Recursion. It generates all the prediction error filters up to a given order from the reflection coefficients.

RLEV
: Reverse of Levinson's Algorithm. From the highest-order prediction-error filter and the corresponding prediction error, it generates all the lower-order prediction-error filters and the autocorrelation lags to that order.

FIRW
: FIR Wiener filter design. It generates the optimal filter weights for both the direct form and the lattice realizations.

YW
: The Yule-Walker, or autocorrelation method of estimating the LPC model parameters from a given block of data.

BURG
: Burg's Method of estimating the LPC model parameters from a given block of data.

SPIKE
: It generates all the spiking filters of a given order and their outputs that reshape a given waveform into a spike.

SCATTER   Direct Scattering Problem. From a given set of reflection coefficients, it generates all the lattice polynomials and a finite segment of the overall reflection response.

DPD   Dynamic Predictive Deconvolution. From a finite segment of the reflection response, it generates all the lattice polynomials and the corresponding reflection coefficients up to a given order.

```
      SUBROUTINE LEV(M,R,L,E)

c  Subroutine inputs: M, R
c  Subroutine outputs: L, E (L must be declared REAL)

c  R is the vector of autocorrelation lags R(0),R(1), . . . ,R(M)

c  The matrix L is unit lower-triangular. Its i-th row holds the i-th
c  prediction-error filter in REVERSE order; that is,
c     A(i,i),A(i,i-1), . . . ,A(i,1),1
c  The first column of L holds the NEGATIVES of the reflection
c  coefficients, GAMMA(i)= -L(i,0)= -A(i,i), for i=1,2, . . . ,M
c  The vector E(0),E(1), . . . ,E(M) holds the prediction errors.
c  The matrix L and the diagonal matrix D formed by the Es define a
c  UL Cholesky factorization of the inverse of the autocorrelation matrix
c     R inverse = L transpose * D inverse * L

      REAL L
      DIMENSION R(0:M),E(0:M),L(0:M,0:M)
      DO 1 I=0,M
      DO 1 J=I+1,M
1     L(I,J)=0
      L(0,0)=1
      L(1,1)=1
      L(1,0)=-R(1)/R(0)
      E(0)=R(0)
      E(1)=E(0)*(1 - L(1,0)**2)
      DO 2 I=2,M
      GAP=0
      DO 3 K=0,I-1
3     GAP=GAP+R(K+1)*L(I-1,K)
      GAMMA=GAP/E(I-1)
      L(I,0)=-GAMMA
      DO 4 K=1, I-1
4     L(I,K)=L(I-1,K-1)-GAMMA*L(I-1,I-1-K)
      L(I,I)=1
2     E(I)=E(I-1)*(1-GAMMA**2)
      RETURN
      END
```

```
      SUBROUTINE BKWLEV(M,A,L)

c   Subroutine inputs: M, A
c   Subroutine outputs: L (L must be declared REAL)

c   Determines the lattice realization of a minimal phase polynomial
c       1,A(1),A(2), . . . ,A(M)
c   by the backward Levinson recursion.
c   The unit lower-triangular matrix L contains all the required
c   information. Its first column holds the NEGATIVES of the reflection
c   coefficients. Its i-th row is the REVERSE of the i-th polynomial
c   L(i,j)=A(i,i-j); that is, A(i,i),A(i,i-1), . . . ,A(i,1),1

      REAL L
      DIMENSION A(0:M),L(0:M,0:M)
      DO 1 I=0,M
      DO 1 J=I+1,M
1     L(I,J)=0
      DO 2 J=0,M
2     L(M,J)=A(M-J)
      DO 3 I=M,1,-1
      G=-L(I,0)
      E=1-G**2
      DO 4 K=0,I-2
4     L(I-1,K)=(L(I,K+1)+G*L(I,I-1-K))/E
3     L(I-1,I-1)=1
      RETURN
      END



      SUBROUTINE FRWLEV(M,G,L)

c   Subroutine inputs: M, G
c   Subroutine outputs: L (L must be declared REAL)

c   Given the reflection coefficients G(1),G(2), . . . ,G(M)
c   it generates all the prediction-error filters, to order M,
c   by the forward Levinson recursion.
c   Their REVERSE polynomials are saved as the rows of L.

      REAL L
      DIMENSION G(1:M),L(0:M,0:M)
      DO 1 I=0,M
      DO 1 J=I+1,M
1     L(I,J)=0
      L(0,0)=1
      L(1,0)=-G(1)
      L(1,1)=1
```

```
      DO 2 I=2,M
      L(I,0)=-G(I)
      L(I,I)=1
      DO 2 K=1,I-1
2     L(I,K)=L(I-1,K-1)-G(I)*L(I-1,I-1-K)
      RETURN
      END
```

```
      SUBROUTINE RLEV(M,A,EM,R,L)
```

```
c  Subroutine inputs: M, A, EM
c  Subroutine outputs: R, L (L must be declared REAL)
```

```
c  This subroutine generates the autocorrelation lags
c  R(0),R(1), . . . ,R(M) of an M-th degree AR process with polynomial
c  1,A(1),A(2), . . . ,A(M), and also it generates all the lower prediction
c  error filters and reflection coefficients.
c  The white noise power EM=E(M) must also be given.
```

```
c  It calls subroutine BKWLEV
```

```
      REAL L
      DIMENSION R(0:M),A(0:M),L(0:M,0:M)
      CALL BKWLEV(M,A,L)
      DO 1 I=M,1,-1
1     EM=EM/(1-L(I,0)**2)
      R(0)=EM
      DO 2 I=1,M
      R(I)=0
      DO 2 K=0,I-1
2     R(I)=R(I)-L(I,K)*R(K)
      RETURN
      END
```

```
      SUBROUTINE FIRW(M,RXY,RYY,L,E,G,H)
```

```
c  Subroutine inputs: M, RXY, RYY
c  Subroutine outputs: L, E, G, H (L must be declared REAL)
```

```
c  The user must supply the desired order M,
c  the cross-correlation lags RXY(0),RXY(1), . . . ,RXY(M)
c  and the autocorrelation lags RYY(0),RYY(1), . . . ,RYY(M).
```

```
c  The outputs of the subroutine are the vectors of filter
```

```
c   weights both for the lattice and direct form realizations; namely,

c       g(0),g(1), . . . ,g(M)      and      h(0),h(1), . . . ,h(M)

c   It calls subroutine LEV

c   The matrix L and vector E are defined in subroutine LEV

        REAL L(0:M,0:M)
        DIMENSION RXY(0:M),RYY(0:M),E(0:M),G(0:M),H(0:M)
        CALL LEV(M,RYY,L,E)
        DO 1 P=0,M
        G(P)=0
        DO 11 I=0,P
        G(P)=G(P)+L(P,I)*RXY(I)
11      CONTINUE
1       G(P)=G(P)/E(P)
        DO 2 P=0,M
        H(P)=0
        DO 2 I=P,M
2       H(P)=H(P)+L(I,P)*G(I)
        RETURN
        END




        SUBROUTINE YW(N,Y,M,L,E)

c   Subroutine inputs: N, Y, M
c   Subroutine outputs: L, E (L must be declared REAL)

c   Yule-Walker method.
c   It computes all the prediction-error filters to order M,
c   the reflection coefficients, and the errors E(0), . . . ,E(M)
c   on the basis of a block of data Y(0),Y(1), . . . ,Y(N)

c   It computes first the sample autocorrelation lags R(0), . . . ,R(M)
c   and then uses Levinson's algorithm.

        DIMENSION Y(0:N),R(0:30),E(0:M)
        REAL L(0:M,0:M)
        DO 1 I=0,M
        DO 1 J=I+1,M
1       L(I,J)=0
        L(0,0)=1
        DO 2 K=0,M
        S=0
        DO 21 J=0,N-K
        S=S+Y(K+J)*Y(J)
```

```
21      CONTINUE
 2      R(K)  = S/(N+1)
        L(1,0)=-R(1)/R(0)
        L(1,1)=1
        E(0)=R(0)
        E(1)=E(0)*(1-L(1,0)**2)
        DO 3 I=2,M
        GAP=0
        DO 4 J=0,I-1
 4      GAP=GAP+R(J+1)*L(I-1,J)
        G=GAP/E(I-1)
        L(I,0)=-G
        L(I,I)=1
        DO 5 J=1,I-1
 5      L(I,J)=L(I-1,J-1)-G*L(I-1,I-1-J)
 3      E(I)=E(I-1)*(1-G**2)
        RETURN
        END



        SUBROUTINE BURG(N,Y,M,L,E)

c   Subroutine inputs: N, Y, M
c   Subroutine outputs: L, E (L must be declared REAL)


c   Burg's method.
c   It computes all prediction-error filters up to order M,
c   the corresponding reflection coefficients, and
c   all the prediction errors E(0),E(1), . . . ,E(M).


c   Must supply the block of data Y(0),Y(1), . . . ,Y(N),
c   and the desired prediction length M.
c   Note N must be less than 400,
c   and M greater than or equal to 2 and less than N.


c   F(j) and B(j) are the forward and backward (actually, the
c   time-reversed) error sequences: For the i-th stage
c   F(i,j)=F(j) and B(i,j)=B(j-i). This definition of B(j)
c   simplifies indexing in do-loop 6: B(i,j)=B(i-1,j-1)-GF(i-1,j)
c   becomes B(j-i)=B(j-i)-GF(j)

        REAL L
        DIMENSION Y(0:N),L(0:M,0:M),E(0:M),F(400),B(0:400)

c   Initialize by computing i=0 and i=1 cases

        E(0)=Y(0)**2
        ANUM=0
```

```
        DENOM=0
        DO 1 J=1,N
        E(0)=E(0)+Y(J)**2
        ANUM=ANUM+Y(J)*Y(J-1)
1       DENOM=DENOM+Y(J)**2+Y(J-1)**2
        E(0)=E(0)/(N+1)
        G=2*ANUM/DENOM
        DO 10 J=1,N
        F(J)=Y(J)-G*Y(J-1)
10      B(J-1)=Y(J-1)-G*Y(J)
        E(1)=E(0)*(1-G**2)
        L(0,0)=1
        L(1,0)=-G
        L(1,1)=1

c   Lower triangular L

        DO 2 I=0,M
        DO 2 J=I+1,M
2       L(I,J)=0

c   Remaining stages i=2 to i=M

        DO 3 I=2,M
        ANUM=0
        DENOM=0
        DO 4 J=I,N
        ANUM=ANUM+F(J)*B(J-I)
4       DENOM=DENOM+F(J)**2+B(J-I)**2

c   Get the i-th reflection coefficient

        G=2*ANUM/DENOM
        E(I)=E(I-1)*(1-G**2)

c   Update L by its i-th row

        L(I,0)=-G
        L(I,I)=1
        DO 5 K=1,I-1
5       L(I,K)=L(I-1,K-1)-G*L(I-1,I-1-K)

c   Update the prediction error sequences

        DO 6 J=I,N
        FJ=F(J)
        F(J)=FJ-G*B(J-I)
6       B(J-I)=B(J-I)-G*FJ
```

```
3     CONTINUE
      RETURN
      END



      SUBROUTINE SPIKE (N,Y,M,R,L,E,P,H,EPS)

c  Subroutine inputs: N, Y, M, EPS
c  Subroutine outputs: R, L, E, P, H (L must be declared REAL)

c     N  =  dimension of the array Y
c     Y  =  the array to be spiked
c     M  =  spiking filter length (actually; M + 1)
c     R  =  sample autocorrelation lags R(0), . . . ,R(M)
c     L  =  Cholesky factor of R-inverse
c     E  =  prediction errors E(0), . . . ,E(M)
c     P  =  performance matrix P = Y Rinverse Ytranspose
c     H  =  all the spiking filters
c   EPS  =  Backus-Gilbert-Oldenburg parameter R(0) → (1 + EPS)R(0)

c     It calls subroutine LEV

      REAL L(0:M,0:M)
      DIMENSION Y(0:N),P(0:N+M,0:N+M),H(0:M,0:N+M),E(0:M)
      DIMENSION R(0:M),YY(0:200,0:70)
      DO 1 I=0,N+M
      DO 1 J=0,M
1     YY(I,J)=0
      DO 2 J=0,M
      DO 2 I=J,N+J
2     YY(I,J)=Y(I-J)
      DO 3 K=0,M
      R(K)=0
      DO 3 I=K,N+M
3     R(K)=R(K)+YY(I,0)*YY(I,K)
      R(0)=(1+EPS)*R(0)
      CALL LEV(M,R,L,E)
      DO 4 I=0,M
      DO 4 J=0,N+M
      H(I,J)=0
      DO 4 K1=I,M
      DO 4 K2=0,K1
4     H(I,J)=H(I,J)+L(K1,I)*L(K1,K2)*YY(J,K2)/E(K1)
      DO 5 I=0,N+M
      DO 5 J=0,N+M
      P(I,J)=0
      DO 5 K=0,M
```

```
5     P(I,J)=P(I,J)+YY(I,K)*H(K,J)
      RETURN
      END




      SUBROUTINE SCATTER(M,RHO,A,B,N,R)

C  Subroutine inputs: M, RHO, N
C  Subroutine outputs: A, B, R

C  Direct Scattering Problem.

C  The subroutine produces all the lattice polynomials
C  A(i,z) and B(i,z), i=0,1, . . . ,M,
C  and N samples of the reflection response R(z)=B(M,z)/A(M,z).

C  Must supply the reflection coefficients RHO(0),RHO(1), . . . ,RHO(M)
C  And a desired number N, defining the number of samples of
C  the reflection response R(0),R(1), . . . ,R(N) to be computed.


      DIMENSION RHO(0:M),A(0:M,0:M),B(0:M,0:M),R(0:N)

C  Forward Lattice Recursion

      A(0,0)=1
      B(0,0)=RHO(0)
      A(1,0)=1
      A(1,1)=RHO(0)*RHO(1)
      B(1,0)=RHO(1)
      B(1,1)=RHO(0)
      DO 1 I=2,M
      A(I,0)=1
      A(I,I)=RHO(0)*RHO(I)
      B(I,0)=RHO(I)
      B(I,I)=RHO(0)
      DO 1 J=1,I-1
      A(I,J)=A(I-1,J)+RHO(I)*B(I-1,J-1)
1     B(I,J)=B(I-1,J-1)+RHO(I)*A(I-1,J)

C  Generate the reflection response R(0),R(1), . . . ,R(N)

      R(0)=B(M,0)
      DO 2 K=1,N
      S=0
      DO 3 J=1,MIN(K,M)
```

```
3     S=S-A(M,J)*R(K-J)
      IF (K.LE.M) THEN
      R(K)=B(M,K)+S
      ELSE
      R(K)=S
      ENDIF
2     CONTINUE
      RETURN
      END



      SUBROUTINE DPD(N,R,M,A,B,PHI,L,E)

c  Subroutine inputs: N, R, M
c  Subroutine outputs: A, B, PHI, L, E

c  Dynamic Predictive Deconvolution.

c  Must supply the reflection response R(0),R(1), . . . ,R(N),
c  and a desired (or, guessed) prediction length M.

c  The subroutine generates all the lattice polynomials
c  A(i,z),B(i,z), i=0,1, . . . ,M,
c  and the reflection coefficients RHO(0),RHO(1), . . . ,RHO(M).

c  The rows of the lower triangular matrices A and B hold the
c  coefficients A(i,0),A(i,1), . . . ,A(i,i), etc.
c  The reflection coefficients are in the first column of B; i.e.
c  RHO(i) = B(i,0), i=0,1, . . . ,M

c  The array PHI (spectral function autocorrelation), the matrix L,
c  and the array E, are all auxiliary to the computation, and are used
c  only to facilitate dimensioning when calling the subroutine LEV.

      DIMENSION R(0:N),A(0:M,0:M),B(0:M,0:M)
      DIMENSION PHI(0:M),E(0:M)
      REAL L(0:M,0:M)
      DO 10 I=0,M
      DO 10 J=I+1,M
      A(I,J)=0
10    B(I,J)=0

c  Compute the spectral function autocorrelation

      DO 1 K=0,M
      S=0
      DO 11 I=0,N-K
```

```
11    S=S+R(I+K)*R(I)
      IF (K.EQ.0) THEN
      PHI(K)=1-S
      ELSE
      PHI(K)=-S
      ENDIF
 1    CONTINUE

c  Get A(M,z) by the Levinson recursion

      CALL LEV(M,PHI,L,E)
      DO 15 J=0,M
15    A(M,J)=L(M,M-J)

c  Convolve A(M,z) with R(z) to get B(M,z)

      DO 4 J=0,M
      B(M,J)=0
      DO 4 K=0,J
 4    B(M,J)=B(M,J)+A(M,K)*R(J-K)

c  Go through the backward lattice recursion

      RHOO=B(M,M)
      DO 5 I=M,1,-1
      RHOI=B(I,0)
      T=1-RHOI**2
      DO 6 J=0,I-1
      A(I-1,J)=(A(I,J)-RHOI*B(I,J))/T
 6    B(I-1,J)=(B(I,J+1)-RHOI*A(I,J+1))/T
      A(I-1,0)=1
 5    B(I-1,I-1)=RHOO
      RETURN
      END
```

# Index